Knowledge-Driven Governance

Lihua Yang

Knowledge-Driven Governance

The Role of Experts and Scholars in Combating Desertification and Other Dilemmas of Collective Action

北京大学出版社
PEKING UNIVERSITY PRESS

Springer

Lihua Yang
School of Government
Peking University
Beijing, China

ISBN 978-981-13-2909-8 ISBN 978-981-13-2910-4 (eBook)
https://doi.org/10.1007/978-981-13-2910-4

Jointly published with Peking University Press, Beijing, China

The print edition is not for sale in China Mainland. Customers from China Mainland please order the print book from: Peking University Press.
ISBN of the China Mainland edition: 978-7-301-25871-2

Library of Congress Control Number: 2018957648

This Springer imprint is published by the registered company Springer Nature Singapore Pte Ltd.
The registered company address is: 152 Beach Road, #21-01/04 Gateway East, Singapore 189721, Singapore

To my wife, Yan Huang,
for her love and help
and my daughters, Ruiqing Yang
and Ruiyuan Yang

Acknowledgements

I would like to thank Prof. G. Zhiyong Lan. He is a very brilliant, patient, and generous advisor, who has helped me far beyond the scope of this research. He taught me how to conduct good research, logically structure my arguments, balance scholarly creativity and professionalism, and work together with others. I am deeply indebted to him.

I would also like to thank Prof. Jianguo (Jingle) Wu for his scholarly input and support, which improved the quality of this research greatly. His contagious enthusiasm for science, scientific debate, and teaching inspired me. Also, I must thank him for letting me share his laboratory with his intelligent doctoral students, and the wonderful experience of playing basketball with him and his colleagues.

Furthermore, I would like to thank Prof. Ronald Perry and Prof. R. F. "Rick" Shangraw, Jr. for serving as my committee. They generously gave their time to review the draft of this research and provided constructive criticism and very helpful suggestions and comments. Also, I must thank them for their trust in me and giving me so much freedom to develop my own ideas and thoughts.

I would also like to thank Profs. John Hall, Barbara McCabe, Chris Herbst, and Edgar Ramirez. As their research associate, I really learned more than I expected. Also, I would like to thank Profs. Thomas Catlaw, Elizabeth Corley, Janet Denhardt, Robert Denhardt, David Guston, and Mary Romero. Through reading their books and taking their wonderful classes, I learned a lot.

Professors Elinor and Vincent Ostrom, as my mentors at the Workshop in Political Theory and Policy Analysis at Indiana University (Bloomington), taught me institutional analysis, introduced me to studies of collective action and common-pool resources, and helped me develop the original ideas for this research.

Lin also reviewed one part of this research and gave me some very helpful suggestions and comments. The talks with Vincent shall be my lifetime asset.

I would also like to thank Prof. Roy J. Gardner in the Department of Economics and Prof. Eric Rasmusen in the Kelley School of Business, as affiliated faculty of the Workshop, who taught me game theory. Further thanks to Profs. David Good, Roger B. Parks, Edwardo L. Rhodes, and Evan J. Ringquist in the School of

Public and Environmental Affairs and Dr. Amos Sawyer in the Workshop, for teaching me various research methods and methodologies and for being kind and friendly to me while I was at IU.

Special thanks should go to Congguo Tang and Jie Zhang for their tremendous help in data collection and data input of the field study results for this research.

Finally, I must thank Xiaoxun Gao, Qian Hu, Bing Li, Wei Liu, Pei Pei, Liang Sun, Mary Reece, Robbie Robichau, Jung-wook Seo, Dan Wang, Jing Wang, Chengbo Yi, Yin Yin, Fei Yuan, Wei Zhong, Dr. Christopher Clark, Dr. Alexander Buyantuyev, and Dr. Yun Ouyang for their criticism, comments, editorial support, and friendship.

Contents

List of Figures

List of Tables

Abstract

This research addresses the central issue of collective action by exploring an alternative to the classical models of collective action: central authority, privatization, and self-governance. This alternative model is temporarily named knowledge-driven governance or expert- and scholar-based governance (simply written as "scholar-based governance").

Through a series of game theoretical analyses of collective decision making, survey and interview studies of combating desertification in multiple counties in northwest China, and reviews of relevant case studies in other regions of the world, the author finds that the scholar-based governance model is an alternative, or at least complementary, solution to the problem of collective action. The findings demonstrate that experts and scholars who have comparative advantages in knowledge, and information over other social actors (such as herders and governments), can help decision makers resolve their collective action dilemmas under certain conditions. Experts and scholars can either be information providers, entrepreneurs, self-interest pursuers, or government agents. In order for the scholar-based governance model to work, seven working rules (or design principles) may apply: (1) sustained participation of field-based experts and scholars; (2) federal organizational structure and concrete and stratified purposes; (3) democratic and collaborative management with the federal mechanism of appropriate awards and sanctions; (4) steady local scholar-entrepreneurship; (5) realization of expected benefits; (6) the experiment-extension method (social actors first do experiments in relatively small experimental areas or zones and then extend them gradually to broader areas after obtaining adequate experience); and (7) reliable external support. The more strongly these rules are abided by, the more successful this model of governance becomes.

In conclusion, this study demonstrates that in addition to Lindblom's observed intellectually guided society and preference-guided/volition-guided society, there may be the possibility of a knowledge-driven society in which knowledge or intellect plays a larger role.

Chapter 1
Scholar-Based Governance as an Alternative Solution to the Collective Action Dilemma

The Philosopher Tsăng (also translated as Tseng-tzu or Zengzi) said, 'The scholar may not be without breadth of mind and vigorous endurance. His burden is heavy and his course is long.

'Perfect virtue is the burden which he considers it is his to sustain; —is it not heavy? Only with death does his course stop; —is it not long? (In Chinese: 曾子曰: "士不可以不弘毅, 任重而道远。仁以为己任, 不亦重乎?死而后已, 不亦远乎?" (《论语·泰伯第八》). The English is revised from *Confucian analects* (1900) translated by James Legge.)

—*Confucian analects* (1900, pp. 74–75).

Collective action is concerned with the provision of public or collective goods through the collaboration of two or more individuals. The collective action dilemma, or sometimes referred to as social dilemma, refers to the situation in which public goods are needed yet they are underprovided (Olson 1971), because, while individuals share a common output as a group, they also have choices for strategies and actions maximizing their own short-term individual benefits. Thus, the rational choice of an individual in a group is to "free-ride" if other individuals are willing to provide, regardless of each person's contribution. This conflict between individual rationality and group interest optimality is the classical problem of collective action (Schelling 1978; Lichbach 1996). Olson once pointed out that even in the smallest groups, "the collective good will not ordinarily be provided on an optimal scale" (1971, p. 34), unless they perfectly arrange marginal cost-sharing.[1] Through voluntary, independent and rational action of one or more of their members, some sufficiently small groups

[1] He said,

The marginal cost of additional units of the collective good must be shared in exactly the same proportion as the additional benefit. Only if this is done well each member find that his own marginal costs and benefits are equal at the same time that the total marginal cost equals the total or aggregate marginal benefit. If marginal costs are shared in any other way, the amount of collective good provided will be *suboptimal* (Olson 1971, pp. 30–31).

L. Yang, *Knowledge-Driven Governance*,
https://doi.org/10.1007/978-981-13-2910-4_1

can provide themselves with some amount of a collective good without "relying on coercion or any positive inducements apart from the collective good itself" (Ibid, p. 33). He, however, argued that "*the larger the group, the farther it will fall short of providing an optimal amount of a collective good*" (Ibid, p. 35). Thus, the core problem of collective action is how the participants in a social activity can avoid the temptation of suboptimal equilibria and move closer to optimal outcomes to gain a cooperators' dividend (Lichbach 1996).

Collective action dilemmas are very serious problems of society and are becoming more and more pressing. For example, sound socioecological governance is essential for building resilient, sustainable, or robust social-ecological systems (SESs) and is currently an active field in political science, environmental policy, resilience research, and sustainability science (Adger 2000; Anderies et al. 2004; Wu 2006).[2] Collective action dilemmas, however, are always an important factor that hinders this endeavor (Crona and Bodin 2006; Janssen and Ahn 2006; Ostrom 1990; Toleubayev et al. 2007). Thus, Professor Ostrom (1998, p. 1) stated that "the theory of collective action is the central subject of political science."

Over the past century, different models and metaphors have been developed to describe the collective action dilemma in SESs, such as the state of nature (Hobbes 1991), the tragedy of the commons (Gardon 1954; Hardin 1968), the tragedy of public land (*Lüshi Chunqiu* 2000; *Shi Jing* 2003; Yang 2007b), the prisoner's dilemma (Pareto 1935; Dawes 1973, 1975; Tucker 1983; Yang 2007b), the free-rider problem (Hume 1978; Pasour 1981), externalities (spillover effects), and public goods (Samuelson 1954, 1955). Various models have also been applied to resolve the dilemma of collective action. These include the central-authority or Leviathan model (Hobbes 1991; Pigou 1932; Olson 1971; Hardin 1978), the privatization model (Buchanan 1965; Coase 1960, 1974; Demsetz 1970; Gordon 1954; Savas 2000; Smith 1981), and the community self-governance model (Lichbach 1996; Ostrom 1990, 2000). The central-authority model argues that governmental control and forceful actions are the major methods to resolve collective action problems and supply public goods, whereas the privatization model claims that privatization is an essential way, if not the only way, to resolve the collective action (or social) dilemma. The community self-governance model emphasizes that community members can realize self-governance of the community's common resources using self-organizing methods under certain conditions.

All these models, however, have limitations. The major problems that undermine the feasibility of the central-authority model include (1) incomplete or inaccurate governmental information, (2) low governmental monitoring capabilities, (3) low governmental sanctioning reliability, (4) high costs of administration, and (5) the possibility of governmental corruption rather than always pursuing the public interests (Ostrom 1990; Yang 2007b). The privatization model also suffers from several problems: (1) some resources cannot be divided; (2) even for resources that can

[2]Other studies include Anderies et al. (2002), Carpenter et al. (2001), Cumming et al. (2005), Dobson (1999) Fiksel (2006), Folke et al. (2002), Fraser et al. (2003), Gallopin (2006), Gottlieb (2001), Kamieniecki et al. (1997); Ostrom (1990), Tompkins and Adger (2004), Young et al. (2006).

be divided, they cannot always be divided fairly because of the heterogeneity and non-stationarity of resources; (3) the division of some resources is not economical and often involves high transaction costs; and (4) numerous clashes exist between private and public interests or between short-term and long-term interests (Ostrom 1990, pp. 12–13; Yang 2007b). Furthermore, there are several basic requirements for Coase theorem[3] in Economics: (1) every player has perfect information; (2) all the game players are price-takers; (3) the cost of court system for enforcing agreement is zero; (4) all players maximize their utility; (5) there are no income and wealth effects; and (6) the transaction cost is zero (Kolstad 2000, p. 108). All these can also be deemed as the requirements of using the polarized model of privatization. Obviously, these conditions cannot be simultaneously held in most cases. For the self-governance model, studies have shown that self-governance tends to be successful in some small-scale and homogeneous settings. In these settings, the number of people who depend heavily upon the CPRs for economic livelihood and who can interfere with each other only within the local community ranges from 50 to 15,000 (Sabatier 1992, p. 249; Yang 2007a, b). Also, self-governance cannot be automatically realized; rather, it often requires high self-organizing capability of community members, a stable and relatively closed inner environment of the community, the nonexistence of external intervention, numerous historical experiments, and a high degree of moral and religious forces (Yang 2007b).

The primary objective of this study is to search for a viable alternative solution, scholar-based governance, to social dilemmas and to achieve an understanding of the fundamental principles of successful scholar-based governance and how to ameliorate effectiveness of scholar-based governance through institutional arrangements and change. In this chapter, I first analyze the assumptions, definitions, and a theoretical framework of this study, and then I address the differences between studying scholars and studying experts. The contributions and limitations of this study are also mentioned. Finally, I briefly sketch the research methods and organization of this study.

1.1 Assumptions, Definitions, and the Fundamental Hypothesis

A conceptual governance tree of collective action is helpful for us to understand why scholar-based governance is possible and necessary. I begin to develop this tree from classifying different possible conceptual models to the solution of the problem of collective action. Classification is always the first step in the scientific process (McCool 1995, p. 174). This is also true of studying social actors and governance. I assume that the major social actors of a community in SESs can be classified into

---

[3]In a perfectly competitive market, when property rights are well-defined and transaction costs are zero, bargaining will lead to an efficient outcome regardless of the initial allocation of property rights, even when there are externalities (Coase 1960).

five groups: the public, firms, governments, scholars, and the fifth sector (including religious groups, clans, and many kinds of NGOs) (Yang 2007a, b). Also, I assume that the major social resources or capital can be grouped into six types: physical and natural, monetary, human, knowledge, social, and organizational and institutional.[4]

[4]Broadly, capital can be defined as a "stock that yields a flow of valuable goods and survives into the future" (Costanza and Daly 1992, p. 38). Coleman (1990) distinguishes four forms of capital: financial, physical, human, and social. Financial capital is comprised by monetary wealth. Physical capital is made up of tools, machinery, and other productive equipment form. Human capital includes people in the production process form. This term in modern neoclassical economic literature can date back to Mincer (1958). The best-known work, however, is supplied by Becker. To him, social capital is "expenditures on education, training, medical care, etc." and social capital differs from physical or financial capital, because "you cannot separate a person from his or her knowledge, skills, health, or values the way it is possible to move financial and physical assets while the owner stays put" (Becker 1993, p. 16). Schultz (1993), however, directly defines human capital investments as enrollment rates multiplied by the cost of education for one individual. Of the human capital, one important dimension is its technical know-how, namely knowledge and information. While knowledge and information are considered two different entities—information is about facts and figures, but knowledge is the understanding of the facts and figures (Frenzel 1987; Morgan and Peha 2003)—they nonetheless represent formless capital that plays an important role in the production and transaction process (e.g., Castells 2002; Bacon 1597; Aghion and Howitt 1992; Lan and Scott 1996; Romer 1990; Sjöberg 1998). Thus, human capital can be improved through an investment in education and training other than medical care (Becker 1993).

The fundamental ideas of social capital can be traced back to the theories raised by Karl Marx, Tocqueville, and Emile Durkheim (Ostrom and Ahn 2003; Carroll and Stanfield 2003: 398). Putnam argues that "the core idea of social capital theory is that social networks have value" (2000, pp. 18–19). To him, social capital in fact refers to "connections among individuals" and "is closely related to what some have called 'civic virtue'" (p. 19). Francis Fukuyama deems social capital "as a set of informal values or norms shared among members of a group that permits cooperation among them" (1999, p. 16). Ostrom and Ahn (2003) stressed three broad forms of social capital: trustworthiness, network, and formal and informal rules or institutions. They view it as "an attribute of individuals and of their relationships that enhances their ability to solve collective action problems" (p. xiv).

Other than the capital identified by Coleman (1990), Prescott and Visscher (1980) argued that organizational capital (or institutional capital) is an important resource from which favored individuals or parties could benefit from realizing their interests. To them, organizational capital mainly refers to accumulated information in the firm. Tomer (1987), however, further defined organizational capital as the procedures and practices that an organization implements in order to get the day's work done. Later, Tomer (1998) also explained how two new types of organizational capital—pre-organizational and linking—are formed. To him, pre-organizational capital "corresponds to an individual's attitudes and understandings regarding what it means to be a good societal citizen," while linking organizational capital is "formed during the joining-up process" (p. 844). Brynjolfsson et al. (2002) even found that investors treat the organizational complements to firms' installed computer capital as intangible assets and in the financial market the organizational assets associated with IT are also treated as importantly as other assets that increase long-term output and profits.

Some scholars even deem ecology, sustainability, and environment as capital (e.g., Harte 1995). In ecological economics, however, a particularly vigorous discourse about this has been generated by the study of natural capital to human being (Harte 1995, p. 158). Based on the study by several scholars (Costanza and Daly 1992; El Serafy 1989, 1991; Pearce and Turner 1990), natural capital is deemed as the environmental stocks that provide environmental goods and services, and natural income is the flows of goods and services (Harte 1990, p. 158). Here I deem physical and natural capital as a joint name of Coleman's physical capital and ecological economists' natural capital. Thus, there at least are six types of capital or resources: physical and natural capital (sometimes,

If I further assume that the collective action dilemma mainly happens among the public, then different models of resolving it can be first grouped into two types: the self-governance model and the other-help governance model in which the public get helps from others.

According to the physical and natural property rights structure, the self-governance model can include four submodels: the community model (Lichbach 1996), the club model (Buchanan 1965), the privileged-group model (Olson 1971), and the privatization model (Coase 1960, 1974; Demsetz 1970; Gordon 1954; Savas 2000; Smith 1981). If property rights are commonly owned by all the community members and cannot be equally divided, they are deemed "common property rights." Then the property can be commonly governed by the community. This is called "community self-governance." If property rights are restrictive and exclusive (i.e., they are owned only by some club members, but others are excluded by tolling), they are deemed as "restrictive property rights." Then the property can be governed by the club members. This is called "restrictive self-governance" or "club self-governance." If property rights are unequally shared by various people and there are privileged groups who have much more property rights than other community members, they are called "privileged-group property rights." Then these privileged groups can provide the public goods for the whole community to resolve the social dilemma. This is called "privileged-group self-governance." If the property is divisible and privately owned by various individuals and there are no privileged groups and club-like arrangements, they are "private property rights." Then the property can be governed through privatization methods. This is called "privatized self-governance."

According to the other four major community actors (government, firms, scholars, and the fifth-sector organizations), the other-help governance model also includes four submodels: the government model, the firm model, the expert and scholar-based model, and the fifth-sector model. In these models, different social actors as helpers participate in collective action to help the public resolve the dilemma. Their different preponderant resources or capital are governments' authority capital (or organizational or institutional capital), firms' financial capital, scholars' knowledge capital, and the fifth sector's social capital (see Fig. 1.1). Obviously, each of the aforementioned classical models which have been developed to resolve the collective action dilemma addresses one part of the whole picture. For example, the self-governance model mainly stresses the important function of the public including some parts of the fifth sector, the Leviathan model emphasizes the importance of government, and the privatization model mainly focuses on the important function of firms. More important, the functions of scholars and the fifth sector, especially the former, are essentially neglected.

The view of how knowledge is owned by individuals influences the view of how the social dilemma can be resolved and why we should pay attention to the function of scholars. Knowledge represents capital that plays an important role in the production and transaction process and provides structure to understanding (e.g., Aghion and

it is also directly abbreviated to physical capital), financial or monetary capital, human capital, knowledge capital, social capital, and organizational or institutional capital.

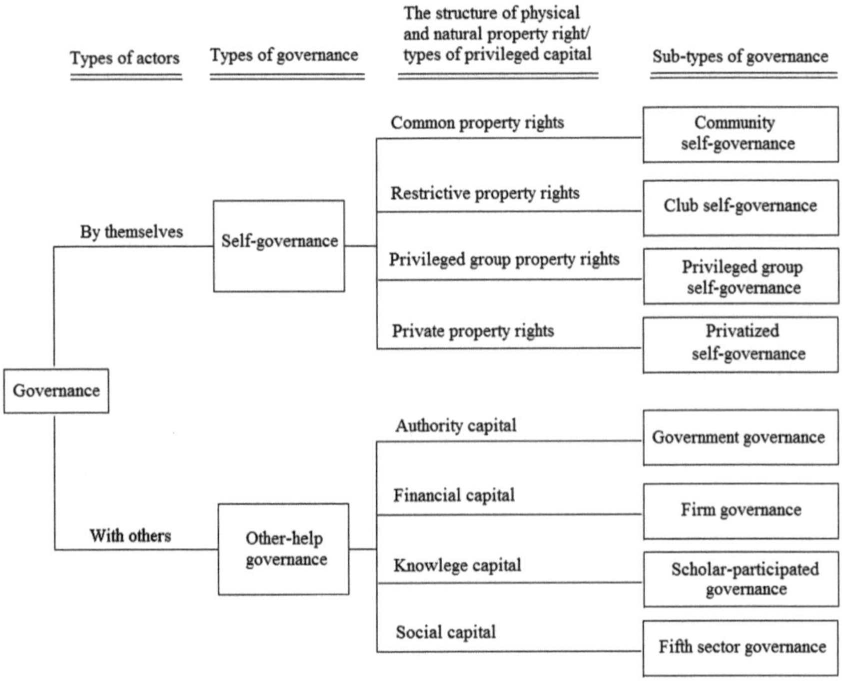

Fig. 1.1 Governance tree of collective action

Howitt 1992; Bacon 1597; Romer 1990). Knowledge is different from information. Information is about facts and figures; however, knowledge is the understanding of the facts and figures (Frenzel 1987; Morgan and Peha 2003). Hayek (1945, p. 519) claims that "the knowledge of the circumstances of which we must make use never exists in concentrated or integrated form, but solely as the dispersed bits of incomplete and frequently contradictory knowledge which all the separate individuals possess." This, however, does not mean that knowledge is equally owned by all the separate individuals. First, unless people are living in an ideal world, the asymmetry of knowledge possession is inevitable. Second, even if some institutional arrangements can be designed to distinctly reduce the degree of the asymmetry of knowledge possession, the division of labor or specialization (Brusoni et al. 2001; Durkheim 1960; Romer 1987; Smith 1965), economies of scale (Christensen and Green 1976; Krugman 1980; Murray and White 1983), economies of scope (Murray and White 1983; Teece 1980), the complex attributes of knowledge, the limits of human nature, and individual heterogeneity often make it impossible.

In this study, I assume that the heterogeneity of knowledge possession exists at the individual level as all individuals possess knowledge that is incomplete and oftentimes incoherent to different degrees. Scholars, in this study, are broadly defined as those who have comparative advantage in knowledge over other social actors such as the general public. In other words, scholars are those who possess more knowledge than others, and their knowledge is gained either from education or experience. Thus, a scholar can be "an intelligent and well-educated person" as defined in Longman Dictionary of Contemporary English (2008) or an intelligent and self-educated person who gains his knowledge from self-study or experience. Specifically, scholars here broadly refer to professors, researchers, experts, technicians, intellectual elites, and any stakeholders who have learned knowledge derived from training or experience.

As we all know, experts are often deemed as those who have "a special skill or special knowledge of a subject, gained from training or experience" (Longman Dictionary of Contemporary English 2008). Particularly, to some researchers, the term "experts" often refers only to technical persons who can use information to construct an expert–client relationship of influence (Rifkin and Martin 1997, p. 30 and 37). Scholars in this study, however, are not necessarily required to have high levels of skills or specific knowledge of great depth that are usually expected of experts. For instance, in the program of People's Planning in the Kerala in India, a new expert is defined to include the "wise farmer" in addition to the civil engineer (Fischer 2000, p. 167). Furthermore, some researchers also argued that scientific knowledge, social knowledge (such as the moral, ethical, cultural, and behavioral dimensions of issues), and particularly the local "non-scientific" knowledge, which should gain legitimacy in policy circles, are important to combine because each one of these is partial knowledge and using only one of them cannot resolve the problem (Eden 1996; Fischer 1999, 2000; Pooley and Wilcox 2000). People who possess scientific knowledge may be called "scientific experts," people who possess local "non-scientific" knowledge may be called "local experts," while people who possess more social knowledge about the moral, ethical, cultural, and behavioral dimensions of issues are often called "scholars" rather than "experts." For example, in Longman Dictionary of Contemporary English (2008) a scholar is also defined as "someone who knows a lot about a particular subject, especially one that is not a science subject." Also, although I broadly define "scholars" on the basis of social actors' comparative advantages in knowledge or information in this study, Yang (2007b) also found that in concrete local communities, local people also recognize scholars on the basis of social norms, prestige, past experiences, social status, and so on, rather than only on the basis of their knowledge and information. This means, in addition to their most important characteristics of comparative advantages in knowledge and information, scholars are more socialized than experts (Yang 2007b).

In summary, the meaning of "scholars" in this study is broader than the traditional meaning of both "scholars" and "experts." I choose to use "scholars" in a broad sense and use it to include all kinds of knowledgeable people mentioned above. The three most important reasons why I decided to choose "scholars" rather than "experts" can be listed here. First, I want to strengthen the important functions of professors and researches in social governance. Although these people are the major scholars to whom we usually refer, their functions are often limited to do research in universities, colleges, or various institutes, but their roles in governance are often negligible. Second, I want to broadly highlight the important functions of knowledge and people who possess knowledge rather than to emphasize the differences between experts and non-experts. Third, considering the situation that this study includes many cases in

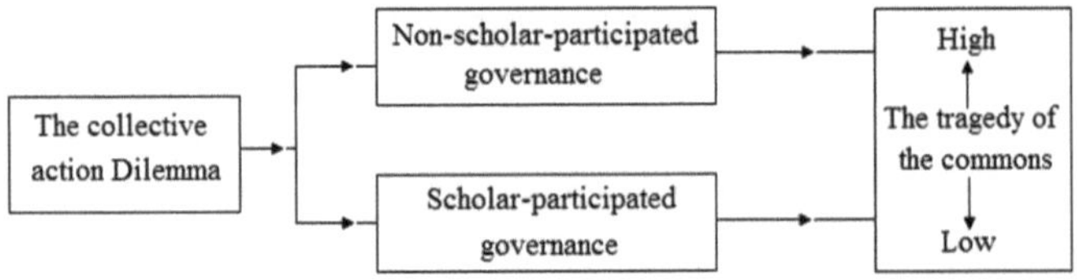

Fig. 1.2 Fundamental hypothesis of the study

China and that in ancient China scholars were broadly defined as those who had more knowledge than other people and were often generalists rather than specialists in a limited field, the word of "scholars" allows me to include some cases in ancient China and connect this study with ancient literature.

I understand that it is very difficult to use only one existing term to include all kinds of knowledgeable people mentioned above and make all readers completely agree with it. In order to simplify my analysis, however, I must choose one concise term. Thus, I hope readers can deem "scholars" in this study as an expedient and flexible term rather than a formal and rigid concept. Certainly, other words or phrases may be used, such as "experts (also in a broad sense)," "scholars and experts," "the knowledgeable," and "knowledgeable people." But all these terms should be defined broadly to include all kinds of knowledgeable individuals investigated in this study, as shown by the aforementioned term "scholars."

Using the criterion whether there is scholars' participation, the governance of the tragedy of the commons can be divided into two types: non-scholar-based governance and scholar-based governance. In regard to non-scholar-based governance, different self-governance models, and other-help governance models, except for the scholar-based model, may be used to resolve the social dilemma. Compared with scholar-based governance, the degree of the social dilemma and the tragedy of the commons in a non-scholar-based world may be relatively strong, because different types of non-scholar-based governance have their own irresolvable problems as stated above. They, however, often can be resolved or partly resolved when certain scholars are involved in them (Yang 2007b). For scholar-based governance, no matter how many kinds of social actors there are and what concrete types[5] of the scholar-based governance they are, scholars' participation may increase the capacity of the public or the public's self-governance, decrease the degree of government intervention failure, or increase the capacity of firms and the fifth-sector organizations' participation in combating the tragedy of the commons (Yang 2007a). Thus, compared with non-scholar-based governance, the degree of the tragedy of the commons may be relatively low when scholar-based governance is applied to resolve the collective action dilemma (Fig. 1.2). This is the fundamental research hypothesis of this study.

[5]Such as scholar-participated self-governance, scholar-participated government governance, scholar-participated firm governance, scholar-participated fifth-sector governance, scholar-participated mixed governance, and pure scholar-participated governance.

Fig. 1.3 Plato's cave. Adapted from Rouse (1999, p. 316)

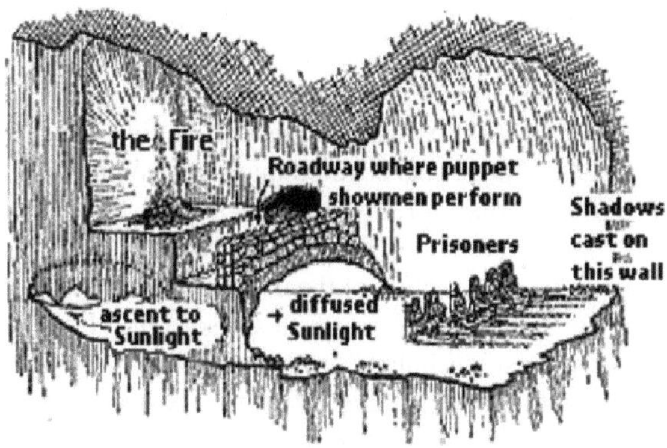

(Adapted from Rouse, 1999, p. 316)

1.2 Differences Between the Study of Experts and the Study of Scholars

Some researchers have studied the functions of experts in policymaking (e.g., Collingridge and Reeve 1986; Guston 2003; Weiss 1991) as well as the interaction between experts and the public (e.g., Barabas 2004; Brown 2001; Fischer 1999; Guston 1999; Rowe and Gammack 2004) for many years. Jasanoff (1990) stressed six types of social actors in policymaking—the public, courts, governmental agencies, different interest groups (especially industrial companies), adversary committees, and experts or expert groups. She further illustrated the tension between politics[6] and science,[7] values and facts,[8] authority (or power) and knowledge (or expertise and information) or truth,[9] science and democracy,[10] interests and legitimacy,[11] and even private interests and public interest. She concluded that the final version of policymaking is the result of a game of negotiation and competition among different social actors, especially among three major actors—the public (or citizens), experts, and officials (or agencies). Guston (1999) studied the relationships among experts, lay citizens, and government agencies in community issues through analyzing consensus

---

[6]Or policy, politicians, agencies, and governments.

[7]Or scientists and research communities. Please also see Guston (2000), Keiper (2005), Morgan and Peha (2003), Rich (2005), The Union of Concerned Scientists, March, July (2004), Weiss (1991), etc.

[8]Please also see Simon (1945) and Waldo (1948). Both authors analyzed the roles of professionals in policymaking, although they debated on other things. Different from Simon (1945), who believed that the administration is relevant to factual judgments but that policymaking is relevant to value judgments, Waldo (1948) did not think that the administration can be value-natural.

[9]Please also see Collingridge and Reeve (1986), Haas (1992), Jasanoff (2003), etc.

[10]Please also see Barabas (2004), Brown (2001), Christoforou (2003), Dahl (1989), Yaron (1990), Fischer (1999), Guston (1999), Jasanoff (2005), Nowotny (2003), Rayner (2003), Rowe and Gammack (2004), Tickner and Wright (2003), etc.

[11]Please also see Weiss (2002).

conferences. There are also a few studies which analyze experts' functions and their interactions with other social actors in combating environmental tragedy. For example, Fischer's book *Citizens, Experts, and the Environment: The Politics of Local Knowledge* (2000) illustrated these functions and interactions well. Jasanoff's studies (1990, 2005) also covered many environmental problems. Professor E. Corell's Ph. D. Dissertation *The Negotiable Desert: Expert Knowledge in the Negotiations of the Convention to Combat Desertification* (1999) even directly analyzed the experts' roles in the negotiation of the convention to combat desertification.

The study of scholars in this research is broader than the traditional study of experts. First, "scholars" in this study are a broader concept than "experts." Second, the traditional studies of experts focus on experts' function as information or knowledge producers or providers. According to the analysis of a five-actor community and scholar-based governance stated above, scholars' role in dealing with collective action dilemmas can be grouped into six types: as information providers for governments and the public; as scholar-entrepreneurs of the public; as agents of governments; as information providers, agents and scholar-entrepreneurs of firms; as information providers, agents and scholar-entrepreneurs of the fifth sector; and as pure game players (for detailed analysis, please see Chap. 2). Third, the traditional study of experts often deems experts as very passive actors who are invited to resolve social affairs by other social actors such as government and the public. Experts seldom actively take part in collective action or initiate a new collective action by themselves. The study of scholars, however, deems scholars as active actors in governments and the public. Finally, the traditional study of experts mainly focuses on the higher levels (such as the national, state or province levels) of policymaking. The study of scholars extends the scholars' research to all levels of policymaking from the local community level to the national and international levels. The study of scholars especially highlights the important functions of scholars at the community, county, and local levels. In summary, to compensate these theoretical gaps between expert and scholar studies, it is really the right time to study the roles of scholars in socioecological governance, policymaking, and collective action.

1.3 Contributions and Limitations

Collective action is central to the studies of public administration as well as other relevant fields such as political science, sociology, economics, and business management. Existing collective decision models, however, all have their own deficiencies. I have sequentially listed the major problems of the three current models—central authority, privatization, and self-governance—in the introduction part of this chapter. One common deficiency shared by all the three current models is that they neglect the important functions of knowledge, information, and scholars' participation in resolving the problem of collective action. This study seeks to identify an alternative governance model, scholar-based governance, to fill this gap. It not only supplies a new tool for us to overcome collective action dilemmas, but also helps researchers exit Plato's cave[12] (Plato 2000) created by currently simplified and polarized models.

In particular, this study focuses on practices of collective action to combat desertification. With global warming, depletion of grassland and rainforest, and global environmental deterioration, combating desertification is becoming a serious global challenge. Every year, millions of dollars and a vast amount of human efforts have been invested in this area, and satisfactory solutions are yet to be found. This study uses a new framework to study the governance model for desertification control. If successful, the research findings will help improve desertification control practices and enlighten us with a new governance model that may help us solve many other governance problems.

This study reviews the most important literature about existing models of collective action, studies of experts and scholars, and methodological approaches for studying collective action, decision making, and scholars' participation. It analyzes their achievements and limitations. Also, it exhaustively searches for relevant cases for studying scholar-based governance in the literature. It finds 43 relevant cases in total, and 23 of them are exclusively on combating desertification. More than one thousand articles, books, archives, etc., in total have been cited and reviewed in this volume. Therefore, this study is also useful to others who are interested in the literature of these problems and can help them engage in studies in these fields.

This study involves both theoretical analysis and empirical examination. Through a game theoretical analysis, its first analyzes the theoretical possibility of scholar-based governance as an alternative solution to the collective action dilemma and the four most important roles of scholars' participation. Game theory can help the author treat the problem rigorously, but the analysis with only a game theoretical method is not enough to study governance, which is often more complicated than the situation described by the simple model of game theory. In order to overcome

[12]In his enormously influential work, *The Republic*, Plato (2000) described a well-known and interesting allegory of the cave (see Fig. 1.3). Through the dialogue between Socrates and Glaucon, he said:

> If we're thinking about the effect of education—or the lack of it—on our nature, there's another comparison we can make. Picture human beings living in some sort of underground cave dwelling, with an entrance which is long, as wide as the cave, and open to the light. Here they live, from earliest childhood, with their legs and necks in chains, so that they have to stay where they are, looking only ahead of them, prevented by the chains from turning their heads. They have light from a distant fire, which is burning behind them and above them. Between the fire and the prisoners, at a higher level than them, is a path along which you must picture a low wall that has been built, like the screen which hides people when they are giving a puppet show, and above which they make the puppets appear. (Plato 2000, p. 220)

If the puppeteers pass along the wall holding up puppet, the prisoners see and hear only shadows cast on the opposite wall of the cave by the fire and echoes but not the puppeteers and puppets—the real objectives. To them, the truth would be only the shadows and echoes, but they would know nothing of the real cause of the shadows. Even if they can converse with each other, they would suppose "they were naming what was actually before them," but in fact they are really talking about a shadow.

this limitation, this study then outlines a framework for product-institutional analysis (PIA), which combines game theoretical analysis and traditional decision theory as well as generalized product analysis (GPA) and institutional analysis (for detailed analysis, please see Chap. 3). This is done not only to provide a map for the theoretical analysis of scholar-based governance (such as the factors influencing scholar-based governance and its institutional change), but also to build a fundamental tool to design an empirical research of scholar-based governance and to analyze empirical data.

A two-step research process is employed in studying scholar-based governance. In the first step, field studies are conducted in seven counties in northwest China. They include 1974 valid responses from surveys and 78 interviews, which together provide first-hand information for the author to study scholar-based governance in combating desertification. In the second step, archives and literature are studied. It involves an analysis of 43 extracted cases to examine and compare their findings to the empirical study results.

The model for collective action revealed in this study is undergirded by (1) the importance of scholars' participation and their play of four important roles, (2) seven rules governing the process of scholar-based governance, and (3) the possibility of a knowledge-driven policy outcome. Also, from a policy perspective, these findings give us some empirical support to promote scholars' participation, to transform unsuccessful or semi-successful participation into more successful participation, to design new institutions for scholars' participation in combating desertification and other collective action dilemmas, and to build a more resilient, sustainable, and robust society.

This study is not without limitations. In order such an important collective action model be supported, a large and more versatile dataset may be needed. The data used for this study are geographically concentrated (32 cases in China and 18 cases in other regions of the world) and may not be representative enough. I currently am collecting information on a large set of empirical cases to determine if the findings from this study are replicated.

1.4 The Organization of the Research

Chapter 2 examines the possibility of scholar-based governance as an alternative solution to the social dilemma using a series of game theoretical analyses. The analysis shows that scholars can help game players to resolve their collective action dilemma under certain conditions, and this positive outcome can be achieved through scholars playing a role as information providers, governmental agents, scholar-entrepreneurs, and pure game players.

Chapter 3 employs a product-institutional analysis (PIA) framework for studying scholar-based governance. This framework is not only a roadmap for theoretically analyzing scholar-based governance in the whole study but also a tool that guides the data collection and analyses in this research.

Chapter 4 focuses on the analyses of empirical evidence. It found that scholars played an important role in combating desertification, but counties in which scholar participation was high did not mean that land amelioration was also high. I further analyze the mechanism of scholars' participation through studying 13 related factors that differentiated effectiveness in combating desertification. It is found that the higher levels of the presence of these factors, the more successful the combating desertification effort is.

Chapter 5 outlines the seven design principles for ensuring successful scholars' participation in collective action through further analysis of the factors that affected the outcome of scholars' participation in combating desertification. It also tested the generalizability of these seven principles by examining the study results from the 43 cases in the second stage of this study. It concludes that the better these rules are applied, the more successful the combat of desertification effort will be. The seven design principles are (1) sustained participation of field-based scholars, (2) federal organizational structure and concrete and stratified purposes, (3) democratic and collaborative management with strictly implemented mechanisms of awards and sanctions, (4) steady local scholar-entrepreneurship, (5) realization of expected benefits, (5) the experiment-extension method, and (6) reliable external support.

Chapter 6 evaluates the institutional performance of the institutions in cases on the basis of a synthetical analysis of governance results, resilience, and sustainability. Then, I explore how to transform fragile and failed institutional arrangements into robust ones through an analysis of three levels of institutional change and knowledge-driven or intellect-driven institutional change. I further argued that a knowledge-driven society may be a third model of a humanitarian society, in addition to Lindblom's intellectually guided society and preference-guided or volition-guided society, especially in China.

References

Adger, W. N. (2000). Social and ecological resilience: Are they related? *Progress in Human Geography*, 24 (3), 347–364.

Aghion, P., & Howitt, P. (1992). A model of growth through creative destruction. *Econometrica*, 60(2), 323–351.

Anderies, J. M., Janssen, M. A. & Walker, B. N. (2002). Grazing management, resilience and the dynamics of a fire driven rangeland system. Ecosytems, 5(1), 23–44.

Anderies, J. M., M. A. Janssen, & E. Ostrom. (2004). A framework to analyze the robustness of social-ecological systems from an institutional perspective. *Ecology and Society*, 9(1): 18.

Bacon, F. (1597). *Meditationes sacrae*. Londres: De Hæresibus.

Barabas, J. (2004). How deliberation affects public opinion. *American Political Science Review*, 98(4), 687–701.

Becker, G. S. (1993). *Human capital: A theoretical and empirical analysis, with special reference to education*. Chicago: University of Chicago Press.

Brown, M. (2001). The civic shaping of technology: California's electric vehicle program. *Science, Technology & Human Values*, 26(1), 56–81.

Brusoni, B., Prencipe, A., Pavitt, K. (2001). Knowledge specialization, organization coupling, and the boundaries of the firm: Why do firms know more than they make? *Administrative Science Quarterly*, 46(4), 597–621.
Brynjolfsson, E., Hitt, L., & Yang, S. (2002). Intangible Assets: Computers and Organizational Capital. *Brookings Papers on Economic Activity*, 1(1), 137–198.
Buchanan, J. M. (1965). The economic theory of club. *Economica*, 32(1), 1–14.
Carpenter, S. R., Walker, B. H., Anderies, M. A., & Abel. N. A. (2001). From metaphor to measurement: resilience of what to what? *Ecosystems*, 4(8), 756–781.
Carroll, M. C., & Stanfield, J. R. (2003). Social capital, Karl Polanyi, and American social institutional economics. *Journal of Economic Issues*, 37(2), 397–404.
Castells, M. (2002). The Internet Galaxy: Reflections on the Internet, Business and Society. London: Oxford University Press.
Christensen, L. R., & Green, W. H. (1976). Economies of scale in US electric power generation source. *The Journal of Political Economy*, 84(4), 655–676.
Christoforou, T. (2003). The precautionary principle and democratizing expertise: A European legal perspective. *Science and Public Policy*, 30 (3), 205–211.
Coase, R. 1960. The problem of social costs. *Journal of Law and Economics*, 3(1), 1–44.
Coase, R. 1974. The lighthouse in economics. *Journal of Law and Economics*, 17(2), 357–376.
Coleman, J. C. (1990). *Foundations of social theory*. Cambridge, Mass.: Harvard University Press.
Collingridge, D., & Reeve, C. (1986). *Science speaks to power: The role of experts in policy making*. New York: St Martin's Press.
Confucian analects. (1900). In *Zhushi jiaozheng huaying sishu* (*Si shu*) (English & Chinese) (Trans. by J. Legge) (pp. 1–218). Shanghai: Shangwu Yinshuguan.
Corell, E. (1999). *The Negotiable Desert: Expert Knowledge in the Negotiations of the Convention to Combat Desertification*. Unpublished doctoral dissertation, Linköping Studies in Arts and Science no. 191, Linköping University, Sweden.
Costanza, R., & Daly, H. E., (1992). Natural capital and sustainable development. *Conservation Biology*, 6(1), 37–46.
Crona, B., & Bodin, Ö. 2006. What you know is who you know? Communication patterns among resource users as a prerequisite for co-management. *Ecology and Society*, 11(2), 7.
Cumming, G. S., Barnes, G., Perez, S., Schmink, M., Sieving, K., Southworth, J., Binford, M., Holt, R. D., Stickler, C., & Van Holt, T. (2005). An exploratory framework for the empirical measurement of resilience. *Ecosystems*, 8(8), 975–987.
Dahl, R. A. (1989). *Democracy and its critics*. New Haven and London: Yale University Press.
Dawes, R. M. (1973). The commons dilemma game: an N-person mixed-motive game with a dominating strategy for defection. *ORI Research Bulletin*, 13(2), 1–12.
Dawes, R. M. (1975). Formal models of dilemmas in social decision making. In M. F. Kaplan & S. Schwartz (Eds.), *Human judgment and decision process: Formal and mathematical approaches* (pp. 87–108). New York: Academic Press.
Demsetz, H. (1970). The private production of public goods. *Journal of Law and Economics*, 10(13), 293–306.
Dobson, A. (1999). *Justice and the environment: Conceptions of environmental sustainability and dimensions of social justice*. Oxford: Oxford University Press.
Durkheim, E. (1960). *The division of labor in society* (Trans. by G. Simpson). New York: The Free Press.
Eden, S. (1996). Public participation in environmental policy: Considering scientific, counter-scientific and non-scientific contributions. *Public Understanding of Science*, 5(3), 183–204.
El Serafy, S., (1989). The proper calculation of income from depletable natural resources. In Y. J. Ahmad, S. El Serafy, E. Lutz (Eds.), *Environmental Accounting for Sustainable Development* (pp. 10–18). Washington, DC: World Bank.
El Serafy, S., (1991). The environment as capital. In:R. Costanza (Ed.), *Ecological Economics: The Science and Management of Sustainability* (pp. 168-175). New York: Columbia University Press.

Fiksel, J. 2006. Sustainability and resilience—toward a systems approach. *Sustainability: Science, Practice, & Policy (SSPP)*, 2(2), 1–8.
Fischer, F. 1999. Technological deliberation in a democratic society: The case for participatory inquiry. *Science and Public Policy*, 26(5), 294–302.
Fischer, F. (2000). *Citizens, experts, and the environment: The politics of local knowledge*. Durham and London: Duke University Press.
Folke, C., Carpenter, S., Elmqvist, T., Gunderson, L. Holling, C. S., & Walker, B. (2002). Resilience and sustainable development: Building adaptive capacity in a world of transformations. *Ambio*, 31(5), 437–440.
Fraser, E. D. G., Mabee, W., & Slaymaker, O. (2003). Mutual vulnerability, mutual dependence: The reflexive relation between human society and the environment. *Global Environmental Change*, 13(2), 137–144.
Frenzel, L. E., Jr. (1987). *Understanding expert systems*. Indianapolis: Howard W. Sams & Company.
Fukuyama, F. (1999). *The great disruption: Human nature and the reconstitution of social order*. London: Profile Books.
Gallopin, G. C. (2006). Linkages between vulnerability, resilience, and adaptive capacity. *Global Environmental Change*, 19(3), 293–303.
Gordon, H. S. (1954). The economic theory of a common-property resource: the fishery. *Journal of Political Economy*, 62(4), 124–42.
Gottlieb, R. (2001). *Environmentalism unbound: Exploring new pathways for change*. Cambridge, MA: The MIT Press.
Guston, D. H. (1999). Evaluating the first U.S. consensus conference: The impact of the 'citizens' panel on telecommunications and the future of democracy. *Science, Technology and Human Values*, 24(4), 451–82.
Guston, D. H. (2000). *Between politics and science: Assuring the integrity and productivity of research*. Cambridge, UK, New York, NY: Cambridge University Press.
Guston, D. H. (2003). Principal-agent theory and the structure of science policy, revisited: 'Science in policy' and the US report on carcinogens. *Science and Public Policy*, 30 (5), 347–357.
Haas, P. M. (1992). Introduction: Epistemic communities and international policy coordination. *International Organization*, 46(1), 1–36.
Hardin, G. (1968). The tragedy of the commons. *Science*, 162(3859), 1243–1248.
Hardin, G. (1978). Political requirements for preserving our common heritage. In H. P. Bokaw (Ed.), *Wildlife and America* (pp. 310–317). Washington, D.C.: Council on Environmental Quality.
Harte, M. J. (1995). Ecology, sustainability, and environment as capital. *Ecological Economics*, 15(2), 157–164.
Hayek, F. A. (1945). The use of knowledge in society. *The American Economic Review*, 35(4), 519–530.
Hobbes, T. (1991). *Leviathan*. Cambridge, UK: Cambridge University Press.
Hume, D. (1978). *A treatise of human nature* (2nd ed.). Oxford: Oxford University Press.
Janssen, M. A., & Ahn, T. K. (2006). Learning, signaling, and social preferences in public-good games. Ecology and Society, 11(2), 21.
Jasanoff, S. (1990). The fifth branch: science advisers as policymakers. London: Harvard University Press.
Jasanoff, S. 2003. (No?) Accounting for Expertise. *Science and Public Policy*, 30(3):157–162.
Jasanoff, S. (2005). *Design on nature: Science and democracy in Europe and the United States*. Princeton and Oxford: Princeton University Press.
Kamieniecki, S., Gonzalez, G., & Vos, R. (Eds). (1997). *Flashpoints in environmental policymaking*. Albany, NY: SUNY Press,
Keiper, A. (2005). Science and congress. In *The New Atlantis-A Journal of Technology and Society*. http://www.thenewatlantis.com/archive/7/keiperprint.htm.
Kolstad, C. D. (2000). *Environmental economics*. Oxford: Oxford University Press.

Krugman, P. (1980). Scale economies, production differentiation, and the pattern of trade. *The American Economic Review*, 70(5), 950–959.
Lan, Z., & Scott, C. R. (1996). The relative importance of computer-mediated information versus conventional non-computer mediated information in public managerial decision making. *Information Resources Management Journal*, 9 (1), 27–37.
Lichbach, M. I. (1996). *The cooperator's dilemma*. Ann Arbor, MI: University of Michigan Press.
Longman Dictionary of Contemporary English. (2008). Retrieved April 28, 2009, from http://www.ldoceonline.com/.
Lüshi Chunqiu. (2000). Retrieved November 3, 2006, from http://chinese.dsturgeon.net/text.pl?node=23107&if=gb.
McCool, D. C. (1995). Public policy theories, models, and concepts: an anthology. New Jersey: Prentice Hall, Inc.
Mincer, J. (1958). Investment in Human Capital and Personal Income Distribution. *Journal of Political Economy*, 66(4), 281–302.
Morgan, M. G., & Peha, J. M. (2003). *Science and technology advice for congress*. Washington, DC: Resources for the Future.
Murray, J. D., & White, R. W. (1983). Economies of scale and economies of scope in multiproduct financial institutions: *A study of British Columbia Credit Unions*. *The Journal of Finance*, 38(3), 887–902.
Nowotny, H. (2003). Democratising expertise and socially robust knowledge. *Science and Public Policy*, 30 (3): 151–156.
Olson, M. (1971). *The logic of collective action: public goods and the theory of groups*. Cambridge, MA: Harvard University Press.
Ostrom, E. (1990). *Governing the commons the evolution of institutions for collective action*. New York: Cambridge University Press.
Ostrom, E. (1998). A behavioral approach to the rational choice theory of collective action: Presidential address, American political science association, 1997. *American Political Science Review*, 92(1): 1–22.
Ostrom, E. (2000). The danger of self-evident truths. *PS: Political Science and Politics*, 33(1), 33–44.
Ostrom, E, and Ahn, T. K. (Eds.). (2003). *Foundations of social capital*. Cheltenham, UK: Edward Elgar Publishing Ltd.
Pareto, V. (1935). *The mind and society*. New York: Harcourt, Brace.
Pasour, E. C. Jr. (1981). The free rider as a basis for government intervention. *The Journal of Libertarian Studies*, 5(4), 453–464.
Pearce, D., & Turner, R. K., (1990). *Economics of natural resources and the environment*. London: Harvester Wheatsheaf.
Pigou, A.C. (1932). The Economics of Welfare, 4th ed. London: Macmillan & Co.
Plato. (2000). *The republic* (Ed. by G. R. F. Ferrari and, & trans. by T. Griffith). Cambridge: Cambridge University Press.
Pooley R, & Wilcox, P. (2000). Distributing decision making using Java Simulation across the World Wide Web. *The Journal of the Operational Research Society*, 51 (4), 395–404.
Prescott, Edward C, & Visscher, Michael. (1980). Organization capital. *Journal of Political Economy*, 88(3): 446–61.
Putnam, R. D. (2000). *Bowling alone: The collapse and revival of american community*. New York: Simon & Schuster Paperbacks.
Rayner, S. (2003). Democracy in the age of assessment: Reflections on the roles of expertise and democracy in public-sector decision making. *Science and Public Policy*, 30 (3), 163–170.
Rich, A. (2005). War of ideas: Why mainstream and liberal foundations and the think tanks they support are losing in the war of ideas in American politics. *Stanford Social Innovation Review*, 18–25 (http://www.ssireview.com/pdf/2005SP_feature_rich.pdf).
Rifkin, W D., & B. Martin. (1997). Negotiating expert status: Who gets taken seriously. *IEEE Technology and Society Magazine*, 16(1), 30–39.

Romer, P. (1987). Growth based on increasing returns due to specialization. *American Economic Review*, 77(2), 56–62.
Romer, P. (1990) Endogenous technological change. *Journal of Political Economy*, 98(5), 71–102.
Rouse, W. H. D. (Trans. & Ed.). (1999). *Great dialogues of Plato: Complete texts of the Republic, Apology, Crito Phaido, Ion, and Meno* (Vol. 1). New York: Signet Classics.
Rowe, G., & Gammack, J. G. (2004). Promise and perils of electronic public engagement. *Science and Public Policy*, 31(1), 39–54.
Sabatier, P. A. (1992). Review: [Untitled]. *The American Political Science Review*, 86 (1), 248–249.
Samuelson, P. A. 1954. The pure theory of public expenditure. *Review of Economics and Statistics*, 36(4), 387–389.
Samuelson, P.A. (1955). Diagrammatic exposition of a theory of public expenditure. Review of Economics and Statistics, 37, 350–356.
Savas, E. S. (2000). *Privatization and public-private partnerships*. New York: Chatham House Publishers/ Seven Bridges Press.
Schelling, T. C. (1978). *Micromotives & macrobehavior*. New York: W. W. Norton.
Schultz, P. T. (1993). The Role of Education and Human Capital in Economic Development: An Empirical Assessment. In H. Siebert (Ed.), *Economic growth in the world economy: Symposium 1992* (pp. 145–164). Tübingen: JCB Mohr.
Shi Jing (*The book of odes*). (2003). Retrieved October 7, 2006, from http://www.tianyabook.com/shijing/index.htm.
Simon, H. A. (1945). Administrative behavior (1st ed.). New York: Free Press.
Sjőberg, L. (1998). Risk perception: Experts and the public. *European Psychologist*, 3(1), 1–12.
Smith, A. (1965). *An inquiry into the nature and causes of the wealth of nations*. New York: Modern Library.
Smith, R. J. (1981). *Resolving the tragedy of the commons by creating private property rights in wildlife. CATO Journal*, 1(2), 439–68.
Teece, D. J. (1980). Economies of scope and the scope of the enterprise. *Journal of Economic Behavior and Organization*, 1(3), 223–247.
Tickner, J. A., & Wright, S. (2003). The precautionary principle and democratizing expertise: A US perspective. *Science and Public Policy*, 30(3), 213–218.
Toleubayev, K., Jansen, K., & Huis, A. Van. (2007). Locust control in transition: the loss and reinvention of collective action in post-Soviet Kazakhstan. *Ecology and Society*, 12(2), 38.
Tomer, J. F. (1987). *Organizational capital: The path to higher productivity and well-being*. New York: Praeger Publishers.
Tomer, J. F. (1998).Organizational capital and joining-up: Linking the individual to the organization and to society. *Human Relations*, 51(6), 825–846.
Tompkins, E. L., & Adger, W. N. (2004). Does adaptive management of natural resources enhance resilience to climate change? *Ecology and Society*, 9(2), 10.
Tucker, A. W. (1983). The mathematics of tucker: a sampler. *The Two-Year College Mathematics Journal*, 14(3): 228–232.
Union of Concerned Scientists. July 2004. *Scientific integrity in policymaking: further investigation of the Bush Administration's misuse of science*. Cambridge: Union of Concerned Scientists. Retrieved November 11, 2006, from http://www.ucsusa.org/assets/documents/scientific_integrity/scientific_integrity_in_policy_making_july_2004_1.pdf.
Waldo, D. (1948). *The administrative state: A study of the political theory of American public administration*. New York: Ronald Press.
Weiss, C. H. (1991). *Organizations for policy analysis: Helping government think*. Newbury Park: Sage Publications.
Weiss, C. (2002). Scientific uncertainty in advising and advocacy. *Technology in Society*, 24(4), 375–386.
Wu, J. (2006). Landscape ecology, cross-disciplinarity, and sustainability science. *Landscape Ecology*, 21(1), 1–4.

Yang, L. (2007a). Scholar-based governance: A fourth model to collective action. *Chinese Public Administration*, 259 (1), 96–103 (In Chinese with English abstract).
Yang, L. (2007b). Building a multi-collaborative community governance system to resolve the dilemma of collective action—a framework of "product-institutional" analysis (PIA). *Journal of Public Management*, 4(2), 6–23 (In Chinese with English abstract).
Yaron, E. (1990). *The Descent of Icarus: Science and the Transformation of Contemporary Democracy*. Cambridge, MA: Harvard University Press.
Young, O. R., Berkhout, F., Gallopin, G. C., Janssen, M. A., Ostrom, E., & van der Leeuw, S. (2006). The globalization of socio-ecological systems: An agenda for scientific research. *Global Environmental Change*, 16(3), 304–316.

Chapter 2
Four Major Roles of Scholars' Participation: A Game Theoretical Analysis

Tsze-hsiâ (also translated as Zi Xia) said, 'Even in inferior studies and employments there is something worth being looked at; but if it be attempted to carry them out to what is remote, there is a danger of their proving inapplicable. Therefore, the superior man does not practice them. (In Chinese: 子夏: "虽小道, 必有可观者焉; 致远恐泥, 是以君子不为也。" (《论语·子张第十九》).)

—CHAP. IV., BOOK XIX. TSZE-CHANG (Zi Zhang), *Confucian analects* (1900, pp. 204–205)

The purpose of this chapter is to use game theoretical analysis to demonstrate why a new "Scholar-based Governance Model" as an alternative or complementary solution to the collective action dilemma in SESs is necessary and possible. In the following sections, I first review a herder game using a common grazing meadow and analyze the underlying assumptions of this research. Then, I explore scholars' four roles in collective action as information providers, agents of government, entrepreneurs, and pure game players in a three-person game.

2.1 A Herder Game and the Assumptions of Research

Ostrom (1990) studied a game of herders sharing a common grazing meadow—a simple example of SES. Suppose that the largest number of grazing animals the meadow can support for an entire season is L, and the "cooperate" strategy is thought

Some of the contents in this chapter was originally published in: Yang and Wu (2009). Reprinted by permission of Elsevier and the authors.

L. Yang, *Knowledge-Driven Governance*,
https://doi.org/10.1007/978-981-13-2910-4_2

of as grazing $L/2$ animals for each herder. That is, a cooperate strategy is a strategy where each herder considers both his and his opponents' benefits in order to move closer to optimal outcomes to gain his cooperators' dividend. Also, a defect strategy is a strategy where each herder tries to graze as many animals as he thinks he can sell at a profit regardless of his opponent's benefits. Suppose both of the herders obtain 10 units of profit when they take the cooperate strategy, whereas they obtain zero profit if they both choose the defect strategy. If one takes the cooperate strategy and the other takes the defect strategy, the "defector" obtains 11 units of profit, and the "sucker" obtains -1 (see the payoff matrix for Game 1 in Fig. 2.1). If each herder chooses strategies independently and has no capacity to engage in a binding contract, both of them will choose their dominant strategy (in the sense that the player is always better-off when choosing this strategy) to defect and both will obtain zero profit. This is the Hardin herder game with the structure of a prisoner's dilemma game, which is a "nonzero-sum" noncooperative game where each game player has a dominant strategy to defect no matter what the other player chooses. Supposing that the central authority has complete information about the carrying capacity of the meadow (L), can always tell which herder is cheating, and has monitoring capabilities, sanctioning determination, and zero cost of administration, the solution to Game 2 becomes (cooperate, cooperate) when the central authority decides to impose a penalty of 2 profit units on the defector (see Fig. 2.1, Game 2). Without complete information, however, a central agency would find it difficult to set the grazing intensity or fine to an optimal level.

Now let's assume that the central government has complete information on the carrying capacity of the meadow (L) but incomplete information on the particular actions of the herders, that the probability of the central agency punishing defections is y, and that the probability of the central agency falsely punishing cooperative actions is x (see Fig. 2.1, Game 3). Several errors made by the central agency include "setting the carrying capacity or the fine too high or too low," "sanctioning herders who cooperate," or "not sanctioning defectors" (Ostrom 1900, p. 10). If $x = 0$ and $y = 1$, Game 3 becomes Game 2. If the central agency does not have complete information about the herders' actions, say $x = 0.3$, $y = 0.7$, Game 3 becomes Game 4 (see Fig. 2.1, Game 4), which makes the herders face a prisoner's dilemma game again. Furthermore, the equilibrium outcome $(-1.4, 1.4)$ for Game 4 is even lower than (1, 1) in Game 1. The method proposed by Ostrom (1990) is one of self-governance. Under the players' self-governance, the players can negotiate various strategies for sharing the meadow and the costs of implementing the agreement, even though a binding contract is still needed to be enforced by an external actor. When both Player 1 and Player 2 agree upon and conform to a contract, the solution to the fifth game becomes (10−e/2, 10−e/2) (here, e is the cost of implementing an agreement). Otherwise, they repeat Game 1 (see Fig. 2.1, Game 5).

The research question of this chapter is how scholars' participation can improve people's capability to resolve the problems analyzed by the above five games and also resolve some problems neglected by them. Our purpose, however, is not to directly supply strict mathematical solutions to the prisoner's dilemma, and the findings we report here focus on how scholars' participation can change the original situation

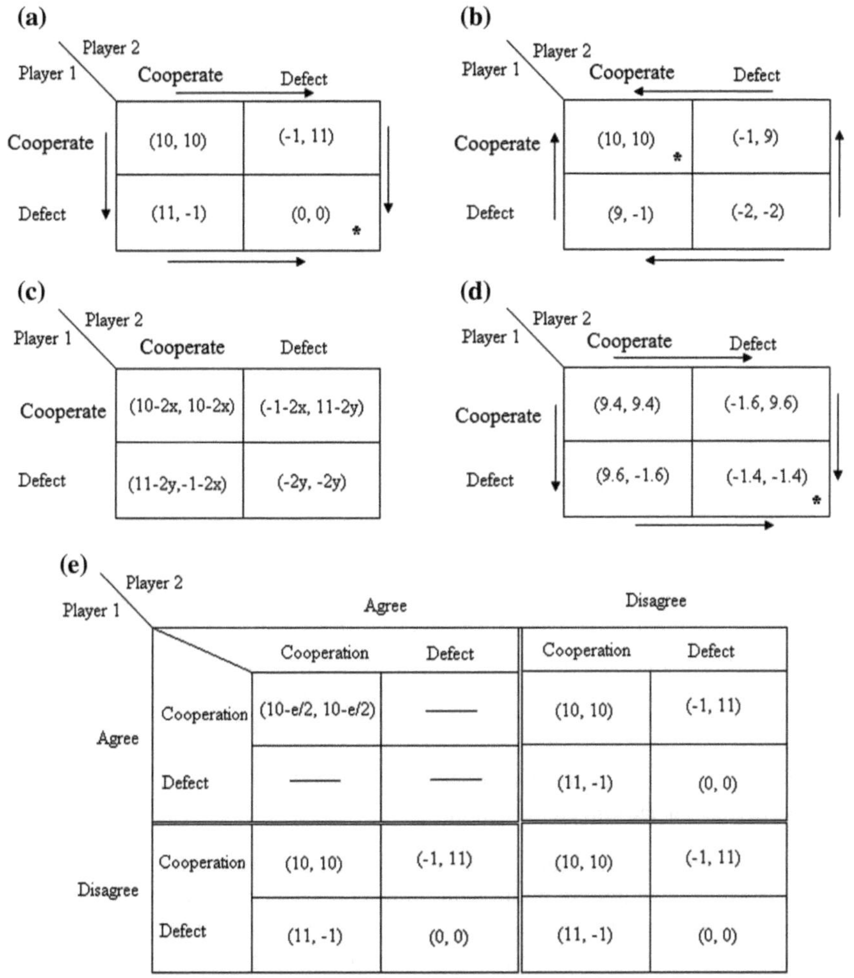

Fig. 2.1 Five games in Ostrom (1990). **a** Game 1—the Hardin herder game. **b** Game 2—the central-authority game with complete information. **c** Game 3—the central-authority game with incomplete information. **d** Game 4—an example of the central-authority game with incomplete information. **e** Game 5—self-financed contract-enforcement game. While all these game are illustrated by Ostrom using extensive forms, we recast them as normal forms

described by the Hardin herder game and the central-authority game with incomplete information and then make cooperative behavior possible. Furthermore, because the situations described by a coordination game and a dictator game are also as important as the prisoner's dilemma in understanding collective action in SESs (reasons will be justified latter), the impact of scholars' participation on these two games will also be considered when necessary. To simplify the analysis, let's first consider a SES with only herders, governments, and scholars, be they resource users or public infrastructure providers (Anderies et al. 2004). There are no firms and the fifth sector. I assume that all these three types of actors are profit maximizers. Herders seek to maximize their returns from herding as an economic activity, governments seek to maximize revenues, and scholars seek to maximize their returns economically or socially (e.g., sometimes they seek reputation rather than economic gains). Also, both herders and governments are assumed to be players with incomplete information. Particularly, incomplete information on actions of herders is also taken into account. Then, from the perspective of a three-person game, in scholar-based governance the scholars can play at least four kinds of roles: as information providers for the herders or the governments, as agents of the governments, as entrepreneurs or leaders for the herders, and as pure game players who pursue their own interests. And thus the game analysis can be fitted into the study of these four roles.

2.2 The Scholar as an Information Provider

Let's first consider a two-player game. Suppose that there is a scholar who has the information about the carrying capacity of the meadow, the particular actions of the two players, and is willing to freely share this information with the public (or sell it as private information) to both of the two players. The costs of this information for Players 1 and 2 are denoted by c_i ($i = 1, 2$), and its values are q_i ($-1 \leq q_i \leq 1, i = 1, 2$). In this case, four situations may be considered as follows.

Situation 1a: The Scholar Has the Information About the Players' Particular Actions When There is the Hardin Herder Game

Let's reconsider Game 1. Suppose that the carrying capacity of a meadow is the common knowledge among both of the players, that the scholar freely shares the information about their opponent's particular actions with each of the two players, and that both players simply think that their opponent has probability p_i ($0 \leq p_i \leq 1, i = 1, 2$) to take the cooperate strategy and probability $1 - p_i$ to take the defect strategy before getting the scholar's information. To the original payoff matrix described by Game 1 in Fig. 2.1, its equilibrium solution is (defect, defect). That is, the two players always take the defect strategy no matter how much p_i is. When the scholar sells the information to the players at cost c_i ($i = 1, 2$) and value q_i ($-1 \leq q_i \leq 1, i = 1, 2$) ($q_i$ as the values of the information for Players 1 and 2 can be deemed as the additional amount for the probability of taking the cooperate strategy by the player's opponent), the equilibrium solution to this game is still (defect, defect) with payoffs $(-c_1, -c_2)$. That is, although both of the players in informed games change (increase or decrease) their expected utility before playing this game, the real payoffs they get are still kept at (0, 0) or become even worse as being $(-c_1, -c_2)$. Thus in the Hardin herder game, both of the players, as the rational individuals, will not buy the information from the scholar. The scholar, then, will not have any actual influence on the players' actions, if he only has the information about the players' particular actions.

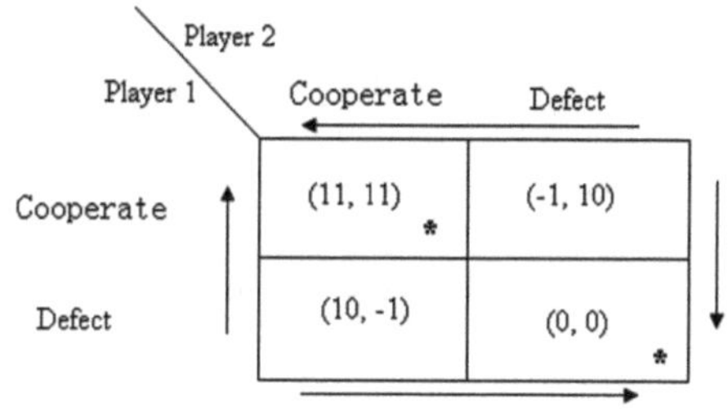

Fig. 2.2 Game 6—a coordination game

Situation 1b: The Scholar Has the Information About the Players' Particular Actions When There is a Coordination Game

In the Hardin herder game, even if both of the players take the cooperate strategy, their payoffs are (10, 10), but if one cooperates and the other defects, the defector will get 11. Let's consider such a game, when both of the players take the cooperate strategy, their payoffs are (11, 11). If one of them takes the defect strategy, then he only gets 10, while the payoff of the other as a cooperator is kept the same (Game 6 in Fig. 2.2).

If neither of the players knows the information about their opponent's particular actions, there are two equilibrium solutions: (cooperate, cooperate) with payoffs (11, 11) and (defect, defect) with payoffs (0, 0). If both of them think that their opponent has probability p_i ($0 \leq p_i \leq 1, i = 1, 2$) to take the cooperate strategy and probability $1 - p_i$ to take the defect strategy, then:

(1) If both p_2 and p_1 are larger than 0.5, then the game solution is (cooperate, cooperate) with payoffs (11, 11).
(2) If both p_2 and p_1 are less than 0.5, then the game solution is (defect, defect) with payoffs (0, 0).
(3) If p_2 is larger than 0.5, but p_1 is less than 0.5, then the game solution is (cooperate, defect) with payoffs (-1, 10).
(4) If p_2 is less than 0.5, but p_1 is large than 0.5, then the game solution is (defect, cooperate) with payoffs (10, -1).

This is a coordination game which has two pure strategy Nash equilibria and a mixed-strategy equilibrium. In such a game, cooperation might fail, because each player has an alternative which is safer. Because the player will fail to coordinate with nonzero probability, the mixed Nash equilibrium is also Pareto dominated by the two pure Nash equilibria. That is, unlike the pure Nash equilibria, the mixed equilibrium is not an evolutionarily stable strategy. This shows that the collective action dilemma still exists under a situation involving coordination games and solving these games is as important as solving the Hardin herder game to combat the collective action dilemma in SESs. Although with communication or repeated play, this type of coordination game can be solved. Here, let's consider how the scholar's information can also change the game solutions. When the scholar sells the information to the players with cost c_i and value q_i, several conclusions are possible:

(1) If both $p_2 + q_2$ and $p_1 + q_1$ are larger than 0.5, then the game solution is (cooperate, cooperate) with payoffs $(11 - c_1, 11 - c_2)$.
(2) If both $p_2 + q_2$ and $p_1 + q_1$ are less than 0.5, then the game solution is (defect, defect) with payoffs $(-c_1, -c_2)$.
(3) If $p_2 + q_2$ is larger than 0.5, but $p_1 + q_1$ is less than 0.5, then the game solution is (cooperate, defect) with payoffs $(-1 - c_1, 10 - c_2)$.
(4) If $p_2 + q_2$ is less than 0.5, but $p_1 + q_1$ is large than 0.5, then the game solution is (defect, cooperate) with payoffs $(10 - c_1, -1 - c_2)$.

In summary, the two players' particular actions and payoffs have been strongly influenced by the scholar's information. This is true especially when $q_2 \geq 0.5 - p_2$, $q_1 \geq 0.5 - p_1$, both of the players take the cooperate strategy, and the game solution is (cooperate, cooperate) with payoffs $(11 - c_1, 11 - c_2)$. When $q_2 \geq 0.5 - p_2$, $q_1 \geq 0.5 - p_1$, $c_1 \leq 11\ (p_2 + q_2) -0.5$, and $c_2 \leq 11\ (p_1 + q_1) -0.5$, both players will buy the scholar's information, which will then drive them to take the (cooperate, cooperate) strategy.

Situation 1c: The Scholar Has Other Valuable Information When There is the Hardin Herder Game

In Situations 1a and 2b, the scholar only has information about the players' particular actions. Now, let's consider the situation where the scholar has no information about the players' particular actions but has other types of valuable information, such as the carrying capability of a meadow, the natural characteristics of the meadow, the technology to properly use the meadow, or the knowledge to organize the two players for cooperation. All these kinds of information or knowledge would influence the payoffs the players get from using the meadow. That is, now the values of scholars' information or knowledge are directly deemed as the players' increased or decreased payoffs rather than the additional amount for the probability of taking the cooperate strategy by the player's opponent. For example, due to the scholar's knowledge of the technology to properly use the meadow, the payoffs for Players 1 and 2 may become (12, 12) in Game 1, when both of them take the cooperate strategy. Here, the values of the scholar's information for both Players 1 and 2 are 2.

Let's use $v_{1-\mathrm{cc}}$, $v_{1-\mathrm{cd}}$, $v_{1-\mathrm{dc}}$, and $v_{1-\mathrm{dd}}$ to represent the values for Player 1 when the players take (cooperate, cooperate), (cooperate, defect), (defect, cooperate), and (defect, defect) strategies, respectively, and $v_{2-\mathrm{cc}}$, $v_{2-\mathrm{cd}}$, $v_{2-\mathrm{dc}}$, and $v_{2-\mathrm{dd}}$ for Player 2, respectively. The costs of this information for Players 1 and 2 are still denoted by c_i ($i = 1, 2$). As a prisoner's dilemma game, in a situation without the scholar, the only solution to Game 1 becomes (defect, defect) with payoffs (0, 0). When the scholar's information is considered, a new payoff matrix is derived (Fig. 2.3a, Game 7).

Now, the payoffs for (cooperate, cooperate) are $(10 + v_{1-\mathrm{cc}} - c_1, 10 + v_{2-\mathrm{cc}} - c_2)$, for (cooperate defect) are $(-1 + v_{1-\mathrm{cd}} - c_1, 11 + v_{2-\mathrm{cd}} - c_2)$, for (defect, cooperate) are $(11 + v_{1-\mathrm{dc}} - c_1, -1 + v_{2-\mathrm{dc}} - c_2)$, and for (defect, defect) are $(v_{1-\mathrm{dd}} - c_1, v_{2-\mathrm{dd}} - c_2)$. Under the four settings as described in Appendix A, the unique solution to this game is (cooperate, cooperate) with payoffs $(10 + v_{1-\mathrm{cc}} - c_1, 10 + v_{2-\mathrm{cc}} - c_2)$

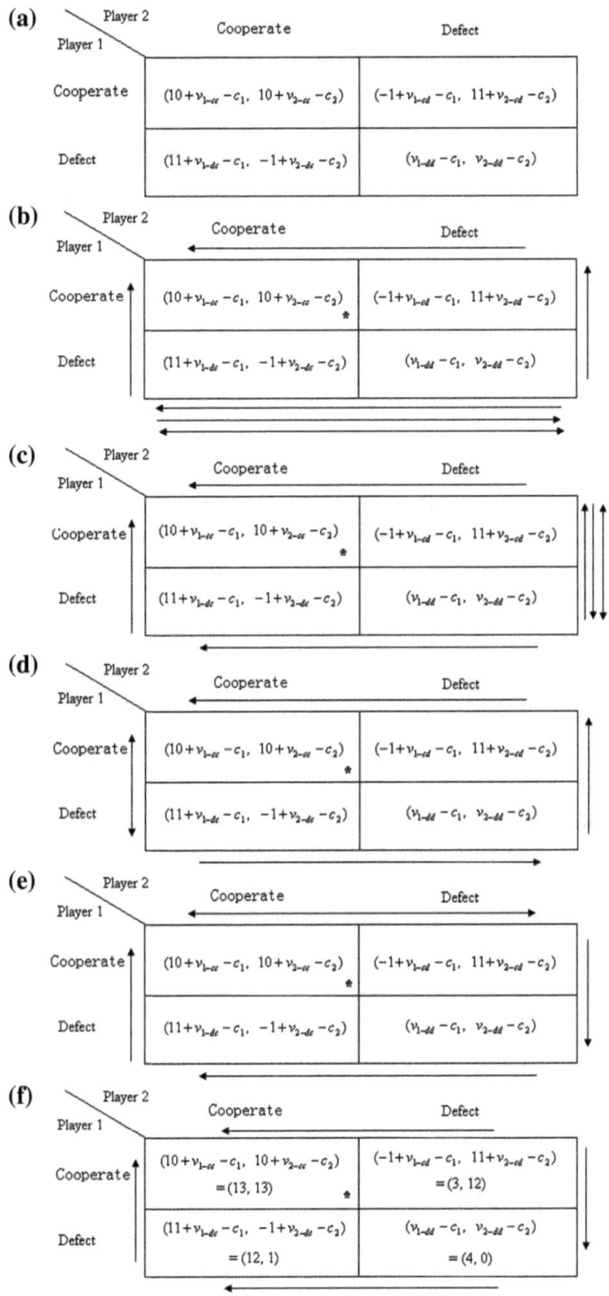

Fig. 2.3 Game 7—the Hardin game with the scholar's information. **a** The Hardin game with the scholar's information. **b** Cooperation between the two players in the Hardin game with the scholar's information—Setting 1. **c** Cooperation between the two players in the Hardin game with the scholar's information—Setting 2. **d** Cooperation between the two players in the Hardin game with the scholar's information—Setting 3. **e** Cooperation between the two players in the Hardin game with the scholar's information—Setting 4. **f** An example of Setting 1

(see Fig. 2.3b, e). For instance, as to Setting 1, if $c_1 = 1$, $c_2 = 1$, $v_{1-cc} = 4 > v_{1-dc} = 2$, $v_{2-cc} = 4 > v_{2-cd} = 2$, $v_{1-cd} = 5 > v_{1-dd} = 3$, and $v_{2-dc} = 3 > v_{2-dd} = 1$, a new game is shown in Fig. 2.3f. The unique equilibrium solution to this game is (cooperate, cooperate) with payoffs (13, 13). Thus, even under the situation as described in Game 1, scholar-based governance can create some conditions for both self-interested players to take the cooperate strategy.

Situation 1d: The Scholar Provides Information for the Government When There is the Central-Authority Game with Incomplete Information

Let's consider a situation where the scholar provides information for the government. In Game 2, the central agency has complete information about the carrying capability of the meadow and the particular actions of the herders; therefore, it can correctly punish defectors. In Game 3 and Game 4, where the central agency only has complete information about the carrying capacity of the meadow but no complete information about the particular actions of the herders, the intervention by the central agency will worsen the situation, unless the probability of correctly imposing sanctions is greater than 0.75. Suppose that the scholar's information helps the central agency improve or reduce its probability to correctly punish defections with q_d ($-1 \leq q_d \leq 1$, and $0 \leq y + q_d \leq$) and its probability to punish cooperative actions with q_c ($-1 \leq q_c \leq 1$, and $0 \leq x + q_c \leq$) and improves the payoffs for both of the players with v. A new payoff matrix is then shown in Fig. 2.4, Game 8, if we use c to denote the cost of this information. For this game, under Settings 1 and 4 as described in Appendix B, its unique solution is (cooperate, cooperate) (see Fig. 2.4b, e). For example, for Settings 1, when $y = 0.5$, $q_d = 0.3$, $x = 0.5$, $q_c = -0.3$, $v = 2$, and $c = 1$, the unique solution to this game is (cooperate, cooperate) with payoffs (10.6, 10.6) (see Fig. 2.4h). The above analysis shows that scholar-based governance strongly influences central-authority governance through providing the scholar's information. When the scholar's information is valuable, the two herder players will take the cooperate strategy under some certain conditions, even if the information government possessed before was partial.

2.3 The Scholar as an Agent of the Government

There are three situations needed to be considered here. First, as an agent, the scholar can be a researcher or consultant within the governmental system, or an information provider of the government (Situation 2a). The analysis of this situation is very similar to Situation 1d, when the scholar acts as an information provider. The only difference is that now the scholar becomes an agent of the government rather than an independent third party. Second, the scholar can use knowledge to directly resolve some problems with the herder on behalf of the government (Situation 2b). The analysis of this situation is also consistent with the analysis of Situation 1d. Third, because the government reduces its intervention cost through delegation, the payoff matrix of games can be changed (Situation 2c). This also improves the possibility of

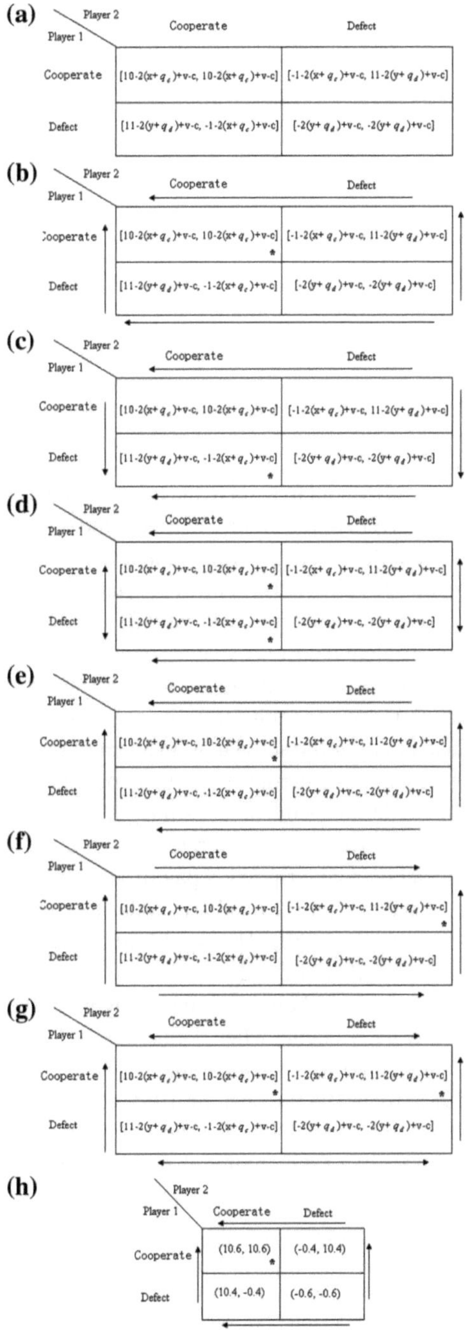

Fig. 2.4 Game 8—the central-authority game with incomplete information but with the scholar's information. **a** A general form of a central-authority game with incomplete information but with the scholar's information. **b** Setting 1. **c** Setting 2. **d** Setting 3. **e** Setting 4. **f** Setting 5. **g** Setting 6. **h** An example of Setting 1

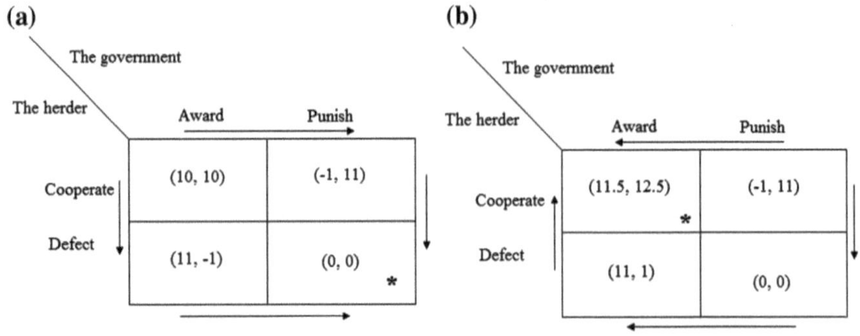

Fig. 2.5 Game 9—comparison between the government and the scholar-government as an agent of the government. **a** The game between the herder and the government without the delegation. **b** The game between the herder and the scholar-government after the delegation

the government to resolve the tragedy of the commons. For example, if we change the two players of the Hardin herder game (Game 1) to a herder and the government, and if we let the two governmental strategies be "award" and "punish," then according to the payoff matrix shown in Game 1, the equilibrium solution to this game is (defect, punishment) with payoffs (0, 0) (see Fig. 2.5a, Game 9). Suppose that the governmental cost is reduced somehow after delegation (e.g., due to the lower cost of seeking information and the scholar's good relationships with the herder—this is the scholar's social capital). As its cost decreases 3 units when the scholar-government (used to refer the scholar as an agent of the government) takes the cooperate strategy (i.e., when the herder takes the cooperate strategy, it gets 13 units of payoffs; when the herder takes the defect strategy, it gets 2), and the scholar-government also gives more awards to the herder, say 0.5 (it must be less than 2 in order to make sure that the scholar-government's payoff is larger than 11 getting from a defect strategy), a new payoff matrix results in order to encourage the herder to take the corporative strategy (Fig. 2.5b). The equilibrium solution to this new game is (cooperate, award) with payoffs (11.5, 12.5).

When the scholar becomes an agent of the government, the principal-agent problem usually occurs when the interests of the agent and those of the principal diverge under conditions of incomplete and asymmetric information (Grossman and Hart 1983; Rees 1985; Rogerson 1985; Ross 1973). This problem is not considered in our study, but we believe that the scholar as an agent can help the government resolve the tragedy of the commons problem.[1]

[1]Scholars played a very important role as the agent of government in ancient China. The local government often depended on them to resolve many local governmental affairs. These affairs include various irrigation works (including building the dams and weirs, dredging rivers, digging and maintaining ditches), works against natural calamities(such as mountain torrents, avalanche, and mud-rock flow), traffic works (like construction and maintenance of the road, bridges), rural economic affairs (such as construction and management of granaries), settling disputes, organizing different charity organizations, acting as guardians, teachers or illustrators of the traditional moral teachings especially Confucianism, organizing different local education schools from elementary

2.4 The Scholar as an Entrepreneur

Scholars can also provide collective goods for unorganized groups as entrepreneurs with their pioneering visions and innovation. Olson (1971) stressed the role of entrepreneur in collective action. According to his analysis, the entrepreneur has two important functions: First, he likes a scholar as an information provider; second, he acts as an ordinary game player to directly pursue his own interest. In addition to be an information provider, Olson said, "The entrepreneur may also get something for himself out of the gains he brings about" (1971, p. 177). To Olson, the methods the entrepreneur can use to influence other game players have two types: (1) bargains and the voluntary cost-sharing method, and (2) the coercion or reward as selected incentives. In light of his analysis, two kinds of games should be analyzed here. The first is a bargaining-voluntary game, in which the scholar as an inactive entrepreneur obtains his gains by providing knowledge and information, and the other two herders can voluntarily reach a agreement, which makes every player better-off in light of the scholar's information and suggestions. The second is a coercion-reward game, in which the scholar acts as an active entrepreneur who is above all an innovator with selective incentives to give either coercion or reward to those who contribute to the collective effort or not, or depending on the different levels of their effort.

---

to high-level schools or even private colleges and free private school, subsidizing students, repairing Confucian temples and different shrines for virtuous people, construction and repair of examinations halls and public libraries, editing local gazetteers as compliers, and privately collecting books (Yang 2007a). For example, Chang pointed out:

> Gentry members were also responsible for the construction and management of granaries in many localities. Decisions on granary policies could affect prices of agricultural products and could save the poor in times of need. Although the managers of granaries were generally appointed by the magistrates from among the local gentry, records reveal that these managers were usually recommended to the magistrates by influential gentry members of their localities. (Chang 1955, p. 62)

Mu-Ling-Shu, a handbook of magistrates in ancient China, even clearly taught the magistrates how to appropriately treat them.

> In administering the affairs of a district, the high families [*shen*] are not to be disturbed. They should be met in a virtuous manner and received courteously and are not to be repressed with power and prestige. Also the scholars [*shih*] are at the head of the people, and since the laws and discipline of the court cannot be exhaustively explained to the people, and since the scholars are close to the people and can easily gain their confidence, and learned and virtuous scholars are exactly the ones to rely upon in persuading the people to follow the instructions of the officials. Therefore, they should be loved and treated with importance. When they happen to come because of public affairs, if they are sincere and self-respecting, they should be consulted on the problems of whether bandits exist in their villages, what the jobs of the villagers are, and whether the customs of their places are praiseworthy.... (Chang 1955, p. 32)

2.4.1 A Bargaining-Voluntary Game

As discussed earlier, when the scholar acts as an information provider in the Hardin herder game, the bargaining-voluntary provision of the collective good cannot be realized. Then, the condition for the bargaining-voluntary provision of the collective good would be that either there is a coordination game or the payoff matrix can be changed in light of the information and knowledge of the scholar. Let's consider two situations.

Situation 3a: Make a Binding Agreement When There is a Coordination Game

Suppose that there are three herders using a meadow. A new coordination game is shown in Fig. 2.6a, Game 10. This game has three pure Nash equilibria: (cooperate, cooperate, cooperate) with payoffs (4, 4, 4), (defect, defect, cooperate) with payoffs (2, 2, −2), and (defect, defect, defect) with payoffs (0, 0, 0). When choosing the cooperate strategy Herder 3 is not sure whether Herder 1 or Herder 2 will take the cooperate or defect strategy. If they take the defect strategy, Herder 3's payoff will be −2, which is less than 0 (when Herder 3 takes the defect strategy). Thus, both of the cooperate and defect strategies can be Herder 3's choices. Now, let one of the three herders become a scholar-entrepreneur, say Herder 3. Suppose that Herder 3 as an entrepreneur has two methods to influence other game players—(1) the bargains and the voluntary cost-sharing method and (2) coercion or reward as selected incentives (for detailed analysis, please see Olson 1971, pp. 169–178)—and does not require any benefit from his information (that is, the payoff matrix of this game is still kept the same). He will suggest this strategy to the other two herders when he realizes that this will make all of them better-off. Because he is generally trusted (or feared) (Ibid), the other two herders voluntarily agree with his suggestion after bargaining. Then, they reach a binding voluntary agreement among them to take (cooperate, cooperate, cooperate) strategy to make all of them better-off. No one has incentives to break this contract, because if he/she does, the result will make him/her worse. Thus, they always take the cooperate strategy and everyone gains 4 (see Fig. 2.6b).

Suppose that Herder 3 should pay some costs to get the information and also wants to "get something for himself out of the gains he brings about" as Olson claimed (Ibid, p. 176). Let's consider the information costs and benefits gains. If the extra gain Herder 3 claims is impartially shared by Herders 1 and 2, the total extra gain he claims to get from Herders 1 and 2 cannot be greater than 2, or finally the payoffs for Herders 1 and 2 are both less than 3 even if all of them take the cooperate strategy. Under bargaining and voluntary arrangement, if the extra gain Herder 3 gets as an entrepreneur is more than the information cost, he will always have the incentives to do so. Suppose that the information cost is 0.5 and the extra gain the scholar-entrepreneur gets from Herders 1 and 2 is 0.33 (that is, the information cost is impartially shared by these three players). If the remaining extra gain from the information is impartially shared by Herders 1 and 2, their payoffs become (3.83, 3.83, 3.83) (see Fig. 2.6c). In this situation, the scholar-entrepreneur gains as much as the other two herders. If he claims that he should gain 1.5 after the bargaining,

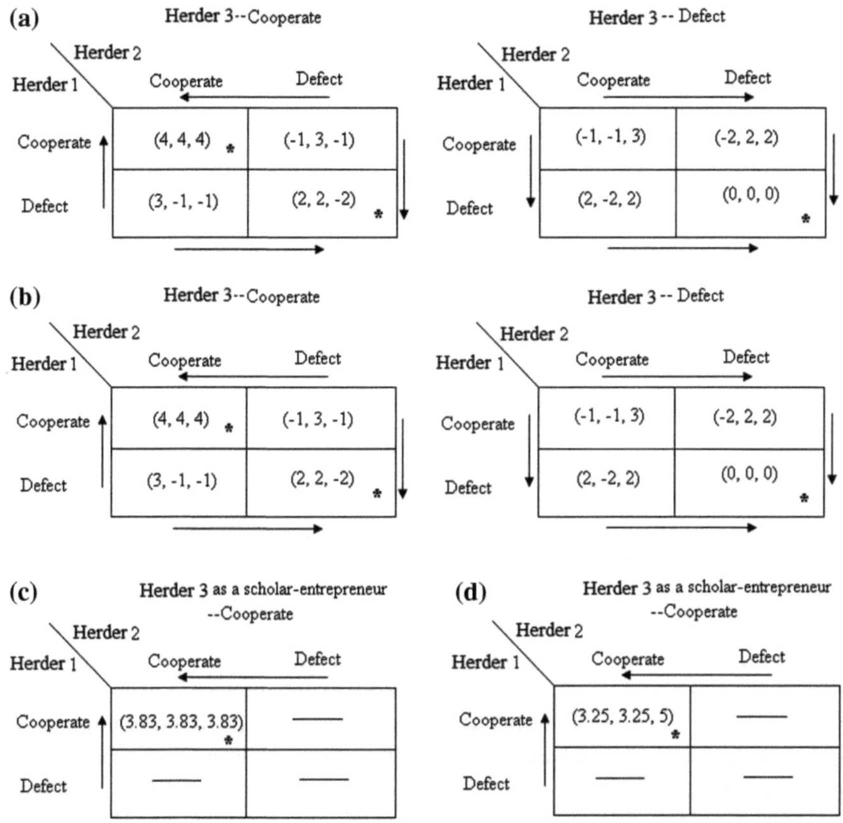

Fig. 2.6 Game 10—the bargaining-voluntary game with a scholar-entrepreneur in a coordination game. **a** A three-herder game without a scholar-entrepreneur. **b** A three-herder game with a scholar-entrepreneur who supplies useful information without any cost and also does not require any benefit of this information. **c** A game in which a scholar-entrepreneur shares the extra gains of the cooperative strategy with the other two herders. **d** A game in which a scholar-entrepreneur gets more extra gains of the cooperative strategy than the other two herders

Herders 1 and 2 should afford the cost 0.75, respectively. Then, his gain is 5, and Herder 1 and Herder 2 gain 3.25, respectively (see Fig. 2.6d). Finally, their payoffs become (3.25, 3.25, 5).

Situation 3b: Make a Binding Agreement When There is the Hardin Herder Game

Given a new payoff matrix shown in Fig. 2.7a, Game 11, the game equilibrium solution is (defect, defect, defect) with payoffs (0, 0, 0). This is a three-person prisoners' dilemma game (or Hardin herder game). Supposing that Herder 3 as a scholar knows that they all take the cooperate strategy and adopt a relative easy new method (e.g., a new technology or designed rule to properly use the meadow) to use the meadow with an affordable cost, their payoffs will be, at least, larger than 4. Let's reconsider this game. For example, if the cost to use this easy method for every player is 1, all of them will take the cooperate strategy; then their gain set becomes (5, 5, 5). After subtracting the cost, their payoffs become (4, 4, 4). If their gain set is (6, 6, 6), their payoffs become (5, 5, 5), which makes everyone better-off (see Fig. 2.7b). That is, in light of the suggestions and information provided by Herder 3 as a scholar-entrepreneur, the bargaining-voluntary cooperation is realized (see Fig. 2.7c). Then, this problem becomes the Situation 3a stated above. The only difference is that Herder 3 now not only provides the same information as in Situation 3a but also provides the information about the physical world and tries to use technology to make all of them better-off. Certainly, the scholar-entrepreneur can require even more residual benefits (e.g., see Fig. 2.7d), and the bargaining-voluntary cooperation can also be realized.

2.4.2 A Coercion-Reward Game

The above analysis shows that if there is a coordination game, or a situation can be easily changed to a coordination game, the bargaining-voluntary game can be realized when the cost is affordable. Now, let's consider even under a situation there is a Hardin herder game, the game also can be ameliorated by a scholar-entrepreneur if he plays a coercion-reward game with the other two herders.

Let's rethink the game shown in Fig. 2.7a, which has only one equilibrium solution (defect, defect, defect). In a bargaining-voluntary game, if the matrix can be changed in light of the scholar-entrepreneur's information, then the cooperation among three game players can be realized. Now, suppose that this payoff matrix cannot be changed, no matter how much information the scholar-entrepreneur has. Then, a bargaining-voluntary game cannot be played. Also suppose that the scholar-entrepreneur not only has information advantage over other two herders, but also has the right and authority to use selective incentives on the other two herders, no matter how he gets this right and authority. There are three classical situations: First, the scholar-entrepreneur only uses punishment but does not use reward. That is, he only uses the negative selective incentives. Second, the scholar-entrepreneur only uses the reward but does not use punishment. That is, he only uses the positive selective incentives. Third, the scholar-entrepreneur uses both punishment and reward. That is, he uses both the positive and negative selective incentives.

Situation 3c: Only Use Negative Selective Incentives

Suppose that penalty is 2, and the scholar-entrepreneur does not award someone using this gain from punishing defectors, that is, this gain is owned by the scholar-entrepreneur himself, then the game shown in Fig. 2.7a becomes Fig. 2.8a. Two solutions of this game are (cooperate, cooperate, cooperate) with payoffs (3, 3, 3) and (defect, defect, cooperate) with payoffs (0, 0, 2).

Let's consider a situation where the scholar-entrepreneur chooses the defect strategy, but also punishes himself. The game becomes the one shown in Fig. 2.8b. Three

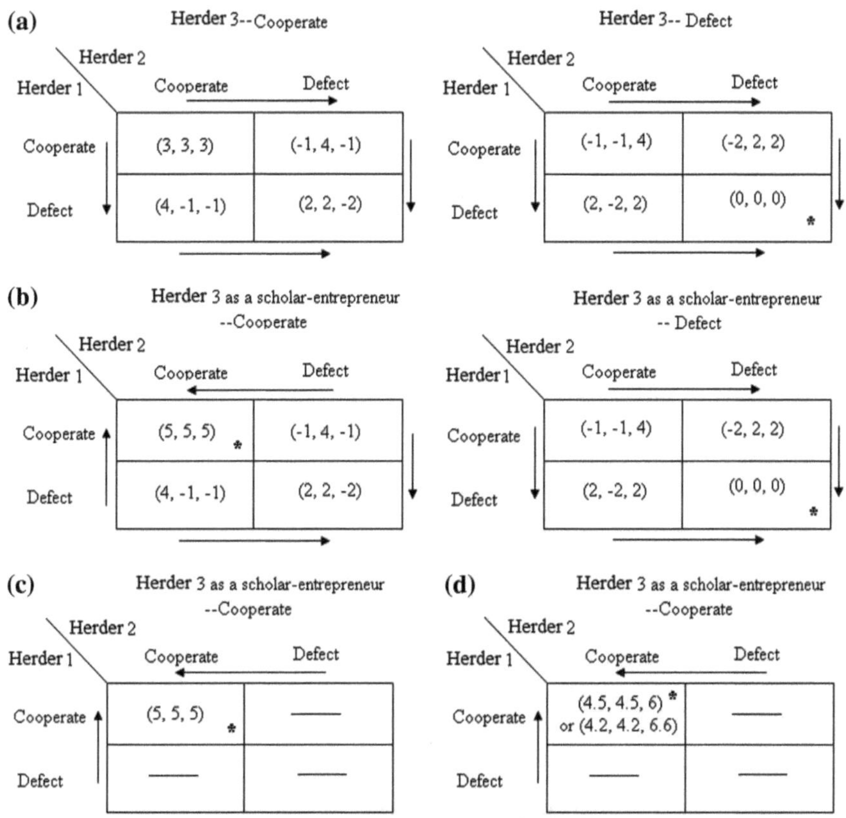

Fig. 2.7 Game 11—the bargaining-voluntary game with a scholar-entrepreneur in a situation where the original payoff matrix can be changed in light of the scholar's information and knowledge. **a** The three-herder game with a scholar-entrepreneur. **b** A three-herder game with a scholar-entrepreneur who supplies the useful information including not only about the game itself but also about the physical world which can make them all better-off when they cooperate. **c** The game in which the scholar-entrepreneur shares the extra gains of the cooperative strategy with the other two herders. **d** The game in which the scholar-entrepreneur gets more extra gains of the cooperative strategy than the other two herders

solutions of this game are (defect, cooperate, defect) with payoffs (0, −2, 0), (cooperate, defect, defect) with payoffs (−2, 0, 0), and (defect, defect, defect) with payoffs (−2, −2, 2). Thus, as a rational game player, the scholar-entrepreneur will never play such a game if he cannot get the benefit from the penalty.

Suppose that the scholar-entrepreneur can choose the defect strategy, but he never needs to pay the penalty even if he takes the defect strategy, and he can punish the other two players if they take the defect strategy. Then, the game becomes the one shown in Fig. 2.8c. Although three are still three solutions—(defect, cooperate, defect) with payoffs (0, −2, 4), (cooperate, defect, defect) with payoffs (−2, 0, 4), and (defect, defect, defect) with payoffs (−2, −2, 4), the scholar-entrepreneur

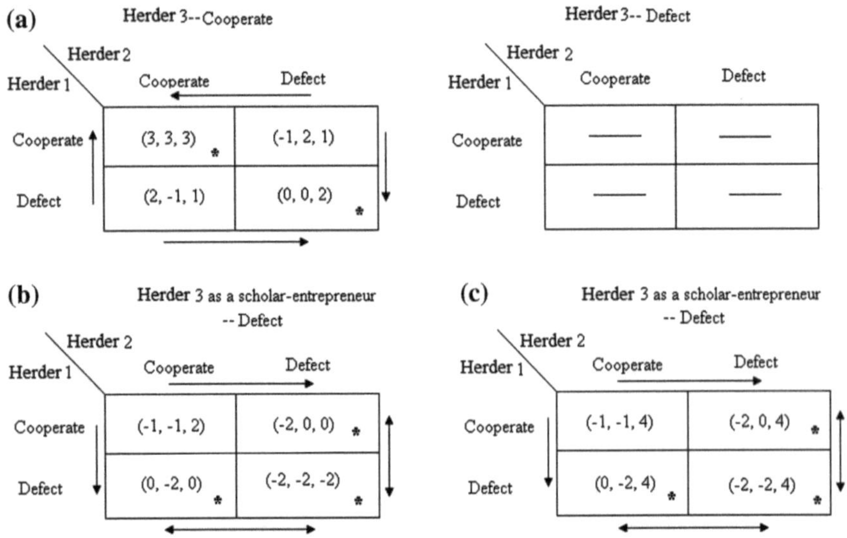

Fig. 2.8 Game 12—the coercion-reward game with a scholar-entrepreneur only using negative selective incentives. **a** A three-person game with a scholar-entrepreneur who only uses negative selective incentives and never uses the defect strategy. **b** A three-person game in which the scholar-entrepreneur chooses the defect strategy but cannot gain the benefit from the penalty. **c** A three-person game in which the scholar-entrepreneur chooses the defect strategy and also gains the benefit from punishing the defectors

always gets 4 which is even more than the amount when all of them take the cooperate strategy. This example is important. It tells us that the rules should be designed and implemented impartially to all of the game players, or it will become a dictator game where the government as the proposer (acting as the dictator) allocates the entire good to itself and gives nothing to the herder as the responder (Bohnet and Frey 1999; Bolton et al. 1998; Hoffman et al. 1996).

Situation 3d: Only Use Positive Selective Incentives

Assume that the value of an award is 1. This means whoever you are, you will be given 1 under whatever situations when you take the cooperate strategy. The payoff matrix can be shown in Fig. 2.9a, Game 13. This game has three solutions: (cooperate, cooperate, cooperate) with payoffs (4, 4, 4), (defect, defect, cooperate) with payoffs (2, 2, -1), and (defect, defect, defect) with payoffs (0, 0, 0). If the scholar-entrepreneur is rational enough, he will never take the defect strategy. Then, the only solution of this game becomes (cooperate, cooperate, cooperate), and every game player will be better-off.

One important problem is from where the award comes. If there is an external agency who can supply some financial support for the herders, then the problem is resolved. If there is another kind of collective action, and all of the players agree to

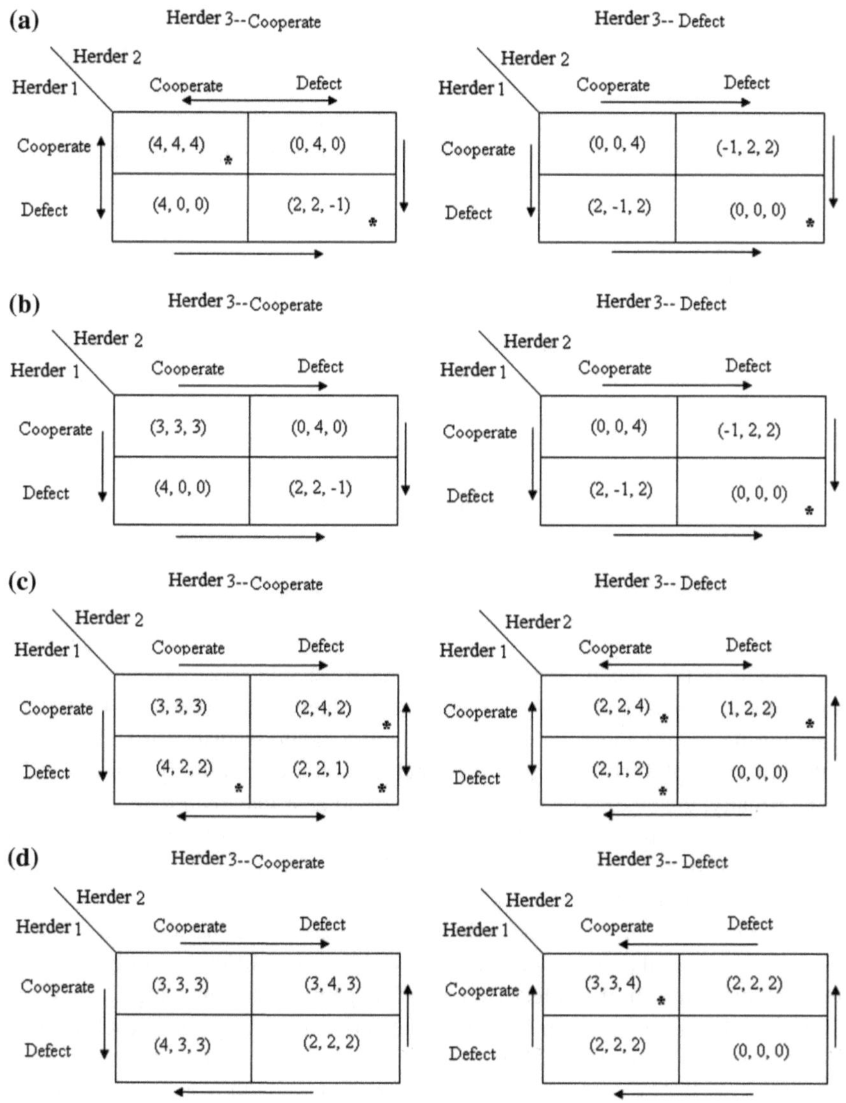

Fig. 2.9 Game 13—the coercion-reward game with a scholar-entrepreneur only using positive selective incentives. **a** A three-person game with a scholar-entrepreneur who only uses positive selective incentives. **b** A three-person game with a scholar-entrepreneur who only uses positive selective incentives but does not award anyone when all of them take the cooperate strategy. **c** A three-person game with a scholar-entrepreneur who only uses positive selective incentives but does not award anyone when all of them take the cooperate strategy and the reward makes the benefit equal to the gain when he takes the defect strategy with another player but one takes the cooperate strategy. **d** A three-person game with a scholar-entrepreneur who only uses positive selective incentives but does not award anyone when all of them take the cooperate strategy and the reward makes the benefit over the gain when he takes the defect strategy with another player but one takes the cooperate strategy

support this from their other businesses, this problem can also be resolved. Anyway, supposing that this problem has been resolved, let's consider another kind of situation.

Assume that the value of reward is still 1, but only award some players who take the cooperate strategy when the others take the defect strategy. This game can be shown in Fig. 2.9b. The solution of this game is (defect, defect, and defect) with payoffs (0, 0, 0).

When the value of award is 3, which makes the player's benefit when he takes the cooperate strategy while at least one takes the defect strategy equal to the benefit when he takes the defect strategy with another game player while one takes the cooperate strategy, the player will take almost all of the strategies except that all the players take the cooperate strategy or the defect strategy (see Fig. 2.9c).

When the value of reward is greater than 3, the player's benefit when he takes the cooperate strategy while at least one takes the defect strategy is larger than the benefit when he takes the defect strategy with another game player while one takes the cooperate strategy. Given the payoffs shown in Fig. 2.9d and the value of reward is 4, the solution of this game is (cooperate, cooperate, defect) with payoffs (3, 3, 4) (see Fig. 2.9d).

Situation 3e: Use Both the Negative and Positive Selective Incentives

Suppose that the penalty is still 2, and the reward depends on the penalty. One basic requirement of the penalty here is that this penalty should make the highest gain of defector less than the gain he gets when all game players take the cooperate strategy. Then for the reward, there are two situations that should be distinguished: The one is that the game players, who take the cooperate strategy, impartially share the value of the penalty; the other is that the scholar-entrepreneur may require more. The first reason is that this arrangement is used to compensate his cost of the information and managements of using selective incentives. The second reason may be that he wants to get more from the residual benefit as his gains as an entrepreneur.

Let's first consider a situation where none of them will be awarded when all the game players take the cooperate strategy. Also, none of them will be punished when all of them take the defect strategy. This game is shown in Fig. 2.10, Game 14. If the scholar-entrepreneur only takes the cooperate strategy and never takes the defect strategy, the solution of this game is (cooperate, cooperate, cooperate) with payoffs (3, 3, 3). Nobody will be punished, and nobody will be awarded. Here, the arrangement reaches a state: Just set the rules and then let them stay there, but never really use them. This is a dream stated by Sun Tze. He suggested defeating your enemy without even fighting.

Let's see whether the solution of the game will be changed, if the scholar-entrepreneur can choose the defect strategy. If he will not be punished, when he takes the defect strategy, there will be a new equilibrium (defect, defect, defect) (see Fig. 2.10b). If he will also be punished, the outcomes of the game are (cooperate, cooperate, defect) with payoffs (0, 0, 2), (defect, cooperate, defect) with payoffs (0, 2, 0), and (cooperate, defect, defect) with payoffs (2, 0, 0) (see Fig. 2.10c). If the scholar-entrepreneur is really knowledgeable as we suppose, he will not do that

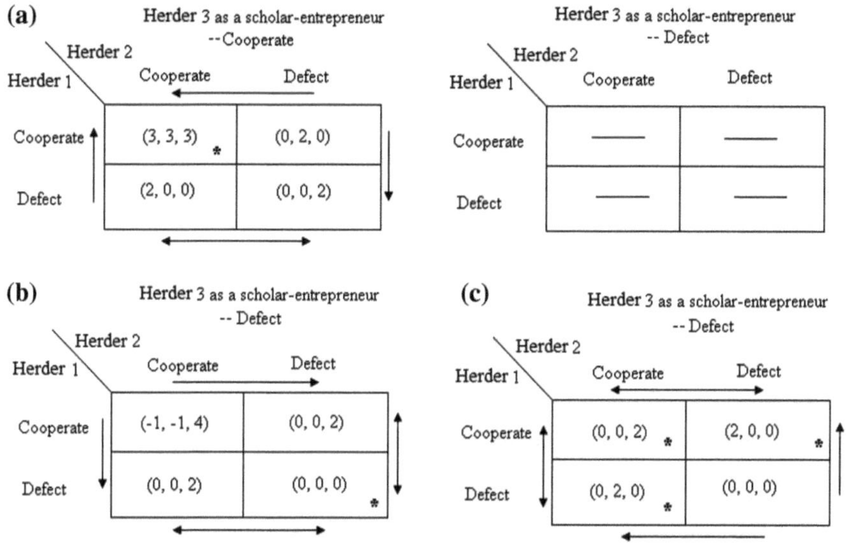

Fig. 2.10 Game 14—a coercion-reward game with a scholar-entrepreneur using both negative and positive selective incentives. **a** A three-person game in which a scholar-entrepreneur uses both negative and positive selective incentives and the gain from the penalty is impartially shared. **b** The scholar-entrepreneur chooses the defect strategy but is never punished. **c** The scholar-entrepreneur chooses the defect strategy but also is punished as to the others

because he already knows that if he did that, his gains would be less than the benefits when he took the cooperate strategy. When the scholar-entrepreneur gets the entire penalty as reward, but does not give any to the other two players, this game will become an aforementioned game, where only the negative selective incentives are used, and the scholar-entrepreneur gets all the values from the penalty.

In summary, when one of individuals becomes a scholar-entrepreneur in a coordination game, the voluntary cooperation among individuals can be reached, based on the suggestions of the scholar-entrepreneur, who is trusted, respected, or even feared sometimes. Even if there is a Hardin herder game, when the payoff matrix can be ameliorated, and everyone's gain from the strategy set (cooperate, cooperate, cooperate) can be at least equal or larger than his largest gain when he takes the defect strategy, the information and knowledge of a scholar-entrepreneur also can help individuals resolve the social dilemma through playing a bargaining-voluntary game.

If the above conditions cannot be all satisfied, the scholar-entrepreneur can use selective incentives. When he only uses the negative incentives, the penalty makes the largest gain of the defector less the gain when all of them take the cooperate strategy, and the scholar-entrepreneur does not play a dictator game, he can help game players resolve the social dilemma. Otherwise, the social dilemma will be remained.

If the external financial support problem can be resolved, when the scholar-entrepreneur takes the reward incentives, he also can help individuals resolve the social dilemma and can do better than when he takes the negative incentives.

If the award can make every player's gain from the strategy set (cooperate, cooperate, cooperate) larger than or at least equal to the largest gain when he takes the defect strategy, the scholar-entrepreneur can do best to help individual resolve the social dilemma as long as he also punishes his own defected actions, when he uses both of the negative and positive incentives.

2.5 The Scholar as a Pure Game Player

Let's consider a situation where the scholar as an independent game player directly competes with the other two players to pursue his own interests. Suppose that the government can have two strategies—intervene or stay out—and that the herder can have two strategies—cooperate or defect—and that the scholar can also have two strategies—accommodate or stay out. Let's consider a game using a meadow among these three actors.

Let's first consider a game between only the government and the herder (i.e., the scholar takes the stay-out strategy). Suppose that the herder gets 2 from using the meadow when he takes the cooperate strategy and gets 3 when takes the defect strategy, the penalty imposed by the government is 3, the intervention cost for government is 1, the government loses 1 due to the damage to its image when it intervenes in the cooperative actions by the herder, and the government gets 0 if it chooses the stay-out strategy. A payoff matrix is shown in Fig. 2.11a, Game 15. The equilibrium solution to this game is (defect, intervene) with payoffs (0, 2). Then, the government takes all the gains and the herder gets nothing. This will make the government richer and the herder poorer when the herder takes the defect strategy repeatedly in order to survive, while the government takes the intervene strategy accordingly to get more benefits. It also likes a dictator game. At first glance, except for transferring the benefit from the herder to the government, this is not different from the (cooperate, stay out) outcome from the perspective of the total benefit. Under this situation, however, the herder has to always take the defect strategy to get 3 from nature. This will definitely lead to the tragedy of the commons, and the government as a proposer also fails to maximize its own expected utility in the long run. Then, in order to get the result (cooperate, stay out) as in the situation of the herder's self-governance, the problems become: (1) how can the herder take the cooperate strategy, and (2) how can the government still take the stay-out strategy when the herder takes the cooperate strategy? Trust building and binding agreements (Fowler 2006; Fowler and Kam 2007; Lubell et al. 2002; Lubell and Scholz 2001; Ostrom 1998; Scholz and Lubell 1998) may be the two ways to solve this dilemma. If the scholar is highly trusted by both the government and the herder because of his good reputation in addition to his comparative advantages in knowledge and can influence their behaviors as a third party

(Yang 2007a, b), he may help them resolve this game through building trust and making a binding agreement. Let's consider seven situations.

Situation 4a: The Scholar is a Peacemaker.

As a peacemaker, the scholar tries to balance the actions and benefits between the government and the herder. For example, when the government punishes the cooperate strategy of the herder, he persuades the government to return 0.5 units of the benefits to the herder; thus, the government only loses 0.5 for the damage to its image. If the scholar gets 0.5 for his benefit, the government gets 0.5. When the government punishes the defect strategy of the herder, he intercedes with the government for the herder and persuades the government to give back 0.5 units of the benefits to the herder. If the scholar requires a benefit of 0.5 units, the government gets 1. When the government stays out and the herder takes the cooperate strategy, he persuades the herder to give 0.5 units of his benefits to the government (like a tax). Or, he may tell the herder that the government will punish him. If he requires 0.5, the herder gets 1. When the government stays out and the herder takes the defect strategy, he persuades the herder to give 1 unit to the government to avoid its punishment. Or he may tell the herder that if the herder does not want to do that, he will tell the government; then, the government will intervene, and the herder can only get 0. If the scholar requires 1 unit for his benefits, the herder, the government, and the scholar all get 1. The scholar himself even can get more as a self-interested individual, as shown in Fig. 2.11b. There are four equilibrium solutions to this game, however. That is, all the outcomes can be possible. Although now the scholar's accommodation does not lead to a uniquely better solution from the perspective of the whole SES, his accommodation actually reduces the possibility of "*the state of war*" between the government and the herder.

Situation 4b: The Scholar Acts as a Proponent of the Herder's Good Behavior and an Opponent of the Herder's Bad Behavior

When the government punishes the cooperate strategy of the herder, the scholar requires the government to return 1 unit of its payoffs to the herder; then the government only loses 0.5 for the damage to its image. If the scholar gets 0.5 for his benefit, the government gets 0. When the government punishes the defect strategy of the herder, and although the scholar dislikes the herder's bad behavior, he intercedes with the government for the herder because of sympathy and persuades the government to give back 0.5 units of its payoffs to the herder. If he also requires a benefit of 0.5 units, the government gets 1, as he is a peacemaker. When the government stays out and the herder takes the cooperate strategy, and although he also persuades the herder to give some benefits to the government, he only suggests a little, say 0.1. If he requires 0.5, the herder gets 1.4. When the government stays out and the herder takes the defect strategy, the scholar may advise the herder to give 1 unit back to the government to avoid punishment. If the scholar also gains 1 unit for his own benefit, the final outcome will be that all parties get 1 (see Fig. 2.11c). The unique equilibrium solution to this game is (cooperate, stay out, accommodate) with payoffs

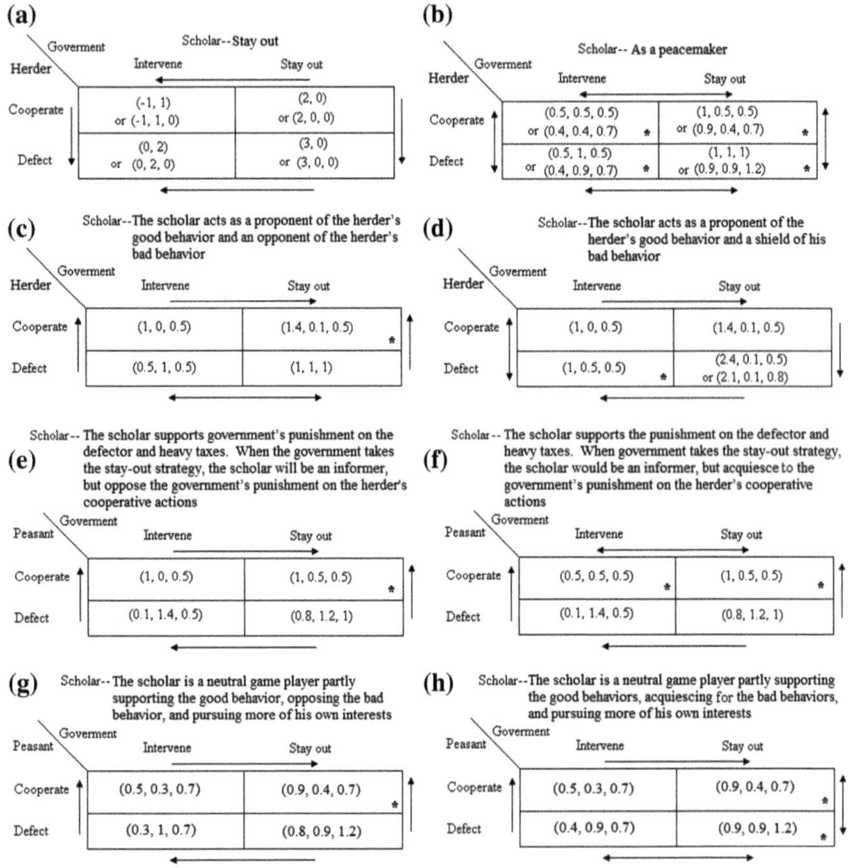

Fig. 2.11 Game 15—a three-player game among the herder, the government, and the scholar taking different roles. **a** A game between the government and the herder when the scholar takes the stay-out strategy. **b** A game among the herder, the government, and the scholar as a peacemaker. **c** A game when the scholar acts as a proponent of the herder's good behavior and an opponent of the herder's bad behavior. **d** A game when the scholar acts as a proponent of the herder's good behavior and a shield of his bad behavior. **e** A game when the scholar supports government's punishment on the defector and heavy taxes. When the government takes the stay-out strategy, the scholar will be an informer, but opposes the government's punishment on the herder's cooperative actions. **f** A game when the scholar supports the punishment on the defector and heavy taxes. When government takes the stay-out strategy, the scholar will be an informer, but acquiesces to the government's punishment on the herder's cooperative actions. **g** A game when the scholar is a neutral game player partly supporting the good behavior, opposing the bad behavior, and pursuing more of his own interests. **h** A game when the scholar is a neutral game player partly supporting the good behaviors, acquiescing for the bad behaviors, and pursuing more of his own interests

(1.4, 0.1, 0.5). That is, the scholar helps the game players resolve the collective action dilemma.

Situation 4c: The Scholar Acts as a Proponent of the Herder's Good Behavior and a Shield of His Bad Behavior

When the herder takes the cooperate strategy, the scholar does the same thing as in Situation 4b. When the herder takes the defect strategy, the scholar tries to cover up this behavior. For instance, when the government punishes the herder's defect strategy, the scholar may ask the government to return 1 unit of its payoffs to the herder. If the scholar also requires 1 unit for his own benefit, the government eventually gets 1. When the herder takes the defect strategy and the government takes the stay-out strategy, the scholar pretends to have known nothing about the herder's behavior and only suggests for the herder to give 0.1 units of his benefits to the government. If the scholar requires 0.5, the herder gets 2.4. The scholar also can get more. For example, he can get 0.8, and then the herder gets 2.1(to keep the herder's gain at least equal to 1.4, the scholar's gain cannot be greater than 1.5) (see Fig. 2.11d). The solution to this game is (defect, intervene, accommodate) with payoffs (1, 0.5, 0.5).

Situation 4d: The Scholar Supports Government's Punishment on the Defector and Heavy Taxes. When the Government Takes the Stay-out Strategy, the Scholar will be an Informer, but Opposes the Government's Punishment on the Herder's Cooperative Actions

A payoff matrix consistent with these statements is shown in Fig. 2.11e. The equilibrium solution to this game is (cooperate, stay out, accommodate) with payoffs (1, 0.5, 0.5), and the collective action dilemma is resolved. Now, the herder gets less than before, while the government gets more taxes.

Situation 4e: The Scholar Supports the Punishment on the Defector and Heavy Taxes. When Government Takes the Stay-Out Strategy, the Scholar will be An Informer, but Acquiesces to the Government's Punishment on the Herder's Cooperative Actions

A payoff matrix consistent with these statements is depicted in Fig. 2.11f. The two equilibrium solutions to this game are (cooperate, intervene, accommodate) with payoffs (0.5, 0.5, 0.5) and (cooperate, stay out, accommodate) with payoffs (1, 0.5, 0.5). This also helps the game players resolve the dilemma. However, because the herder always takes the cooperate strategy when the government punishes his cooperative actions, the total social benefit is less than that when the government takes the stay-out strategy.

Situation 4f: The Scholar is a Neutral Game Player Partly Supporting the Good Behavior, Opposing the Bad Behavior, and Pursuing More of His Own Interests

A payoff matrix consistent with these statements is shown in Fig. 2.11g. The solution to this game is (cooperate, stay out, accommodate) with payoffs (0.9, 0.4, 0.7). Under this situation, the collective action dilemma is also resolved, although scholar himself gets more, while the herder gets less (see Fig. 2.11g).

Table 2.1 Success situations of the scholars' four roles

The four roles	Success situations
Information providers	Situations 1b, 1c, and 1d
Agents of the government	Situations 2a, 2b, and 2c
Entrepreneurs	Situations 3a, 3b, 3c, 3d, and 3e
Pure game players	Situations 4b, 4d, 4e, and 4f (under Situations 4a, 4c, and 4 g, although the collective action dilemma is not resolved, its situation is ameliorated)

Situation 4g: The Scholar is a Neutral Game Player Partly Supporting the Good Behaviors, Acquiescing for the Bad Behaviors, and Pursuing More of His Own Interests

A payoff matrix consistent with these statements is depicted in Fig. 2.11h. The two solutions to this game are (cooperate, stay out, accommodate) with payoffs (0.9, 0.4, 0.7) and (defect, stay out, accommodate) with payoffs (0.9, 0.9, 1.2). Although the collective action dilemma is not completely resolved, this situation at least makes the solution (cooperate, stay out, accommodate) possible.

Certainly, there are other situations which can be analyzed here when the scholar takes the accommodating strategy. The situations stated above are only some special cases. Nevertheless, a clear conclusion emerges that the scholar's accommodation can help the government and the herder build trust or reach a binding agreement between them, thus solving the collective action dilemma under some certain situations (e.g., Situations 4b, 4d, 4e, and 4f). The mechanism in Situation 4b is the best for the herder, the mechanism in Situation 4d is the best for the government, and the mechanism in Situation 4f is the best for the scholar. These findings are also consistent with modern pluralism that claims that a society or political system is at least polyarchical, if not democratic, because multiple actors take part in the process of policymaking (Dahl 1961; Lindblom 1968, 1977).

2.6 Conclusion

Based on a series of game theoretical analyses, this chapter shows that scholar-based governance can help game players resolve their collective action dilemma in a three-party (herder, government, and scholar) SES under some certain situations when scholars act as information providers, governmental agents, scholar-entrepreneurs, or pure game players. I have summarized the successful situations (in the sense that the collective action dilemma is resolved) of the scholars' four roles in Table 2.1.

Thus, different from the three classical models, scholar-based governance is an alternative solution to collective action in SESs. Different models are better suited for different situations, and they may be used together for some complex problems of collective actions.

When other actors such as firms and the fifth sector are also included, the above analysis can be extended to five-person games which would be quite complex to analyze. To simplify the analysis, one may focus only on how scholars can help firms and the fifth sector resolve the collective action dilemma together with other game players. Just as the way they work with governments and herders, scholars also can play at least four types of roles when they work with firms and the fifth sector. They can be information and knowledge providers to the firms and the fifth sector, agents of the firms and the fifth sector, scholar-entrepreneurs (or leaders and organizers) of the firms and the fifth sector, and pure game players competing with the firms and the fifth sector. Thus, through scholars' participation, many collective action dilemmas among these five social actors in SESs can be resolved under certain conditions.

References

Anderies, J. M., M. A. Janssen, & E. Ostrom. (2004). A framework to analyze the robustness of social-ecological systems from an institutional perspective. *Ecology and Society*, 9(1): 18.

Bohnet I., & Frey B. S. (1999). Social distance and other-regarding behavior in dictator games: Comment. *The American Economic Review*, 89(1), 335–339.

Bolton, G. E., Katok, E., & Zwick, R. (1998). Dictator game giving: Rules of fairness versus acts of kindness. *International Journal of Game Theory*, 27(2), 269–299.

Chang, C. (1955). *The Chinese gentry: Studies on their role in nineteenth-century Chinese society*. Washington: University of Washington Press.

Confucian analects. (1900). In *Zhushi jiaozheng huaying sishu* (*Si shu*) (English & Chinese) (Trans. by J. Legge) (pp. 1–218). Shanghai: Shangwu Yinshuguan.

Dahl, R. A. (1961). *Who governs? Democracy and power in the American city*. New Heaven: Yale University Press.

Fowler, J. H. (2006). Altruism and turnout. *Journal of Politics*, 68 (3), 674–683.

Fowler, J. H., & Kam C. D. (2007). Beyond the self: Altruism, social identity, and political participation. *Journal of Politics*, 69 (3), 811–825.

Grossman, S., & Hart, O. (1983). An analysis of the principal-agent problem. *Econometrica*, 51(1), 7–45.

Hoffman, E., McCabe, K., & Smith V. L. (1996). Social distance and other-regarding behavior in dictator games. *The American Economic Review*, 86(3), 653–660.

Lindblom, C. E. (1968). *The policy-making process*. Englewood Cliffs, NJ: Prentice-Hall, Inc.

Lindblom, C. E. (1977). *Politics and markets: The world's political-economic systems*. New York: Basic Books, Inc., Publishers.

Lubell, M., & Scholz, J. T. (2001). Cooperation, reciprocity, and the collective-action heuristic. *American Journal of Political Science*, 45(1), 160–178.

Lubell, M., Schneider, M., Scholz, J. T., & Mete, M. (2002). Watershed partnerships and the emergence of collective action institutions. *American Journal of Political Science*, 46(1): 148–163.

Olson, M. (1971). *The logic of collective action: public goods and the theory of groups*. Cambridge, MA: Harvard University Press.

Ostrom, E. (1990). *Governing the commons the evolution of institutions for collective action*. New York: Cambridge University Press.

Ostrom, E. (1998). A behavioral approach to the rational choice theory of collective action: Presidential address, American political science association, 1997. *American Political Science Review*, 92(1): 1–22.
Rees, R. (1985). The theory of principal and agent: Part I. *Bulletin of Economic Research*, 37(1): 3–26.
Rogerson, W. (1985). The first-order approach to principal-agent problems. *Econometrica*, 53(6), 1357–1368.
Ross, S. (1973). The economic theory of agency: The principal's problem. *American Economic Review*, 63(2), 134–139.
Scholz, J. T., & Lubell, M. (1998). Trust and taxpaying: Testing the heuristic approach to collective action. *American Journal of Political Science*, 42(2), 398–417.
Yang, L. (2007a). Scholar-based governance: A fourth model to collective action. *Chinese Public Administration*, 259 (1), 96–103 (In Chinese with English abstract).
Yang, L. (2007b). Building a multi-collaborative community governance system to resolve the dilemma of collective action—a framework of "product-institutional" analysis (PIA). *Journal of Public Management*, 4(2), 6–23 (In Chinese with English abstract).
Yang, L., & Wu, J. (2009). Scholar-participated governance as an alternative solution to the problem of collective action in social–ecological systems. Ecological Economics, 68, 2412–2425.

Chapter 3
A Product-Institutional Analysis (PIA) Framework to the Study of Scholar-Based Governance

The Master said, 'The mechanic, who wishes to do his work well, must first sharpen his tools.' (In Chinese: 子曰: "工欲善其事, 必先利其器。" (《论语·卫灵公第十五》)).
—*Confucian analects* (1900, p. 161).

In Chap. 2, based on a game theoretical analysis, I concluded that scholar-based governance is a possible alternative solution to the problems of the social dilemma. Governance, however, is often more complicated than the situation described by the simple model of game theory. Frederickson and Smith summarized three distinct conceptions of governance:

> (1) Governance is simply a surrogate word for public administration and policy implementation. Thus governance theory is an intellectual project attempting to unify the various intellectual threads running through a multidisciplinary literature into a framework that covers this broad area of government activity. This, essentially, is the position staked by Lynn et al. (2000, 2001). (2) Governance equates to the managerialist or NPM movement. This is particularly evident in nations associated with the Westminster model, where NPM followed from serious attempts to reform the public sector by defining and justifying what government should and should not do, and to reshape public service provision by attacking the pathologies of bureaucracy (Kettl 2000). (3) Governance is a body of theory that comprehends lateral relations, interinstitutional relations, the decline of sovereignty, the diminishing importance of jurisdictional borders, and a general institutional fragmentation. (Frederickson and Smith 2003, pp. 225–226)

The common ideas shared by these three conceptions are as follows: First, governance is a broader concept than government; second, governance describes the connections between government and society or other social actors such as scholars and citizens; third, governance stresses the importance of other non-governmental social actors in collective action and public affairs; fourth, governance highlights collaboration among multiple social actors through networks, shared power, shared resources and decisions collectively decided. In summary, governance is a new method of public administration. Simon (1997, p. 7) argued that "the administrative processes are decisional processes." Thus, the processes of governance can also be deemed as

L. Yang, *Knowledge-Driven Governance*,
https://doi.org/10.1007/978-981-13-2910-4_3

decisional processes which involve multiple interactive social actors rather than only the government.

The primary objective of this study is to develop a basic road map—a product-institutional analysis (PIA) framework—for the study of governance involving multiple actors. This framework is a fundamental tool not only to theoretically analyze governance, but also to design empirical studies and analyze empirical data. After analyzing the limitations of the analysis using game theory, I first develop a basic game theoretical model of decision making through combining game theory and decision theory. Then, I analyze the concrete structure and several major problems of the product-institutional analysis (PIA) framework. And finally, an empirical research design and its data collection for studying scholar-based governance in field settings are described.

3.1 A Basic Game Theoretical Model of Decision Making

Game theory is concerned with the interactions of decision makers. Its paradigm is: "the modeler assigns payoff functions and strategy sets to his players and sees what happens when they pick strategies to maximize their payoffs" (Rasmusen 2001, p. 2). The basic building blocks of modern game theory include eight elements: players, the players' strategy, an action or move, information, the player's payoff, the outcome of the game, the strategy combination, and an equilibrium. Among these, players, actions, payoffs, and information form the four essential elements of a game and are collectively known as the rules of a game.

The difference between traditional decision theory and game theory is that in decision theory, there is only one player and the interactions among different actors cannot be considered. For example, based on the complete rationality model, the stages heuristic (Sabatier and Jenkins-Smith 1993) or the textbook approach (Nakamura 1987) assumed that there was a single group of decision makers and a single decision or policy cycle (Sabatier 2007), and often divided decision- or policymaking processes into a series of stages such as agenda setting, policy formulation and legitimating, implementation, and evaluation (Sabatier 2007). The major contributors of the textbook approach included Lasswell (1956), Jones (1970), Anderson (1975), and Brewer and deLeon (1983). Some researchers even traced their analyses back to Dewey (1927). Although Simon's bounded rationality model of decision making (1997) moved the focus of decision-making studies to the operational level from the policy level (Denhardt 2004), he still assumed that there was a single decision maker and argued that the decision-making process, at whatever level, consists of three sections: intelligence by which the individual makes a decision through scanning the environment and identifying occasions, design which refers to finding or developing alternative courses of action, and choice which means the selection of the alternative with the best chance of success (Simon 1967). Lindblom further examined the tension analyzed by Simon in greater detail (Denhardt 2004). He deemed the rational-comprehensive method and the "successive limited comparisons" or incremental method as two models of policymaking or decision making and suggested that the second was much more likely to be used in practice although the first had received the greatest attention in the literature. When using the incremental model, the policy maker only needs settle on a limited goal achieved by the policy, outline a limited number of alternatives that were immediately available, and make a choice that combined objectives or values in different ways (Lindblom 1959). Thus, to Lindblom, policy making is often a nonlinear, incremental, and evolutionary, rather than revolutionary, process that takes "baby-steps" and "muddles through," policymaking and policy practice often influence one another, and all these actually hold definite advantages for policy formulation in a democratic society.

Modern decision models that involve more and more social actors and interactions among different decision actors have become the core problem to explain decisions. Pluralists and group theorists (see Allison 1971; Arnold 1990; Dahl 1961; Downs 1972; Hirschman 1970; Kingdon 2003; Latham 1951; Lindblom 1968, 1977; Lowi 1979; Olson 1971; Schattschneider 1960) argued that the decision is not made by a single group of decision makers or dominated only by managers, supervisors, politicians, or bureaucrats, but there is a wide range of actor groups which can influence the process and outcomes of decision making. Following several criteria, such as meeting the criteria of a scientific theory, being the subject of a fair amount of recent conceptual development and/or empirical testing, being a positive theory, and addressing the broad sets of factors, Sabatier (2007) selected eight as the more promising number of theoretical frameworks for contemporary approaches of policymaking. They are the institutional analysis and development (IAD) framework (Ostrom 2007), the multiple-stream framework (Kingdon 2003), the social construction and policy design framework (Ingram et al. 2007), the network approach (Adam and Kriesi 2007), the punctuated-equilibrium framework (Baumgartner and Jones 1993; True et al. 2007), the advocacy coalition framework (Sabatier and Jenkins-Smith 1988, 1993), the innovation and diffusion models (Berry and Berry 1990, 1992), and a related set of frameworks used in large-N comparative studies (Blomquist 2007). Compared with the aforementioned approaches, one of the most important characteristics of the contemporary approaches is that some of these models began to develop simple and theoretical frameworks to directly study interactions of different actors or groups in the decision- and policymaking process. Furthermore, policymakers in these contemporary approaches were no longer only interest groups as claimed by the pluralist models and might be individuals, groups, or governments, etc. The typical frameworks among them are the network approach, the advocacy coalition framework, and the IAD framework.

Based on the above analysis, if we combine decision theory and game theory, a new game theoretical model of decision making for furthering the study of governance can be built. In addition to the variables considered in game theory, some other important variables should also be deemed as endogenous variables in a game theoretical model of decision making. They can be analyzed from six aspects.

1. In game theory, the type of the player is deemed as "the strategy set, information partition, and payoff function which nature chooses for him at the start of a

game of incomplete information" (Rasmusen 2001, p. 53). This understanding is still very important in a game theoretical model of decision making, but the player's type in production process (as a producer, a consumer, a person of pure transaction) and his social position[1] should also be considered (Yang 2002, 2007b).

2. For the actor or player resources, game theory only considers the importance of information. The other resources such as physical, human, and financial capital are only blurredly considered when it analyzes the player's strategies and utility function. All types of resources—physical, financial, human, information and knowledge, social, and organization or institutional capital (Yang 2004, 2007b; Yang and Lan 2008), however, influence decision making. Furthermore, information is different from knowledge. Information is about facts and figures; knowledge is the understanding of those facts and figures (Frenzel 1987; Morgan and Peha 2003). Thus, all the six types of resources should be considered in a game theoretical model of decision making.
3. Although game theory should not be misunderstood to say it "ignores nonrational and nonmonetary motivations" and to some extent emotions are gauged to "determine exactly how the actions and outcomes affect the players' utility" (Rasmusen 2001, p. 101), the careful analyses of the player's motivation and preferences are often not necessary in economic games. In a game theoretical framework of decision making, however, motivations, beliefs, ideologies, and preferences sometimes should be carefully considered as shown by the contemporary approaches of decision making (Sabatier 2007).
4. As a modeling tool, game theory always tries to treat "unimportant subcomponents of a model in a cursory way" (Rasmusen 2001, p. 4), and this is called "black boxing" by Kreps and Spence (1985). Decision analysis sometimes, however, has to open some black boxes to see their inner structure. For example, as stated above, only the information capital is considered in economic game theory, other kinds of the players' recourses or capital are put into a black box. In decision theory, however, all these resources should be considered. Furthermore, in economic game theory, when the players choose the actions, its analysis goes directly to the game outcomes. In decision analysis, however, the inner box between the actions and outcomes should also be carefully studied. It is impossible for a researcher to include all possible variables in his small model. Thus, whether some variables can be put into black boxes or not should be decided depending on at least three factors. The first is the nature of the problem which determines whether some factors are significant; the second is the particular research purpose; and the third is the researcher's personal preferences (Yang 2007b, 2008).

[1] Social position is different from social status. Social status often refers to the honor or prestige which is attached to someone's position in society. Social position influences social status. Although someone can have different social positions simultaneously, he only can have one social status which is a representative of his total honors or prestige. Honors and prestige are the important part of his social capital.

5. As stated above, the game outcomes are the set of values of whichever variables are considered interesting. They may be strategy combinations, different players' actual payoffs, etc. In economic game theory, after the outcomes, a new game will begin, and the analysis of the previous game has been completed. For decision analysis, however, the game outcome does not mean that a final decision has been made, so it is necessary to continue analyzing some new products of the effects of game outcomes on the research problems. There may be a final decision, or only some *intermediate products* of a final decision. If the latter, a new game will begin (Yang 2007b).
6. Although the four essential elements of a game—players, actions, payoffs, and information—are collectively known as the rules of a game, there are also some more micro-rules which have been considered in game theory analysis, for example, the order of play in a sequential game (Rasmusen 2001). In decision analysis, more interaction rules among the players such as the rules which describe that who should do what kinds of things at a certain moment should be considered (Yang 2007b).

The nature of human beings is complex. In addition to rationality, behaviors can also be influenced by instincts, emotion, faith, belief, ideology, or other irrational factors as shown by the aforementioned contemporary approaches of decision or policymaking (Sabatier 2007). Furthermore, as Simon and Lindblom analyzed, even rationality is bounded, nonlinear, incremental, and evolutionary (Simon 1967, 1997; Lindblom 1959). Some researchers further argued that the bounded rationality and satisficing models are still too rational. Especially, when it is moved from the individual level to the social level, Arrow's impossibility theorem (1963) argued that it is impossible to formulate a social preference ordering that satisfies the five conditions—unrestricted domain, transitive property, independence of irrelevant alternatives, weak Pareto principle, and non-dictatorship. The Allais paradox or problem (1953) even highlighted an inconsistency of actual observed choices with the predictions of expected utility theory. Because individuals often pay more attention to the choice with high certainty in the decision-making process (this is called the certain effect), the expected utility theory is not robust enough to capture such "bounded rationality" choice. All these irrational factors should be considered to determine how different strategies, actions, or game outcomes, etc., affect their personal utility in a game theoretical model of decision making.

Based the above extension of game theory in decision analysis, a new game theoretical model of decision making can be shown in Fig. 3.1.

The above model shows various types of decision makers with different production and social positions taking part in the game arena of decision making. They build their different strategy sets depending on their own resources or capital, motivations and preferences, and formal and informal rules, and choose different actions according to their owned information and knowledge and their utility functions to pursue (maximize or satisfy) their payoffs through different concrete processes shown by the game's inner box. Then, the collective decision or other intermediate products of the collective decision are finally produced. The actors in this model may be indi-

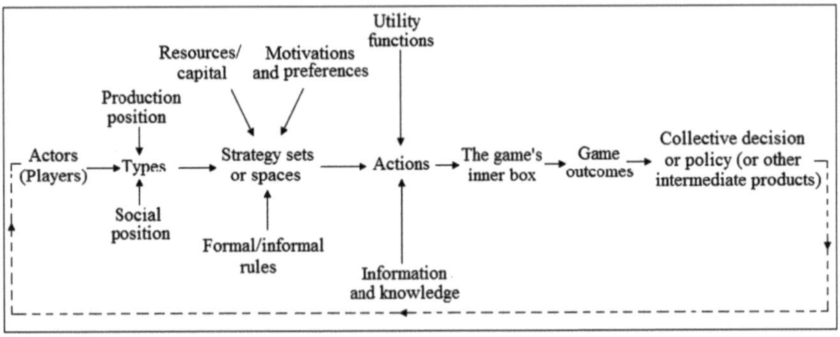

Fig. 3.1 Basic game theoretical model of decision making. *Source* Adapted from Yang (2007b)

viduals, groups, coalitions, or other organizations. Their actions, however, should be studied through their interactions with other social actors. Furthermore, this process is also a continuous process which can be characterized by long period of incremental change punctuated by brief period of major decision change, just as analyzed by Baumgartner and Jones (1993) through studying the policymaking process.

3.2 A Product-Institutional Analysis (PIA) Framework

3.2.1 A Basic Product-Institutional Analysis (PIA) Framework

Olson (1971) argued that the purpose of collective action, groups, or organizations is to produce collective, common, or public goods. He said, "Since most (though by no means all) of action taken by or on behalf of groups of individuals is taken through organizations, it will be helpful to consider organizations in a general or theoretical way" (p. 5); "*The achievement of any common goal or the satisfaction of any common interest means that a public or collective good has been provided for that group*" (p. 15, italics in original). If the benefit that can be derived from the collective good is fixed and limited in the market situation where the members of a market group attempt to reduce the size of their group, this sort of collective goods can be called an "exclusive collective good"; if the supply of collective goods in non-market situation automatically expands when group expands, it can be called an "inclusive collective good" (p. 38). Then, he analyzed the relationship of group size and the production of collective action by the group. He pointed out, "The larger the group the farther it will fall short of providing an optimal supply of a collective good, and very large groups normally will not, in the absence of coercion or separate, outside incentives, provide themselves with even minimal amounts of a collective good" (p. 48). He furthermore analyzed three types of organizations:

> A "privileged" group is a group such that each of its members, or at least some one of them, has an incentive to see that the collective good is provided, even if he has to bear the full burden of providing it himself. In such a group there is a presumption that the collective good will be obtained, and it may be obtained without any group organization or coordination whatever. An "intermediate" group is a group in which no single member gets a share of the benefit sufficient to give him an incentive to provide the good himself, but which does not has so many members that no one member will notice whether any other member is or is not helping to provide the collective good. In such a collective good may, or equally well may not, be obtained, but no collective good may ever be obtained without some group coordination or organization. It is distinguished by the fact that, if one member does or does not help provide the collective good, no other one member will be significantly affected and therefore none has any reason to react. Thus an individual in a "latent" group, by definition, cannot make a noticeable contribution to any group effort, and since no one in the group will react if he makes no contribution, he has no incentive to contribute. Accordingly, large or "latent" groups have no incentive to act to obtain a collective good because, however valuable the collective good might be to the group of a whole, it does not offer the individual any incentive to pay due to any organization working in the latent group's interest, or to bear in any other way any of the costs of the necessary collective action. (Olson 1971, pp. 49–51)

To summarize, the nature and other characteristics of products decide which kinds of groups or organizations should be chosen to provide the products; the forms and other characteristics of groups or organizations also decide what kinds of products they can produce (Yang 2002, 2004, 2006b, 2008). Thus, Olson's analysis already links the theory of public goods or the theory of *Generalized Product Analysis* (GPA) (Yang 2000, 2002, 2004, 2006a, b, 2007b, 2008; Yang and Yang 2004) and *Institutional Analysis* (*IA*) together. I call this analytical method *Product-Institutional Analysis* (*PIA*). The nature and other characteristics of the product first define the particular problem that the decision maker needs to consider and then influence all of the variables which are analyzed in the above basic game theoretical model. For example, De Groot, Wilson, and Boumans (2002) divided ecosystem functions, goods, services into 23 kinds. Of these ecosystem products, in addition to their nature as public assumption goods, their complexity also requires the involvement of scholars, and then the scholars' participation and scholar-based governance should be considered (Yang 2007a, b). Furthermore, the nature of the product and its other characteristics (including utility characters or other more concrete characters such as price and quality) often can be analyzed depending on three parts of a whole production process—production, transaction, and consumption (Yang 2002, 2004, 2008). Thus, through combining the game theoretical framework of decision making and product-institutional analysis stated above, a basic *Product-Institutional Analysis* (*PIA*) framework can be shown in Fig. 3.2.

3.2.2 *Internal and External Worlds*

The environmental or exogenous variables which influence the endogenous variables involved in a game arena of decision making have been analyzed by various frameworks. For example, the IAD framework stresses three groups of variables—phys-

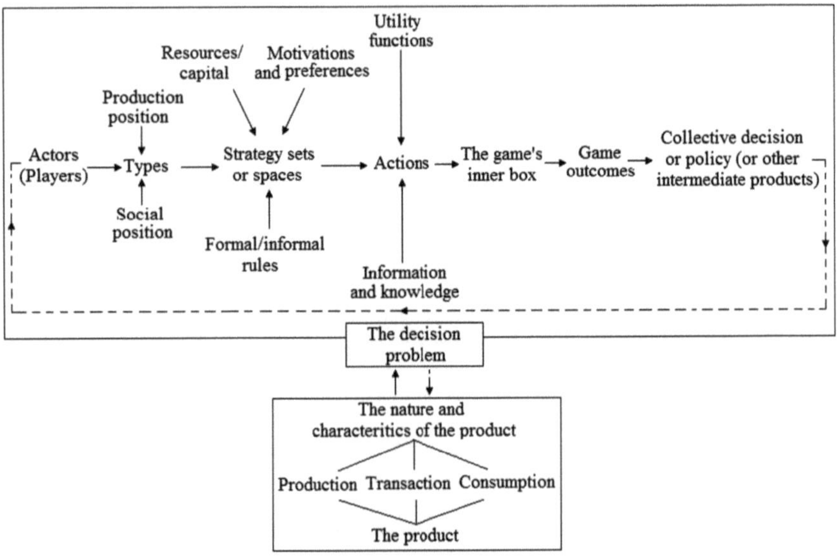

Fig. 3.2 Basic framework for product-institutional analysis (PIA). *Source* Adapted from Yang (2007b)

ical/material conditions, attributes of community, and rules in use (Ostrom 2005). The network approach highlights transnational context, national context, and policy domain-specific context (Adam and Kreisi 2007). The diagram of the advocacy coalition framework even involves more, such as relatively stable parameters (including basic attributes of the problem area, basic distribution of natural resources, fundamental social–cultural values and social structure, and basic constitutional structure), external system events (including changes in socioeconomic conditions, changes in public opinion, changes in systemic governing coalition, policy decisions, and impacts from other subsystems), long-term coalition opportunity structure (including degree of consensus needed for major policy change, and openness of political system), and short-term constraints and resources of subsystem actors (Sabatier and Weible 2007). Based on these precursory studies and the analyses stated above, to a certain game arena of decision making, its exogenous or environmental variables can be first divided into two levels: The first are the exogenous or environmental variables in the inner world of the game arena (the micro-level); the second are the ones in the external world of the game arena (the macro-level). For different research problems, this external world also can be divided into different levels or scales. For example, if a game arena is focused on the community level, its external world can be divided into county, state, national, or international levels. Whether it should be divided into different levels or which levels should be considered can be decided by concrete problems. No matter what they are, these variables can be grouped into three streams: attributes of the arena and its environment (including natural, population, economic, cultural, social, and political or governance environments), resources or capital (including physical, human, financial, knowledge, social, and organizational or institutional capital), and formal or informal rules (including nature, population, economy, culture, society, and governance rules). Through comparing their concrete analytical contents as stated above, it can be found that the first stream and the second stream, in fact, analyze the same things through different perspectives; thus, the researcher can choose either one when using this framework. In this manuscript, I prefer to use the second stream. Thus, the above three streams, in fact, are two steams. The whole game theoretical framework for decision making can be demonstrated in Fig. 3.3. This figure shows that decision making is a collective and successive game process among various actors influenced by many internal and external factors of the decision game arena (e.g., a community).

3.2.3 Defining a Particular Research

The intended scope of the above model is to give a general game framework to study interactions among different social actors in the decision- or policymaking process influenced by many internal and external variables of a decision or policy arena in governance. To a certain problem, for example, scholar-based governance, this framework could also be specialized depending on the certain unit of analysis (such as the community, the county, the city, the state, and the nation), the certain level of analysis (such as the constitutional level, the collective level, and the individual level from a perspective of institutional analysis; or the individual, family, community, county, city, state, national, international, and all-over-the-world levels from a scale perspective) (North 1981; Ostrom 2007; Yang 2004, 2007b), the temporal span, and the type of the problem (Yang 2007b). All these dimensions should be carefully considered when we design an empirical study of governance and analyze the data obtained from archives, surveys, interviews, observations, focus groups, filed studies, and so on. Furthermore, even to a purely theoretical analysis, the consideration of all these dimensions is very important. Thus, certain research can be defined using a dynamic framework shown in Fig. 3.4.

3.3 Remodeling the PIA Framework: Black Boxing and Opening the Black box

To any particular research problem, in addition to defining the particular research as stated above, the PIA framework needs to be remodeled, in order to choose which particular endogenous or exogenous variables should be considered in a particular research. Thus, if some variables stated above are not very important, they may be treated in a cursory way (Rasmusen 2001), and this is a process of black boxing; by contrast, if some variables stated above are particularly important, they may need

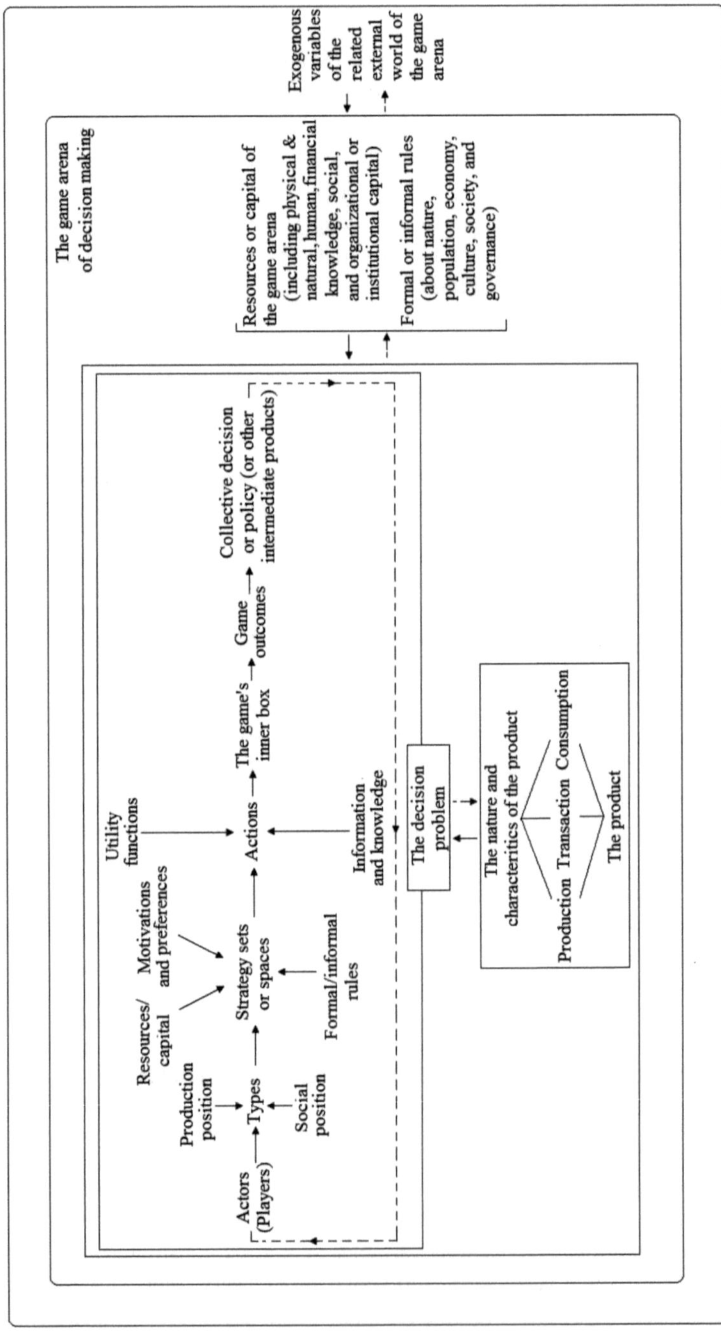

Fig. 3.3 A game theoretical framework for decision making. *Source* Adapted from Yang (2007b)

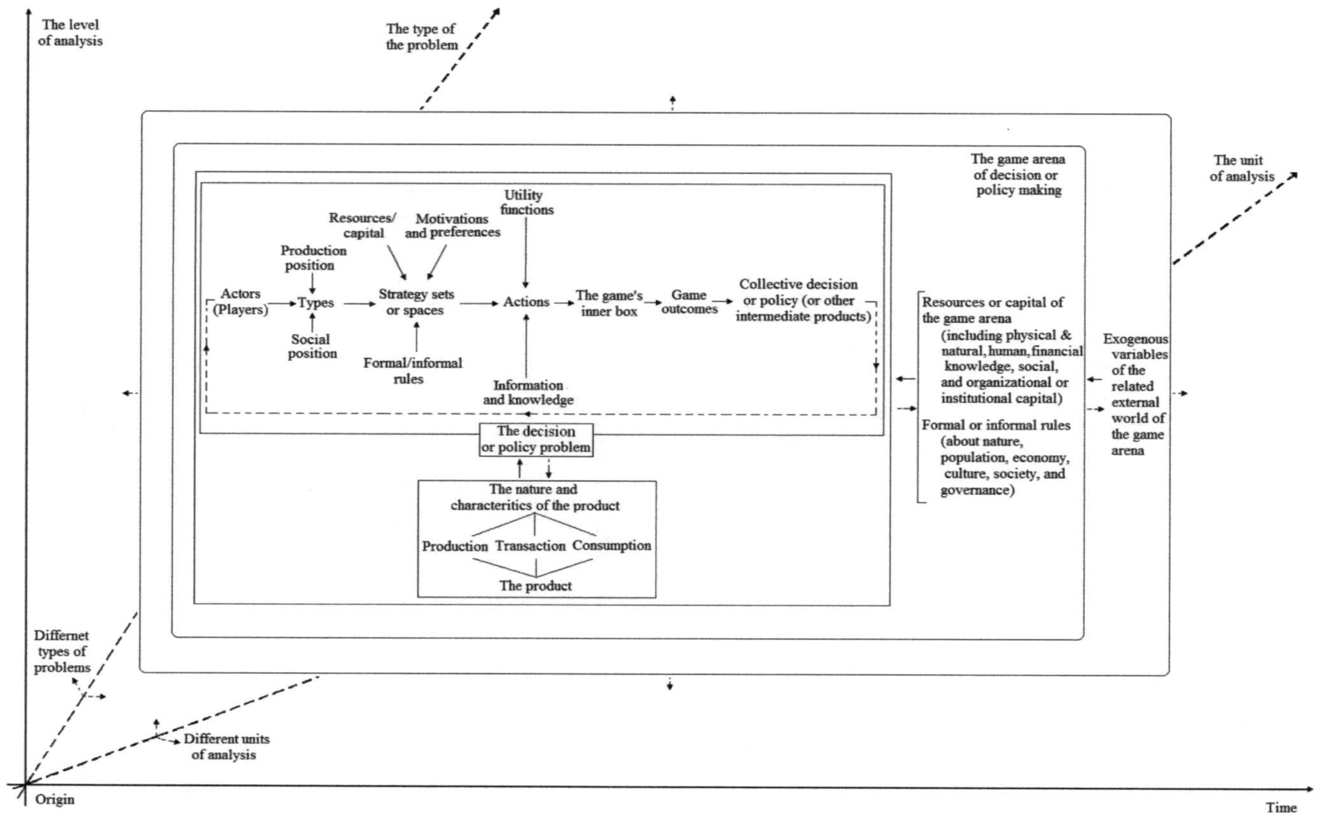

Fig. 3.4 A dynamic game theoretical framework for decision or policymaking for defining a particular research. *Source* Adapted from Yang (2007b)

not only to be just involved in the particular framework but also to be particularly considered through further opening their own black boxes. A good particular framework should make a balance between black boxing and opening new black boxes. Using this strategy, the danger of explaining nothing by including everything can be avoided. For example, to the study of scholar-based governance, the variables which are directly related to scholars should be particularly considered through further opening their own black boxes; by contrast, some variables about other social actors such as citizens, governments, firms, and the fifth sector (Yang 2007b), no matter what they are (endogenous or exogenous variables), should be treated in a cursory way, if they are not very important for the study of scholar-based governance.

3.4 Studying Scholar-Based Governance in Field Settings

Game theory is a rigorous tool to theoretically examine how scholar-based governance can help game players reach a stable equilibrium through changing incentive structure. This rigor, however, also weakens its real-world appliance and relevance for policymaking. To examine the validity of the findings from the game theoretical analysis and study scholar-based governance in more details, we conducted a field study about desertification.

3.4.1 Desertification and Its Studies

Desertification usually refers to land degradation in arid, semiarid, and dry subhumid regions due to human activities and climate variations,[2] which may lead to the permanent loss of land productivity (UN 1992; Wu 2001). These dryland regions together cover about 41% of the global land surface and are home to more than 38% of the world's human population (Reynolds et al. 2007). Desertification has serious

[2]What has caused desertification? Some researchers deem human effects such as over-cultivation, overgrazing, land abuse, and water mismanagements as the major reasons of desertification (e.g., Chao 1984; Zhu 1989; Zhu et al. 1988); some attribute desertification to the natural factors such as climate change, the geographical movement, and other physical variations(e.g., Wang and Dong 1994; Zhu and Liu 1983). Although the factors and their casual relationships of desertification are still debated and vague, more and more researchers deem desertification as a joint product of human and natural factors (Berry et al. 1977; Bryson 1973; Chouhan 1992; Johnson 1977; Kates et al. 1977; Unitized Nations 1992; Wang and Wu 2005; Wang et al. 2001, 2003; Zhu 1990, 1998). Many researchers believe that although natural factors first formed the Chinese desert, human behaviors contributed to their enlargement (e.g., Bureau of the Ministry of Forestry 1990; He 1991; Zhu 1990, 1998). For instance, the officials within the Ministry of Forestry estimated that only one-third (or 500,000 km^2) of current total desert area was formed by nature, and the rest had been making of human activities (Bureau of the Ministry of Forestry 1990, p. 20). Zhu (1990) even claimed that sand dune encroachment by natural factors account for only a tiny percentage of the current ecological problem, but human affects account for at least 94.5%.

consequences and affects[3] directly about 250 million people in the developing counties (Reynolds et al. 2007). United Nations Environment Programme (UNEP 2003) pointed out that appropriately 5 million square kilometers of global land (including two-thirds of its productive agricultural land) are affected by land degradation. According to the Millennium Ecosystem Assessment (MEA), an area of 10–20 million square kilometers, or about 10–20% of the global drylands, has been degraded (MEA 2005). GTZ (2005) also estimated that about 50–60 thousand square kilometer global lands are permanently lost to agriculture each year because of human-induced soil degradation. In China over the past several decades, this process is even more pronounced. Its annual invading speed of desertification process was 1560 km^2 per year in the 1950s, but in 2002 this speed was expanded to 10,400 km^2 per year (CCICCD 2000, 2002). Understanding the mechanisms of desertification and developing effective management and mitigation plans become increasingly important for maintaining the ecological and socioeconomic sustainability as well as political stability of both the dryland areas and the entire world (Wu 2001, 2005; Reynolds et al. 2007).

The above analysis shows that desertification can be deemed as a public bad; combating desertification, on the contrary, can be deemed as a public good. As a public good, it will be underprovided, and this is the dilemma of collective action. Many researchers highlighted important roles of government (CSC 1994; Sheehy 1992), firms (Skuras et al. 2000), local people, or communities (Reynolds et al. 2007), various NGOs or international organizations (Betsill and Corell 2008; Thomas 1997; UNCCD 1994), and even science in combating desertification (SACDS 1997; Seely 1998; Thomas 1997), the functions of scholars as a whole group of independent social actors, however, have received little attention in the mainstream discourse concerning contemporary desertification control.

[3]Desertification has serious consequences and affects on land degradation (Zhu and Liu 1983), ecological systems (CCICCD 2000; Ci 2001; Wang et al. 2001; Yang et al. 2001), people's living environment and economy (CCICCD 2000; Ci 2001; Dong 2002; Wheaton et al. 1992), and even politics. For instance, during the calamity of the dust storms (desertification is its one of the major conditions) on May 5, 1993, in the northwest of China, 4412 houses were destroyed, 118.2 thousand domestic animals died, about 373,333 million hectares of crops were destroyed, over 2000 km of ditches were buried, 200 thousand of diversiform-leaved poplar (Huyang in Chinese) trees and 130 thousand hectares of sacsaoul trees (Suosuo in Chinese) were destroyed, about 10 thousand hectares of forest were buried by over 1 m thickness of the dust, and 270 km of road surface were eroded by wind. Also, communication was disrupted for 4–6 days, and the production and living condition of one fourth the herdsman were destroyed. Totally, the direct economic damage caused by this event over Alashan, an Alliance of Inner Mongolia, was up to 436 million RMB (55 million US dollar), the indirect lost even went up to 1370 million RMB, according to statistics (Ci 2001, p. 218; Dong 2002, p. 21). Furthermore, desertification often results in cross-regional even national problems. For example, North Africa is a source of dust for southern European dust deposition (Yaalon 1996), and the arid and semiarid regions around the Sahara region are one of the principal sources of global dust (Mattsson and Nilsen 1996; Pease et al. 1998).

3.4.2 A Typology of Scholars in Combating Desertification

According to the criteria whether they are living in an area, scholars in combating desertification can be divided into local scholars and outside scholars.

According to the standard whether they can be easily recognized, they can be classified into easily recognizable and evident scholars (such as professors or experts in universities or research institutes, researchers in local desert research stations and their substations and technicians in the county, townships, or villages), and relative and latent scholars (e.g., some knowledgeable farmers or herders).

Furthermore, according to their different degrees of intimacy with the activities of combating desertification, they can be grouped into three parts—the core scholars (such as the researchers or technicians in desert research stations and their substations), the directly related scholars (such as experts or technicians related to the affairs of environmental protection, water resources, forestry, and land and resources), and the indirectly related scholars (such as experts or technicians related to the economic development, political affairs, and population control). Together, these related scholars can be divided into 12 types (Fig. 3.5).

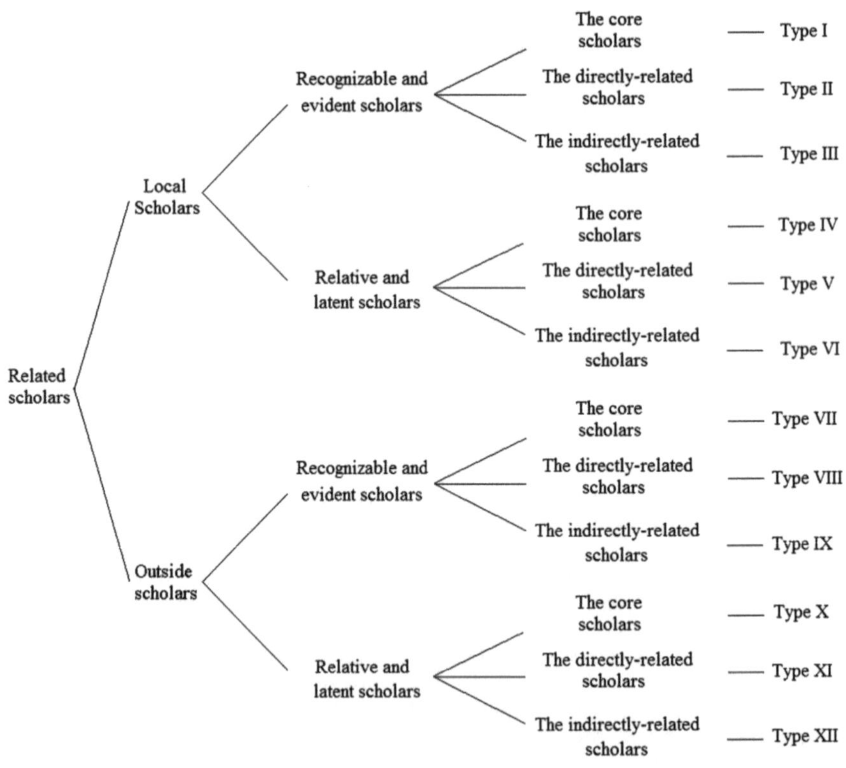

Fig. 3.5 Typology of related scholars

My study tried very hard to cover all these types of scholars, but focused on local scholars. Among local scholars, I paid much attention on the easily recognizable and evident scholars. Furthermore, among these easily recognizable and evident scholars, I paid the most attention to the core scholars, then the directly related scholars, and finally the indirectly related scholars.

Also, according to the knowledge types, the scholars can be divided into groups mainly with technical knowledge or groups mainly with social and policy knowledge. This is also important for the study of combating desertification and is concerned in the field study.

3.4.3 A Two-Step Study

The main objectives of this study are twofold: (1) to explore whether scholars' participation plays an important role in combating desertification, and (2) if so, how scholar-based governance can be most effective to improve the outcomes of combating desertification.

According to the PIA framework, I conducted a two-step research project. In the first step of research, a field study was carried out in seven counties of northwest China (see Fig. 3.6) to gauge the significance of scholars' participation in combating desertification and their different roles from 1949 to 2008. Among these seven representative counties with various results of combating desertification and scholars' participation (King et al. 1994), Zhongwei in Ningxia Hui Autonomous Region, known as the "capital of desert," is deemed as the most successful case of combating desertification in China. The other counties, including Minqin, Jingtai, Linze, Jinta, Guazhou, and Dunhuang, are all important oases in the Hexi Corridor in Gansu Province whose capital is Lanzhou, the geographic center of China. Minqin currently suffers the most serious desertification. The situation in Jinta is very similar. Also, these counties have very similar geological, geographical, climatic, and ecological conditions and original environments as physical and natural capital or resources.[4]

[4]Concretely, in terms of geological and geographic conditions, the continental height of seven research sites is all between 1000 and 2000 m. Their geological time mainly belongs to Quaternary and Paleozoic; their rock types are mainly Tishan-Mts Hinggan Ling lectono-magmatic domain and unmetamorphosed strata. The degree of the geological hazards in these counties falls into the slight or very slight domains. Regarding the hydrogeological attributes, they mainly locate in the desert pore water area. Their environment landforms chiefly belong to the Aeolian desert-basin region. All seven research sites are located in the arid area of middle latitude temperate zone, the annual temperature is between 4 and 12 °C, and the annual precipitation is mainly between 100 and 200 mm. In summer, they are far from the atmospheric depression region in Asia; but in winter, they are in main area of the high atmospheric pressure region. As to ecology, their vegetation zone belongs to the semi-desert and desert zone of vegetation, their zoogeographical zone is Inner Mongolia and Xinjiang region, their population density is less than 100 persons per square kilometer, and the percentage in cultivation is less 10%. For detailed information, please see AXBW (1992), DSBW (1994), JDBW (1992), JXBW (1996), LXBW (2001), MXBW (1994), ZXBW (1995).

In summary, these selections are consistent with Przeworski and Teune's most similar/most different research design (1970).

Following a random sampling strategy, a field survey was carried out from June 26, 2006, to February 12, 2008. Considering that some farmers might not be able to read Chinese characters, the questionnaires were randomly distributed to the students in different high schools, who were carefully trained to teach and help their family members. The reason why I do this is that many old farmers now in villages cannot read Chinese characters. I received 1974 valid responses in total, with a response rate of over 90% (Table 3.1a). In order to get more detailed information, 78 people (including farmers, herders, scholars, government officials, businessmen, members in religious groups or NGOs) were also interviewed from June 6 to July 31, 2007 (Table 3.1b). For the interviewees in the county seat, I contacted the offices of the bureaus of the county, the research institutes, schools, etc., to let them recommend interviewees. For the interviewees in rural areas, I directly went to villages to find volunteers. The questions I asked interviewees are consistent with the questions in survey questionnaires (for the recruitment script and the questionnaire of the interview, please see Appendix C; for the recruitment script, the outline of the training content for high school students, and the questionnaire of the survey, please see Appendix D; for the approval document by the Institutional Review Board, please see Appendix E).

According to the PIA framework, a particular study is defined from four dimensions—the units of analysis, the levels of analysis, the temporal span, and the particular research question, and the generalization of the findings from a particular study means whether these findings still hold when the above four dimensions are changed. Changing coverage and spatial extent will change the units of analysis. When the hierarchical level of an organization as a unit is changed, the level of analysis is also changed. Also, changing fields will change the particular research question. Coverage, also called sampling density or intensity, means "proportion of the study area or duration actually sampled" (Wu 2007, p. 118). Spatial or temporal extent refers to spatial or temporal span of a phenomenon or study or the study area or study duration (Wu 2007). Culture also plays an important role in forming institutions of governance (Avruch et al. 1991; Held et al. 1999). Ostrom (1997, p. 254) ever said, "The aspiration to achieve greater complementarities among the societies of mankind requires us to understand common features in our respective civilizations that might serve as common foundations." Through changing cultural contexts, I want to study whether the findings in the seven Chinese counties still hold in other countries with different cultures. Furthermore, I want to study whether the findings from the seven counties still hold when scholars participate in other types of collective action rather than combating desertification. I call this the generalization through changing fields. Thus, in the second step of the study, in order to test the causal generalization as extrapolation of concerns about "whether a causal relationship holds over variations in persons, settings, treatments, and outcomes" (Shadish et al. 2002, p. 21), I also analyze forty-three cases (Fig. 3.1) based on changing coverage, spatial and temporal extent, cultural contexts, and fields.

Table 3.1 Survey and interview distribution in the seven counties

(a) *Survey distribution*

Counties	The number of sent copies	Response rates	The number of valid copies
Zhongwei	300	0.95	280
Minqin	370	0.92	322
Jingtai	280	0.93	236
Linze	250	0.96	239
Jinta	300	0.93	260
Guazhou	260	0.93	237
Dunhuang	450	0.92	400
Total	2210	0.93	1974

(b) *Interview distribution*

Counties and time	T	F	S	G	B	R
Minqin (6/6-7/13)	29	6	11	11	0	1
Zhongwei (7/20-7/22)	10	5	4	1	0	0
Jingtai (7/23-7/23)	9	3	3	3	0	0
Linze (7/24-7/26)	8	4	3	1	0	0
Jinta (7/27-7/28)	6	4	2	0	0	0
Guazhou (7/29-7/30)	8	6	2	0	0	0
Dunhuang (7/30-7/31)	8	1	2	1	4	0
Total	78	29	27	17	4	1

T total interviewees in this county; *F* farmers, herders, other citizens; *S* scholars; *G* government officials; *B* businessmen; *R* members of religious groups or NGOs

For coverage, five counties were chosen. Their locations, annual precipitation, representations (what they represent), and related literature are shown in Table 3.2. For spatial extent, one village called Songhe in Minqin County and two prefectures (between the county level and the provincial level)—Yulin Prefecture in Shaanxi Province and Zhangye Prefecture (Linze is under its jurisdiction) in Gansu Province—were chosen. For cultural context, fifteen cases from other 13 counties including Australia (two cases), Israel, Pakistan, Spain, Turkmenistan, USA (two cases), Chile, India, Iran, Iraq, Niger, Tunisia, Uzbekistan in six continents including Africa, Asia, Australia, Europe, North America, and South America were chosen. For fields and temporal extent, I chose a total of 20 cases—17 cases in China and three cases from India, South Korea, and Sri Lanka. Their distribution is shown in Table 3.3 (Fig. 3.6).

All the case studies in the second step of the research were based on the existing literature. Many archives, published articles, books, and reports were used to measure problem severity and the results of problem resolving, and to evaluate whether the findings from the seven counties could be generalized.

Furthermore, based on a meta-analysis of all the Chinese cases used in this study, I studied the problem of the casual generalization as representation which concerns about "how to go from the particular units, treatments, observations, and settings on which data are collected to the higher order constructs these instances represent" (Shadish et al. 2002, p. 20). The purpose was to see whether the findings from scholar-based governance in combating desertification as a representative is best described the role and its design principles of scholar-based governance in the whole Chinese society.

Table 3.2 Five counties chosen by changing coverage

Counties	Location	Precipitation (mm)	Representation	Literature
Hotan	Xinjiang uyghur autonomous region	34.8	The extreme arid area in Taklimakan Desert	Lu et al. (2004), Zhao et al. (2003)
Yanchi	Ningxia Hui autonomous region	300	The desert steppe area in Mu Us sandy land	Zhao et al. (2003)
Ejin Horo	Inner Mongolia autonomous region	340–400	The transitional belt from semiarid to arid areas in Mu Us sandy land	Zhao et al. (2004)
Aohan	Inner Mongolia autonomous region	310–460	The semiarid pasturing area in Horqin sandy land	Teng and Dang (2003)
Naimai	Inner Mongolia autonomous region	362	The semiarid farming-pasturing area in Horqin sandy land	Duffy et al. (2003), Zhao et al. (2003)

Table 3.3 Case distribution for generalization analysis

	Coverage	Spatial extent	Cultural contexts	Fields with temporal extent
Distribution	5	3	15	20

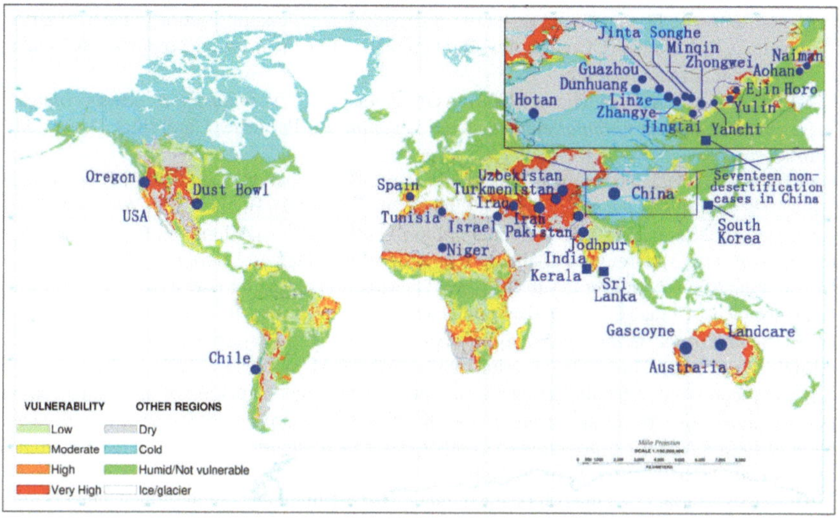

Fig. 3.6 Seven cases in northwest China and forty-three cases used for generalization analysis. *Note* (1) This figure was adapted from the desertification vulnerability map by U.S. Department of Agriculture, Nature Resources Conservation Service, Soil Survey Division, & World Soil Resources. 1998. Washington, DC. http://soils.usda.gov/use/worldsoils/mapindex/desert.html. (2) County boundaries are not authoritative. (3) Seventeen non-desertification cases in China, the South Korea case, the Sri Lanka case, and the Kerala case in India are non-desertification cases. All other cases are about combating desertification

References

Adam, S., & Kriesi, H. (2007). The network approach. In P. A. Sabatier (Ed.), *The theories of the policy process* (2nd ed., pp. 129–154). Boulder, CO: Westview Press.

Allais, M. (1953). Le comportement de l'homme rationnel devant le risque: critique des postulats et axiomes de l'école Américaine. *Econometrica,* 21(4), 503–546.

Allison, G. T. (1971). *Essence of decision: Explaining the Cuban missile crisis*. Boston: Little, Brown.

Anderson, J. (1975). *Public policy-making*. New York: Praeger.

AXBW, Anxixian Xianzhi Bianzuan Weiyuanhui. (1992). *Anxi xianzhi* (*County annals of Anxi*). Beijing: Zhishi Chubanshe (In Chinese).

Arnold, R. D. (1990). *The logic of congressional action*. New Haven: Yale University Press.

Arrow, K. (1963). Social choice and individual values (2nd ed.). New York: Wiley and Sons.

Avruch K., Black, P. W., Scimecca, J. A. (1991). *Conflict resolution: Cross-cultural perspectives*. Westport, CT: Praeger.

Baumgartner, F. R., & Jones, B. D. (1993). *Agendas and instability in American politics*. Chicago and London: The University of Chicago Press.

Berry, F. S., & Berry, W. (1990). State lottery adoptions as policy innovations: An event history analysis. *American Political Science Review*, 84 (2), 397–415.

Berry, F. S., & Berry, W. (1992). Tax innovation in the states: Capitalizing on political opportunity. *American Journal of Political Science*, 36 (3), 715–742.

Berry, L., Campbell, D. J., and Emker, I. (1977). Trends in man-land interaction in the west african Sahel. In D. R. J. Dalby, H. Church, & F. Bezzaz (Eds.), *Drought in Africa 2* (pp. 83–92). London: International African Institute.

Betsill, M. M., & Corell, E. (Eds). (2008). *NGO diplomacy: The influence of nongovernmental organizations in international environmental negotiations*. Publisher Cambridge, Mass.: MIT Press.

Blomquist, W. (2007). The Policy Process and Large-N Comparative Studies. In P. A. Sabatier (Ed.), *The theories of the policy process* (2nd ed., pp. 261–289). Boulder, CO: Westview Press.

Brewer, G., & deLeon, P. (1983). *The foundations of policy analysis*. Monterey, CA: Brooks/Cole.

Bryson, R. A. (1973). Drought in Sahelia: Who or what is to blame? *Ecologist*, 3(10), 366–371.

Bureau of the Ministry of Forestry in Charge of Construction of Shelter Forest in North, Northeast, and Northwest China [Ed.]. (1990). Spring has come to the desert. In *A rising green great wall: Construction of the sanbei shelter-forest system* (pp. 22–23). Beijing: Dadi Publishing House. (English version of Chinese manuscript, quoted from D. M. Williams, 1997).

CCICCD (Secretariat of China National Committee for the implementation of the United Nations Convention to Combat Desertification). (April 2000). *China national report on the implementation of United Nations Convention to Combat Desertification and National Action Programme to Combat Desertification*. Beijing: Author.

CCICCD (Secretariat of China National Committee for the implementation of the United Nations Convention to Combat Desertification). (April 2002). *China national report to implement the United Nations Convention to Combat Desertification* (*UNCCD*). Beijing: Author.

Chao, S. (1984). The sandy deserts and the Gobi of China. In F. El-Baz (Ed.). *Desert and Arid Lands* (pp. 95–113). The Hague: Martinus Nijhoff Publishers.

Confucian analects. (1900). In *Zhushi jiaozheng huaying sishu* (*Si shu*) (English & Chinese) (Trans. by J. Legge) (pp. 1–218). Shanghai: Shangwu Yinshuguan.

CSC, China State Council. (1994). *China's agenda 21: White paper on China's population, environment, and development in the 21st century*. (Adopted at the 16th Executive Meeting of the State Council on 25 Mar.). Beijing: China Environmental Science Press (English and Chinese versions).

Chouhan, T. S. (1992). *Desertification in the world and its control*. Jodhpur: Scientific Publishers.

Ci, L. (2001). China's experience with calamitous sand-dust storms. In Y. Yang, V. Squire, & Q. Lu (Eds.), *Global Alarm: Dust and Sandstorms from the World's Drylands* (pp. 109–124). Haus Carstanjen: UNCCD. Retrieved October 11, 2005, from http://www.unccd.int/publicinfo/duststorms/menu.php.

Dahl, R. A. (1961). *Who governs? Democracy and power in the American city*. New Heaven: Yale University Press.

De Groot, R. S., Wilson, M. A., and Boumans, R. M. J. (2002). A typology for the classification, description and valuation of ecosystem functions, goods and services. *Ecological Economics*, 41(3), 393–408.

Denhardt, R. 2004. *Theories of Public Organization* (4th ed.). Belmont, CA: Wadsworth/Thompson Learning.

Dewey, John. 1927. The Public and its Problems. New York: Holt.

Dong, J. (2002). *Sandstorms in Beijing-Occurrence, protection and control*. TRITA-LWR Master Thesis. Stockholm: Department of Land and Water Resources Engineering Royal Institute of Technology (KTH).

Downs, A. (1972). Up and down with ecology: The issue attention cycle. *Public Interest*, 28 (summer), 38–50.

Duffy, P., & Migongo-Bake, E. (2003). The application of EIA in the evaluation of sustainable desertification control activity: examples from Namibia and Inner Mongolia. *Impact Assessment and Project Appraisal*, 21(1), 65–73.

DSBW, Dunhuang Shizhi Bianzuan Weiyuanhui. 1994. *Dunhuang Shizhi* (*City Annals of Dunhuang*). Beijing: Xinhua Chubanshe (In Chinese).

Frederickson, H. G, & Smith, K. B. (2003). *The public administration theory primer*. Boulder and Cumnor Hill: Westview Press.
Frenzel, L. E., Jr. (1987). *Understanding expert systems*. Indianapolis: Howard W. Sams & Company.
GTZ. (2005). *Face sheet desertification: Africa. Convention project to combat desertification*. Retrieved December 2, 2005, from www2.gtz.de/desert/download/basicfactssheets/factsheet_africa.pdf.
Held, D., Mcgrew, A., Goldblatt, D., Perraton, J. (1999). *Global transformations: Politics, economics and culture*. Stanford: Stanford University Press.
He, B. (1991). *China on the edge: The crisis of ecology and development*. San Francisco: China Books and Periodicals.
Hirschman, A. O. (1970). *Exit, voice, and loyalty*. Cambridge, Massachusetts: Harvard University Press.
Ingram, H., Schneider A. L., & Deleon P. (2007). Social Construction and Policy Design. In P. A. Sabatier (Ed.), The theories of the policy process (2nd ed., pp. 93-126). Boulder, CO: Westview Press.
JXBW, Jingtai Xianzhi Bianzuan Weiyuanhui. (1996). *Jingtai xianzhi* (*County annals of Jingtai*). Lanzhou: Lanzhou University Press (In Chinese).
JDBW, Jintaxian Difangzhi Bianzuan Weiyuanhui. (1992). *Jinta xianzhi* (*County annals of Jinta*). Lanzhou: Gansu Renmin Chubanshe (In Chinese).
Johnson, D. L. (1977). *The human face of desertification. Economic Geography*, 53(4), 317–432.
Jones C. (1970). *An introduction to the study of public policy*. Belmont, CA: Wadsworth.
Kates, R. W., Johnson, D. L., & Johnson-Haring, K. (1977). *Population, society and desertification*. Nairobi: UN Conference on Desertification.
Kettl, D. (2000). *The global public management revolution: A report on the transformation of governance*. Washington, D. C.: Brookings Institution.
King, G., Keohane, R. O., & Verba, S. (1994). *Designing social inquiry: Scientific inference in qualitative research*. Princeton: Princeton University Press.
Kingdon, J. W. (2003). *Agendas, alternatives, and public policies* (2nd ed.). New York: Longman.
Kreps, D., & A.M. Spence (1985). Modelling the role of history in industrial organization and competition. In G. Feiwel (Ed.), Issues in Contemporary Microeconomics and Welfare (pp. 340–378). London: Macmillan.
Lasswell H. (1956). *The decision process*. College Par, MD: University of Maryland Press.
Latham, E. (1951). *The governmental process*. New York: Knopf.
Lindblom, C. E. (1959).The science of muddling through. *Public Administration Review*, 19(2), 79–88.
Lindblom, C. E. (1968). *The policy-making process*. Englewood Cliffs, NJ: Prentice-Hall, Inc.
Lindblom, C. E. (1977). *Politics and markets: The world's political-economic systems*. New York: Basic Books, Inc., Publishers.
LXBW, Linzexian Xianzhi Bianzuan Weiyuanhui. (2001). *Linze xianzhi* (*County annals of Linze*). Lanzhou: Gansu Renmin Chubanshe (In Chinese).
Lowi, T J. (1979). *The end of literalism: The second republic of the United States* (2nd ed.). New York and London: W.W. Norton & Company.
Lu, Q., Yang, Y., Wang, S., Liu, T., etc. (Eds.). (2004). *Zhongguo zhisha qishilu*. Beijing: Kexue Chubanshe (In Chinese).
Lynn, L. Jr., Heinrich, C., & Hill, C. (2000). Studying governance and public administration: Why? How? In C. Heinrich, & L. E. Lynn Jr. (Eds.), *Governance and performance: New perspectives* (pp. 1–33). Washington, DC: Georgetown University Press.
Lynn, L. Jr., Heinrich, C., & Hill, C. (2001). *Improving governance: A new logic for empirical research*. Washington, DC: Georgetown University Press.
Mattsson, J.O., & Nilsén, J. (1996). The transport of Saharan dust to southern Europe: a scenario. *Journal of Arid Environments*, 32(2), 111– 119.

MEA, Millennium Ecosystems Assessment. (2005). *Ecosystems and human well-being: Desertification synthesis*. Washington, DC: World Resources Institute. Retrieved December 10, 2005, from www.maweb.org//proxy/document.355.aspx.

MXBW, Minqin Xianzhi Bianzuan Weiyuanhui. (1994). *Minqin xianzhi* (*County annals of Minqin*). Lanzhou: Lanzhou University Press (In Chinese).

Morgan, M. G., & Peha, J. M. (2003). *Science and technology advice for congress*. Washington, DC: Resources for the Future.

Nakamura, R. (1987). The textbook process and implementation research. *Policy Studies Review*, 7(1), 142–154.

North, D. C. (1981). *Structure and change in economic history*. New York: W. W. Norton & Company, Inc.

Olson, M. (1971). *The logic of collective action: public goods and the theory of groups*. Cambridge, MA: Harvard University Press.

Ostrom, V. (1997). *The meaning of democracy and the vulnerability of democracies a response to Tocqueville's challenge*. Ann Arbor, MI: The University of Michigan Press.

Ostrom, E. (2005). *Understanding institutional diversity*. Princeton, NJ: Princeton University Press.

Ostrom, E. (2007). Institutional rational choice: An assessment of the institutional analysis and development framework." In P. A. Sabatier (Ed.), *Theories of the policy process* (2nd ed.) (pp. 21–64). Boulder, CO: Westview Press.

Pease, P. P., Tchakerian, V. P., & Tindale, N. W. (1998). Aerosols over the Arabian Sea: geochemistry and source areas for aeolian desert dust. *Journal of Arid Environments*, 39(3), 477–496.

Przeworski, A., & Teune, H. (1970). *The logic of comparative social inquiry*. New York: Wiley-Interscience.

Rasmusen, E. (2001). *Game and information: An introduction to game theory* (3rd ed.). Oxford: Blackwell Publishers Ltd.

Reynolds, J.F., Smith, D.M.S., Lambin, E.F., Turner, B.L., Mortimore, M., Batterbury, S.P.J., Downing, T.E., Dowlatabadi, H. (2007). Global desertification: building a science for dryland development. Science, 316, 847–851.

Sabatier, P. A. (ed.).2007. *The theories of the policy process* (2nd ed.). Boulder, CO: Westview Press.

Sabatier, P. A., & Jenkins-Smith, H. C. (Eds). (1988). Policy change and policy-oriented learning: exploring an advocacy coalition framework. *Policy Sciences*, 21 (2–3), 123-272.

Sabatier, P. A., & Jenkins-Smith, H. C. (Eds.). (1993). *Policy change and learning: An advocacy coalition approach*. Boulder, CO: Westview Press.

Sabatier, P. A., & Weible, C. M. (2007). The advocacy coalition framework: Innovations and clarifications. In P. A. Sabatier (Ed.), *The theories of the policy process* (2nd ed., pp. 189–220). Boulder, CO: Westview Press.

SACDS. (1977). Science Associations to Convene Desertification Seminar. *Science*, New Series, 197(4303), 548–549.

Schattschneider, E. E. (1960). *The semisovereign people: A realistic view of democracy in America*. Belmont, CA: Wadsworth/Thompson Learning.

Seely, M. K. (1998). Can science and Community Action Connect to Combat Desertification? *Journal of Arid Environments*, 39(2), 267–277.

Shadish, W. R., Cook, T. D., & Campbell, D. T. (2002). *Experimental and quasi-experimental designs for generalized causal inference*. Boston: Houghton Mifflin.

Sheehy, D. P. (1992). A perspective on desertification of grazingland ecosystems in North China. *Ambio*, 21(4), 303–307.

Simon, H. A. (1967). *The shape of automation for men and management*. New York: Harper & Row.

Simon, H. A. (1997). *Administration behavior: A study of decision-making processes in administrative organizations* (4th ed.). New York: The Free Press.

Skuras, D., Dimara, E., & Vakrou, A. (2000). The day after grant-aid: Business development schemes for small rural firms in lagging areas of Greece. *Small Business Economics*, 14 (2), 125–136.

Teng, X, & Dang, Y. (2003). Technology on transforming low-yield poplars forest in loess hill and sandy land. *Inner Mongolia Forestry Investigation and Design*, 3, 60–62 (In Chinese with English Abstract).

Thomas, D. S. G. (1997). Science and the desertification debate. *Journal of Arid Environments*, 37(4), 599–608.

True J. L., Jones B. D., & Baumgartner F. R. (2007). Punctuated-equilibrium theory: explaining stability and change in public policymaking. In P. A. Sabatier (Ed.), The theories of the policy process (2nd ed., pp. 155–188). Boulder, CO: Westview Press.

UNEP, United Nations Environmental Programme. (2003). *Global environmental outlook 3*. Retrieved October 3, 2008, from www.unep.org/geo/geo3/english.

UN, United Nations. (1992). *Managing fragile ecosystems: Combating desertification and drought, chapter 12 of agenda 2*. New York: United Nations.

UNCCD, United Nations Convention to Combat Desertification. (1994). *Elaboration of an international convention to combat desertification in countries experiencing serious drought and/or desertification, particularly in Africa* (U.N. Doc. A/AC.241/27, 33 I.L.M. 1328). New York: United Nations.

Wang, Y., & Dong, G. (1994). Sand sea history of the Taklimakan for the past 30,000 years. *Geografiska Annaler. Series A, Physical Geography*, 76A (3), 131–141.

Wang, T., & Wu, W. (2005). Sandy desertification in northern China. In K. A. Day (Ed.), *China's environment and the challenge of sustainable development* (pp. 233–247). New York: M. E. Sharpe.

Wang, T, & Zhu, Z. (2001). Some problems of desertification in northern China. *Quaternary Sciences (Disiji Yanjiu)*, 1, 56–65 (In Chinese with English abstract).

Wang, S., Dong, G., Shang, K. &Chen, H. (2001). Progress of research on understanding sand and dust storms in the world. In Y. Yang, V. Squire, & Q. Lu (Eds.), *Global alarm: Dust and sandstorms from the world's drylands* (pp. 29–48). New York: United Nations. Retrieved September 15, 2005, from http://www.unccd.int/publicinfo/duststorms/menu.php.

Wang, R., Zhang, H., & Ma, Y. (2003).The reason for ecological fragility in Hotan river basin. *Journal of Northwest University* (Natural Science Edition), 33(1), 106–110.

Wheaton, E. E., Arthur, L.M., Chorney, B., Shewchuk, S., Thorpe, J., Whiting, J., & Wittrock, V. (1992). The prairie drought of 1988. *Climatological Bulletin*, 26(3), 188–205.

Wu, J. (2001). Desertification. In R, Robinson (Ed), *Plant sciences for students* (pp. 70–73). New York: Macmillan Reference.

Wu, J. (2005) A new paradigm for a transdisciplinary science of desertification. *Diversity and Distributions*, 11(4), 362–363.

Wu, J. (2007). Scale and scaling: A cross-disciplinary perspective. In J. Wu, & R J. Hobbs. (Eds.), *Key topics in landscape ecology* (pp. 115–142). Cambridge: Cambridge University Press.

Yaalon, D. (1996). Comments on the source, transportation and deposition of Saharan Dust to Southern Europe. *Journal of Arid Environments*, (36), 193–196.

Yang, L. (2000). A new speculation on the problems of government functions: From a product perspective. *Seeker*, 1, 32–38 (In Chinese).

Yang, L. (2002). *Product classification and a formal model of identifying product classification*. Master thesis, Peking University, Beijing (In Chinese).

Yang, L. (2004). *Chaoyue zhengfu yu chaoyue qiye: Zhengfu lilun yu qiye lilun de da shehui kexue he chanpin yanjiu fa (Beyond government and firms: A study on the theory of the firm and the theory of government using the big-social-science and product analysis approaches)*. Beijing: China Economy Press (In Chinese).

Yang, L. (2006a). *The impact of social capital of qiaoxiang ties on local economic development: Overseas Chinese merchants in Gongdong and Fujian Provinces in China*. Master thesis, Indiana University, Bloomington.

Yang, L. (2006b). Product approach to product classification within a framework of product approach. *Chinese Public Administration Review*, 1(1), 30–69.

Yang, L. (2007a). Scholar-based governance: A fourth model to collective action. *Chinese Public Administration*, 259 (1), 96–103 (In Chinese with English abstract).

Yang, L. (2007b). Building a multi-collaborative community governance system to resolve the dilemma of collective action—a framework of "product-institutional" analysis (PIA). *Journal of Public Management*, 4(2), 6–23 (In Chinese with English abstract).

Yang, L. (2008). *Wanmei quanmian chanpin guanli: Shehui jiegou he guanli de chanpin fenxi (Perfect total product management: A product analysis of social structure and management)*. Beijing: Beijing University Press, Beihang University Press (In Chinese).

Yang, L., & Lan, Z. (2008). The impact of the internet on expert-citizen interactions in public policymaking. In *the annual meeting of the Midwest Political Science Association 67th Annual National Conference.* Chicago: The Palmer House Hilton. Retrieved from March 12, 2008, from http://www.allacademic.com/meta/p360906_index.html.

Yang, L., & Yang, A. (2004). On the concepts of institutions from three perspectives. *Journal of Renmin University of China*, 2, 115–121 (In Chinese with English Abstract).

Yang, Y., Squires, V. R., & Lu, Q. (2001). *Global alarm: Dust and sandstorms from the World's Drylands*. Haus Carstanjen: UNCCD. Retrieved October 11, 2005, from http://www.unccd.int/publicinfo/duststorms/menu.php.

Zhao, C., Chen, G, Zhang, T, & He, X. 2003. Chapter 19. In T. Wang (Ed.), *Desert and Desertification in China* (pp. 816–876). Shijiazhuang: Hebei Science and Technology Publishing House (In Chinese).

Zhao, L., Chen, H., & Bai, X. (2004). Evaluation of ecological service value of soil and water conservation based on GIS technology in Yijinhuoluo County. *Soil and Water Conservation in China*, 7, 20–22(In Chinese).

Zhu, Z. (1990). Desertification in the northern territory of China: Present status and trend of development. *Journal of Chinese Geography*, 1(1) 61 –70.

Zhu, Z. (1998). Concept, cause and control of desertification in China. *Quaternary Sciences (Disiji Yanjiu)*, 18(2), 145–155 (In Chinese).

Zhu, Zhenda. 1989. Advances in desertification research in China. *Journal of Desert Research*, 9(1), 1–13 (In Chinese with English abstract).

Zhu, Z. & Liu, S. (1983). *Combating Desertification in Arid and Semi-arid Zones in China*. Lanzhou: Institute of Desert Research, Academia Sinica.

Zhu, Z., Liu, S. & Di, X. (1988). *Desertification and rehabilitation in China*. Lanzhou: The International Centre for Education and Research on Desertification Control.

ZXBW, Zhongweixian Xianzhi Bianzuan Weiyuanhui. (1995). Zhongwei xianzhi (County annals of Zhongwei). Yinchuang: Ningxia Renmin Chubanshe.

Chapter 4
Significance and Mechanisms of Scholars' Participation

Boldly hypothesize and carefully validate (In Chinese: "大胆假设, 小心求证。").

—Hu (1928)

Based on field survey and interview data from seven counties in northwest China, this chapter studies the significance of scholars' participation in combating desertification, its relationship with the degree of land amelioration, and the key factors and mechanisms that were most crucial to the effectiveness of scholars' participation. This chapter indicates that scholars played an important role in combating desertification in general, although a higher level of scholar participation did not always correspond to a higher level of land amelioration. Thirteen related, but different, factors were responsible for the effectiveness and underlying mechanisms of scholars' participation in combating desertification. The first seven are related to the characteristics of scholars' behaviors, while the others are germane to the attributes of scholars' organizations and external support in combating desertification. Furthermore, these factors were positively correlated with the success rate of land amelioration efforts.

4.1 Significance of Scholars' Participation and Its Relationship with Land Amelioration

4.1.1 Significance of Scholars' Participation

Scholars' participation is evaluated by surveys regarding three aspects: on recognizable and evident scholars, on both evident scholars and latent scholars, and on scholars' organizations. Six criteria were given to evaluate scholars' participation: very important, important, moderately important, dispensable, negative, and unknown. Except for Minqin County, more than 50% of the respondents indicated that recog-

L. Yang, *Knowledge-Driven Governance*,
https://doi.org/10.1007/978-981-13-2910-4_4

Table 4.1 Evaluation of scholars' participation

Counties options	Zhongwei	Minqin	Jingtai	Linze	Jinta	Guazhou	Dunhuang
a. Recognizable and evident scholars							
Very important	0.43	0.23	0.40	0.34	0.29	0.46	0.40
Important	0.26	0.24	0.28	0.44	0.28	0.24	0.39
Total	0.69	0.47	0.68	0.78	0.57	0.70	0.79
b. All kinds of scholars							
Very important	0.38	0.28	0.37	0.33	0.20	0.29	0.39
Important	0.34	0.36	0.36	0.46	0.47	0.31	0.42
Total	0.72	0.64	0.73	0.79	0.67	0.60	0.81
c. Scholars' organizations							
Very important	0.48	0.27	0.56	0.33	0.20	0.40	0.51
Important	0.28	0.29	0.21	0.46	0.47	0.27	0.33
Total	0.76	0.57	0.77	0.79	0.67	0.67	0.84
d. Average of all scholars							
Average	0.72	0.56	0.73	0.79	0.64	0.66	0.81

nizable and evident scholars' participation in combating desertification was "very important" or "important" (Table 4.1a). That is, most respondents agreed that the recognizable and evident scholars played an important role in combating desertification.

When evaluating the impact of both kinds of scholars in combating desertification, more than 60% of the respondents indicated that they were "very important" or "important" in all seven counties (Table 4.1b). In Dunhuang, even more than 80% of the respondents indicated this. That is, when both kinds of scholars were considered, the scholars' total impact in combating desertification was increased.

For the evaluation of the impact of the scholars' organizations, more than 50% of the respondents in all seven counties indicated that they were "very important" or "important" (Table 4.1c). In Dunhuang, this number surpassed even 80%. That is, most respondents agreed that the scholars' organizations played an important role in combating desertification.

To evaluate the scholars' total effects, all these three aforementioned aspects were simultaneously considered to calculate their averages (Table 4.1d). In all the seven counties, their averages are larger than 56%. In Dunhuang, it is even larger than 80%. That is, most respondents agreed that the scholars played an important role in combating desertification, and scholar-based governance was an important governance mechanism.

Table 4.2 Evaluation of desertification severity and comparison between the land amelioration degree and the desertificated area

Counties options	Zhongwei	Minqin	Jingtai	Linze	Jinta	Guazhou	Dunhuang
a. Evaluation of desertification severity							
Very serious	0.09	0.82	0.15	0.11	0.15	0.22	0.33
Serious	0.22	0.16	0.49	0.27	0.45	0.36	0.51
Moderately serious	0.33	0.01	0.19	0.33	0.15	0.11	0.10
Not obvious	0.06	0.01	0.04	0.05	0.04	0.02	0.01
Greatly ameliorated	0.29	0.00	0.13	0.18	0.18	0.30	0.05
b. Comparison between the land amelioration degree and the desertificated area							
Amelioration	0.29	0.00	0.13	0.18	0.18	0.30	0.05
Desertificated areas	No data	55.18	21.08	14.53	13.24	9.39	34.45

4.1.2 Degree of Land Amelioration

Five criteria were given to evaluate desertification severity in the survey: very serious, serious, moderately serious, not obvious, and greatly ameliorated (Table 4.2a). Percentages of the criterion "greatly ameliorated" are used to evaluate the degree of land amelioration. The respondents indicated that the results of combating desertification are different in the seven counties (Table 4.2b). Furthermore, the order, from the least to the highest, of respondents in the six counties that thought land had been ameliorated is consistent with the order, from the highest to the least, of the percentage of their desertified area (see Li et al. 2003; Zhongwei was not included in this study).

4.1.2.1 The Relationship Between Scholars' Participation and Land Amelioration

Counties in which scholars' participation was high did not necessarily have land amelioration that was also high. For example, in Dunhuang about 81% of the respondents indicated that scholar' participation was important, but only about 5% of the respondents agreed that desertification had been ameliorated (see Table 4.3a).

In order to clarify, I divided both the land amelioration degree and the significance of scholars' participation into three levels: high (larger than 0.29 for the amelioration degree, and 0.72 for the significance of the scholars' impact), middle (between 0.13 and 0.19 for the amelioration degree, and 0.6–0.7 for the scholars' impact), and low (lower than 0.1 for the amelioration degree, and 0.6 for the scholars' impact)

Table 4.3 Comparison between land amelioration and scholars' participation

Counties	Zhongwei	Linze	Jingtai	Guazhou	Jinta	Dunhuang	Minqin
a. Comparison of percentages							
Land amelioration	0.29	0.18	0.13	0.30	0.1	0.05	0.00
Scholars' participation	0.72	0.79	0.73	0.66	0.64	0.81	0.56
b. Comparison of three levels							
Levels of land amelioration	High	Middle	Middle	High	Middle	Low	Low
Levels of participation	High	High	High	Middle	Middle	High	Low
Results	Successful	Semi-successful	Unsuccessful				

(Table 4.3b). One exception is Guazhou. Before 1966, the old county seat was heavily influenced by sand dunes and desertification, and in 1966 the county seat was moved to the current place. After 1949, people were constantly organized to combat desertification, and really in some places desertification was ameliorated. People especially in the county seat in fact overestimated the degree of land amelioration, and its problem of desertification is still very serious (Dong 2007; Zhang et al. 2003). Thus, I characterized its level of land amelioration as "middle." Furthermore, I deemed the cases with high levels of land amelioration as successful cases of combating desertification, with middle levels of land amelioration as semi-successful, and with low levels as unsuccessful (also see Table 4.3b). Thus, the nonlinear relationship between land amelioration and scholars' participation is shown clearer. For example, Zhongwei is a successful case with a high level of scholars' participation, and Linze and Jingtai are two semi-successful cases with high levels of scholars' participation. Dunhuang, however, is an unsuccessful case with a high level of scholars' participation.

4.1.3 Comparison with the Results from the Interview Data

The aforementioned findings from the survey data are consistent with the findings from the interview data. For example, scholars' impact in Zhongwei was praised by interviewees, especially the scholars from the 1950s to the 1970s. One scholar recalled stories of several eminent Chinese scientists (e.g., Kezhen Zhu, Shen'e Liu, Yin Liu, Minggang Li, and Junzhao Ji) who made important contributions to combating desertification. One researcher argued that contemporary scholars' impact accounted for only about 30% of desertification control, while they accounted for at least 40–60% from the 1950s to the 1970s. All ten interviewees mentioned that present scholars did not help as much as their predecessors. Also, all ten interviewees mentioned the activities of combating desertification from the 1950s to the 1970s and said the condition has been greatly ameliorated.

Scholars' participation in Jingtai was also mentioned. For example, two farmers praised their important roles. Several government officials, however, argued that they only focused on technical instructions. Also, four interviewees (including one scholar, two farmers, and one official) indicated that their functions were heavily influenced by financial support. The importance of the cooperation among scholars was also mentioned. For example, one scholar told me the story about their cooperation with some scholars in the Desert Control Station in Wuwei City. Furthermore, all nine interviewees agreed that the trend of desertification had been partly controlled, particularly after the 1990s.

In Minqin, several interviewed government officials and most scholars agreed that scholars played some roles in desertification control. One official argued that their work increased the name recognition of Minqin, leading the central government to dramatically increase investment. All of them, however, agreed that scholars' functions were limited by political institutions. Scholars complained that their words carry little weight in policymaking, and officials were too bureaucratic and did not respect scholars. Officials, however, complained that scholars were not likely to cooperate with governments and had no practical knowledge or suggestions. One official argued that scholars did not have their own definite view and often changed their ideas according to government officials' likes and dislikes. Several citizens and farmers agreed that scholars could have played more important roles, if they did better. They also pointed out that their functions were limited because of little financial support, low salary, and selfish behaviors of scholars. Furthermore, all 29 interviewees mentioned the severity of desertification, and no interviewees mentioned amelioration. A few scholars, however, indicated that the severity of desertification in fact was not as serious as described by local officials, who were likely to aggrandize its severity in order to get more financial support from the central government.

Scholars' impact in Linze was also mentioned by the interviewees. One official argued that in a society with five kinds of members as stated in Chap. 1, scholars' functions accounted for at least 20% in combating desertification. He also suggested that scholars should increase cooperation with governments and supply more practical methods for desertification control. Two researchers agreed that scholars played an important role in desertification control, but one researcher in the Desert Control Station pointed out that it was very difficult to evaluate scholars' impact, and governments should cooperate more with scholars. Other interviewees, however, did not answer this question. Furthermore, all eight interviewees agreed that the condition of desertification control had been relatively ameliorated. Also, they pointed out that the problem of desertification in Linze originally was not very serious, as compared with Minqin, Guazhou, and Dunhuang.

Interviewees in Jinta agreed that scholars played an important role in combating desertification. For example, among four farmers, except for one who did not answer this question, all three other farmers praised the scholars' functions in desertification control. One scholar, however, pointed out that scholars' impact on local problems was limited if they focused on pure science, while their impact was large when they focused on practical techniques. Furthermore, all six interviewees agreed that desertification control had been ameliorated in Jinta and was better than Minqin and Dunhuang, although it did not do as well as in Zhongwei.

All eight interviewees in Guazhou mentioned the roles of scholars in combating desertification, but only two of them thought their impact was large, and the rest thought their impact was very limited. Furthermore, all the interviewees agreed that desertification control had been ameliorated. Also, all of them mentioned serious desertification and dust storms during the 1950s and 1960s. An old woman even recalled some dangerous and miserable experiences before 1949.

In Dunhuang, several interviewees mentioned the significance of scholars' participation, but they pointed out that their impact was mainly on agricultural techniques rather than directly on desertification control. For example, one farmer indicated that he had good relationships with a technician in the Center of Forestry Technology who often directly went to his home to teach him how to plant grapes. Several businessmen also mentioned scholars' technical instructions. Furthermore, all nine interviewees mentioned desertification severity, especially the danger of desertification to Yueya Spring, a famous scenic location in Dunhuang.

4.2 Mechanisms and Thirteen Related Factors of Scholars' Participation

I also paid close attention to the mechanisms that enable effective and efficient scholar participation. According to the PIA framework, some factors such as scholars' production position, their social position, their resources/capital, their motivations and preferences, some formal/informal rules as individual or organizational factors,[1] and some external resource factors affecting their strategy sets to participate in desertification control should be considered. Also, this study found that these factors differentiated scholars' participation in combating desertification in the seven counties. That is, the empirical findings are consistent with the theoretical analysis according to the PIA framework.

As for production position, this study found that whether scholars as producers of good desertification control significantly participated in activities of combating desertification (called F1) should be analyzed. Furthermore, two major factors—how long scholars stayed in the county (time) and where they came from (place)—heavily decided their roles in combating desertification.

[1]Scholars' utility function was considered when I studied the rule of the combination between economic development and environmental protection.

Regarding social position, two problems were important. The first was whether scholars had independent social identity, and the second was what their social status was. Social identity is an individual-based perception; it describes the individual's self-concept derived from perceived membership of social groups (Hogg and Vaughan 2002). Social position affected scholars' participation in desertification control through influencing social status. Thus, scholars' independent social identity and social status (called F2) should be considered.

Among the six kinds of resources—physical and natural, human, financial, information and knowledge, social, and organizational or institutional capital—this study found that scholars' human capital (especially their enthusiasm and capability of participation) (F3), knowledge capital (F4), and social capital (F5) were the main types affecting their participation in desertification control. For their motivations and preferences, their social responsibility and spirit of practice (F6) were the two most important factors in combating desertification. For formal and informal rules, this study found that respect for ideas of other social actors (F7) was an important informal rule which influenced scholars' strategy sets of combating desertification.

Furthermore, this study found that federal organizational structure and concrete organizational purposes (F8), democratic, collaborative management, and the federal mechanism of award and sanction (F9), scholar-entrepreneurship (F10), the combination between economic development and environmental protection (F11), and experiment-extension governance method (F12) as formal and informal rules were the most important organizational factors differentiating scholars' participation in the seven counties. Also, scholars' participation was influenced by some external resource factors. This study indicated that external resource support (F13), especially technical, financial, and institutional support, should be considered.

The objective of this section is to explore the mechanisms of scholars' participation in combating desertification as well as analyze how these 13 factors differentiated scholars' participation in combating desertification in the seven counties from 1949 to 2007. Roughly, among these 13 groups of mechanisms and factors, the first seven (F1–F7) analyze the characteristics of scholars' individual behaviors, and the other six (F8–F13) analyze the mechanisms of organizing scholars' participation in combating desertification.

4.2.1 Characteristics of Scholars' Individual Behaviors and Its Seven Related Factors

F1. Significance of scholars' participation, time they stayed in the county, and places from where they came

The aforementioned analysis already showed that the significance of scholars' participation influenced scholars' impact on combating desertification, but high scholars' participation did not mean high scholars' impact on desertification control. Furthermore, the differences in the significance of scholars' participation (denoted as Factor F1.1) in the seven counties have been shown in Table 4.1.

Ten choices in the survey were given to evaluate how long scholars stayed in the county (F1.2): several hours, several days, several weeks, about one month, over two months, over a half year, over two years, and permanently live here. Percentages of the respondents to the choices over a half year (including over a half year, over a year, over two years, and permanently live here) in the seven counties were different and are divided into three groups: long (greater than 0.6), middle (between 0.2 and 0.6), and short (less than 0.2) (Table 4.4a).

Six choices in the survey were given to collect the information about places from where these scholars came: the same village, the same township, the same county, the same province, other provinces, Beijing, and foreign countries. Percentages of the respondents in the seven counties on coming from the same village, township, and county (F1.3) were different and were divided into three groups: high (greater than 0.6), middle (between 0.5 and 0.6), and low (less than 0.5) (Table 4.4b).

F2. Independent social identity and social status

Two problems were important to social position: the independence of scholars' social identity (F2.1) and the level of scholars' status (F2.2). The interviewees indicated that the scholars in Zhongwei, Jingtai, and Minqin were more independent than they were in the other counties, and they had more freedom and desire to do something they wanted to do (Table 4.4c). Their experimental research stations in particular were relatively independent from the county governments. The research stations in the other counties, however, were governmental bureaucracies. Also, the interviewees indicated that although independence, especially from the government, was a necessary condition for successful scholars' participation, some dependence allowed the scholars to cooperate with other social actors, particularly with the governmental officials as in Linze.

The survey data showed that the scholars in every county had relatively high social status (Table 4.4c). Also, the interview data indicated that high social status was a necessary condition for scholars' participation, but it did not guarantee successful scholars' participation.

F3. Enthusiasm and capability of scholars' participation

Eleven choices in the survey were given to evaluate disadvantages of scholars' participation. The percentages of the respondents lacking participatory enthusiasm, participatory capacity, and leadership and organizing capacity were different. Based on their total percentages, levels of scholars' enthusiasm and participatory capacity were broken down into three groups: high (less than 0.28), middle (between 0.29 and 0.43), and low (larger than 0.44) (Table 4.4d).

Participatory capacity included scholars' ability to learn new knowledge and to teach other social actors. The survey indicated that scholars in Jingtai had the highest participatory capacity while scholars in Linze had the lowest. These were different than the findings from the interview data and the direct observations. According to

Table 4.4 Differences in scholars' characteristics

Counties	Zhongwei	Linze	Jingtai	Guazhou	Jinta	Dunhuang	Minqin
a. Total percentages which scholars stayed in the seven counties over a half year							
Total evaluation	0.793	0.452	0.523	0.533	0.404	0.558	0.162
Classification	Long	Middle	Middle	Middle	Middle	Middle	Short
b. Total percentages for coming from the same village, township, and county							
Total evaluation	0.507	0.565	0.713	0.660	0.631	0.578	0.432
Classification	Middle	Middle	High	High	High	Middle	Low
c. Levels of scholars' independence and their social status							
Independence	High	Middle	High	Middle	Middle	Middle	High
Social status	High	High	High	High	High	High	High
d. Percentages of lacked resources and the levels of scholars' participation							
Enthusiasm	0.068	0.138	0.085	0.113	0.092	0.113	0.171
Capacity	0.096	0.234	0.062	0.060	0.142	0.145	0.165
Leadership	0.107	0.167	0.093	0.087	0.100	0.188	0.124
Total	0.271	0.540	0.240	0.260	0.335	0.445	0.460
Levels	High	Low	High	High	Middle	Low	Low
e. Total percentages of the three kinds of the most important knowledge							
Desertification	0.375	0.352	0.364	0.307	0.400	0.265	0.370
Local desertification	0.300	0.372	0.287	0.253	0.246	0.290	0.335
Environ-governance	0.168	0.155	0.128	0.150	0.177	0.203	0.245
f. Total percentages of unknowing local conditions							
Percentages	0.325	0.289	0.275	0.320	0.254	0.245	0.326
g. Total percentages of unknowing local social relationships and webs							
Percentages	0.057	0.167	0.070	0.037	0.100	0.096	0.103
h. Degrees of the scholars' three kinds of knowledge							
Scientific knowledge	High	S-high	S-high	S-middle	S-middle	S-middle	High
Local knowledge	High	S-high	S-high	Middle	Middle	S-middle	Middle
Social relationships	S-High	High	Middle	Middle	Middle	High	Low

(continued)

Table 4.4 (continued)

Counties	Zhongwei	Linze	Jingtai	Guazhou	Jinta	Dunhuang	Minqin
i. Levels of social capital							
Social capital	High	High	Middle	Middle	Middle	High	Low
j. Percentages of social capital							
With governments	0.071*	0.155*	0.105*	0.073*	0.131*	0.115*	0.112*
With farmers	0.032	0.050	0.012	0.050	0.004	0.020	0.087
With firms	0.046	0.046	0.016	0.033	0.023	0.030	0.093
With the fifth sector	0.032	0.017	0.016	0.017	0.019	0.008	0.112*
k. Percentages of strong social responsibility							
Percentages	0.15 [3]	0.10 [6]	0.081 [6]	0.103 [4]	0.158 [3]	0.103 [4]	0.17 [3]
l. Percentages of having no strong ability to practice and take actions							
Percentages	0.064	0.151	0.085	0.117	0.173	0.090	0.171
Levels	High	Low	High	Middle	Low	High	Low
Amended levels	High	S-high	High	Middle	Low	High	Low
m. Total percentages of the scholars' different roles							
Role 1	0.50 [1]	0.44 [2]	0.38 [2]	0.43 [1]	0.45 [2]	0.50 [1]	0.41 [1]
Role 2	0.47 [2]	0.54 [1]	0.43 [1]	0.38 [2]	0.47 [1]	0.41 [2]	0.51 [2]
Role 3	0.15 [3]	0.21 [3]	0.17 [3]	0.15 [3]	0.13 [3]	0.07 [3]	0.24 [3]
Role 4	0.054 [4]	0.05 [6]	0.03 [4]	0.04 [4]	0.04 [4]	0.02 [5]	0.05 [5]
Role 5	0.046 [5]	0.12 [4]	0.03 [5]	0.03 [6]	0.01 [6]	0.03 [4]	0.04 [6]
Role 6	0.03 [6]	0.08 [5]	0.004 [6]	0.04 [5]	0.015 [5]	0.01 [6]	0.09 [4]
n. Percentages of respecting and not respecting the other social actors' ideas							
Respecting	0.129*	0.113	0.089	0.070	0.058	0.088	0.099
Not respecting	0.096	0.113	0.070	0.037	0.073	0.098	0.121*
Differences	0.032	0.000	0.019	0.033	−0.015	−0.010	−0.021
Levels	High	Middle	Middle	High	Low	Low	Low

Note S- = Semi; for example, S-high = Semi-high; *means the highest; and [1]–[6] refer to the rank

them, Linze should be the second highest county because there were many scholars who directly took part in the governmental affairs. For example, the head of the Experimental Station of Combating Desertification often was the vice supervisor of the county. This was not found in the other counties, but most respondents did not know this.

F4. Three kinds of knowledge

The interviewees indicated that three kinds of knowledge for scholars were important. The first was related knowledge about combating desertification and environmental governance (F4.1). This was also echoed by the survey data. Ten choices in the survey were given to evaluate the importance of different kinds of knowledge in combating desertification: general climatic knowledge, scientific knowledge about desertification and dust storms, environmental governance knowledge, social governance knowledge, knowledge about laws and regulations, knowledge about local desertification and dust storms, knowledge about various local social relationships, local knowledge and local methods, irrigation knowledge, and agricultural and forestry knowledge. The respondents indicated that the three most important subtypes of knowledge about combating desertification and environmental governance were (1) general scientific knowledge of desertification and dust storms, (2) particular knowledge of understanding the local problems of desertification and dust storms, and (3) knowledge of environmental governance (evaluated by the percentages of the respondents) (Table 4.4e). The respondents in Linze, however, also indicated that the knowledge of laws and regulations was as important as the knowledge of environmental governance.

The second was knowledge about local conditions (F4.2). Many interviewees indicated that understanding local conditions and having some knowledge of local conditions were very important for scholars' participation in combating desertification. The survey data also showed this. For example, more than 24% of the respondents in the seven counties indicated that lack of regarding local conditions was the most important disadvantage of the scholars among the 14 choices (Table 4.4f). On the other hand, this meant that having enough knowledge of local conditions was very important for successful scholar-based governance.

The third was knowledge about local social relationships and webs (F4.3). This was also shown by the survey data. Among the 14 choices of scholars' disadvantages, the percentages of the respondents to being unaware of local social relationships and webs in Linze, Jinta, Dunhuang, and Minze were above the average (0.071) (Table 4.4g). In Jintai, it was close to the average (0.070).

The interviewees indicated that many scientists worked in Zhongwei to combat and study desertification, particularly during the 1950s, 1960s, and 1970s. Until 2007, many scientists were still living and working in Zhongwei as members of the *Shapotou* Desert Experimental Research Station and the Forestry Center of Sand Fixation. Thus, the degrees of scholars' scientific knowledge and local knowledge in Zhongwei were high. Scholars in Zhongwei had much knowledge on local social relationships from the 1950s to the 1970s. Current scholars, however, often did not have much knowledge about this, because many of them only focused on their own academic research and selfish interests rather than on practical knowledge and helping the local people to combat desertification. Because of their experience in building the National Station of Desert Control and implementing national programs, the degrees of scholars' scientific and local knowledge in Linze and Jingtai were also high. Compared with Zhongwei, however, their degree is at the "semi-high" level (between the high level and the middle level). Many farmers praised the work of scientists in Linze. This showed that these scholars had more knowledge on social relationships than the scholars in Jingtai. In Guazhou and Jinta, because only the county-owned stations of desert control were established, their degree of scholars' scientific knowledge was at the "semi-middle" level (between the middle level and the low level), while the degree of scholars' local knowledge was at the "middle." The interview data also indicated that scholars' knowledge on social relationships in these two counties was at the middle level. Although the scholars in Dunhuang had much knowledge on social relationships, their scientific and local knowledge only focused on grape planting and other technology of agriculture rather than directly on desertification control. Thus, their degree of scientific and local knowledge about desertification control was at the "semi-middle" level. Scientific and academic studies of scholars in Minqin were very famous, and the degree of scholars' scientific knowledge was at the "high" level. These scholars, however, did not have much knowledge on social relationships (at the middle level), and their local knowledge was limited (at the low level). Together, based on both the interview data and the survey data, the levels of the scholars' three kinds of knowledge in the seven counties are shown in Table 4.4h.

F5. Social capital

The interviewees pointed out that whether scholars had good relationships with other social actors (i.e., high social capital) played an important role in scholars' successful participation in combating desertification. Although scholars in Zhongwei, Linze, and Dunhuang all had good relationships with government officials, firms, and the farmers, they paid particular attention to different groups. For example, in Zhongwei, because firms played an important role in combating desertification, scholars often had very complex relationships with them. Scholars in Linze as stated above had very special relationships with governmental officials. Scholars in Dunhuang, however, had very good relationships with farmers, although their relationships were often based on economic relationships (e.g., researchers helped farmers to plant grapes, and so on) rather than on combating desertification. In Minqin, the scholars in the Experimental Station of Desertification almost had no important relationships with government, firms, and farmers. Together, the levels of scholars' social capital in the seven counties are shown in Table 4.4i.

Furthermore, the interviewees in all the seven counties indicated that scholars always had stronger relationships with government than with the other social actors. This was reflected in the survey findings. Among 14 choices of scholars' disadvantages, there were four choices considering their relationships with governments, farmers, firms, and the fifth sector. The respondents indicated that in all the seven counties the relationship with government was the most important one (Table 4.4j).

F6. Social responsibility, spirit of practice, and four major roles

The interviewees indicated that scholars' social responsibility and practical spirit strongly influenced their effectiveness in combating desertification. Sixteen choices in the survey were given to evaluate scholars' advantages in combating desertification. More than 10% of the respondents in the seven counties indicated that strong social responsibility was important and its rank was greater than six among 14 choices (Table 4.4k).

One criticism of scholars is that they do not have enough spirit of practice. This was included in the 14 choices of scholars' disadvantages in the survey. Except for in Zhongwei, more than 8.5% of the respondents indicated that lack of spirit of practice influenced the scholars' participation in combating desertification, and this number was greater than the average 7% among the 14 choices (Table 4.4l). Based on these percentages, the degrees of the scholars' responsibility and spirit of practice in the seven counties were divided into three groups: high (less than 0.06), middle (between 0.1 and 0.15), and low (greater than 0.15) (Table 4.4l). According to this table, Linze was at the low level. The interviewees, however, indicated that the scholars in Linze often directly participated in the activities organized by government, and the leader of the Experimental Station of Combating Desertification was the vice supervisor of the county. The respondents in the survey might deem these activities only as governmental activities and lowered the scholars' functions. Considering this, I amended the level in Linze to "semi-high" (Table 4.4l).

Based on the game theoretical analysis, scholars at least have six kinds of roles in combating desertification: as information providers for governments and farmers (Role 1), as scholar-entrepreneurs of farmers (Role 2), as pure game players (Role 3), as agents of governments (Role 4), as information providers, agents and scholar-entrepreneurs of firms (Role 5), and as information providers, agents, and scholar-entrepreneurs of the fifth sector (Role 6). A multiple-choice question in the survey was designed to evaluate this problem (Table 4.4m). The results showed that the two most important roles for scholars were information providers for governments or farmers and scholar-entrepreneurs of farmers (supported by over 37% of responders). In the four counties—Zhongwei, Guazhou, Dunhuang, and Minqin—the role as information providers was the most important. In the other three counties—Linze, Jingtai, and Jinta—the role as scholar-entrepreneurs of farmers is the first. When all counties are considered, acting as pure game players is the third most important role. The fourth important role is acting as agents of governments, which get four votes of the 4th rank, two votes of the 5th rank, and one vote of the 6th rank. These findings are consistent with their four roles stressed in my three-person game theoretical analysis.

F7. Respect for the ideas of the other social actors

The interviewees indicated that whether scholars respected the ideas of the other social actors, especially the farmers', influenced not only their communication with these other social actors but also the effectiveness of their participation in combating desertification. Among the 14 choices of scholars' advantages and the 14 choices of their disadvantages in the survey, except for in Jinta and Guazhou, both the percentages of "respecting" and "not respecting" other social actors' ideas were equal or greater than the average 7% (Table 4.4n). Especially, the percentage of "respecting" in Zhongwei was the highest, and the percentage of "not respecting" in Minqin was the lowest. This table, however, shows that the percentages of "respecting" and "not respecting" are not completely consistent with each other. Thus, their net values or differences ("respecting" subtracts "not respecting") were calculated to evaluate the degrees of scholars' respect for the other social actors. The values were broken down into three groups: high (greater than 0.03), middle (0.03–0), and low (less than −0.011) (Table 4.4n).

4.2.2 Mechanisms of Organizing Scholars' Participation

F8. Federal organizational structure and concrete organizational purposes

The interviewees argued that two problems were important to the scholars' organizations: organizational structure and purposes. Especially bottom-up federal organizations (F8.1) and concrete organizational purposes (F8.2) were important for successful scholars' participation in combating desertification. Four types of structure—hierarchical organizations and vertical control (Type 1), fragmented organizations without much communication (Type 2), fragmented organizations with much communication (Type 3), and bottom-up federal organizations (Type 4)—were evaluated in the survey (Table 4.5a). Except for in Jingtai and Minqin, more than 30% of the respondents indicated that bottom-up federal organizations were the most important organizations in these counties. The degrees of bottom-up federal organizations in the seven counties were broken down into three levels: high (larger than 0.35), middle (between 0.35 and 0.30), and low (0.29–2.75) (Table 4.5b).

A concrete organizational purpose refers to a purpose which is more concrete than so-called combating desertification. For example, in Zhongwei, the concrete purpose of combating desertification was to protect Yellow River and Lanzhou–Xinjiang railroad; in Jingtai, was to protect the Jingdian Pumping Irrigation Project; and in Guazhou, was to protect the county seat. Because the Jingdian Pumping Irrigation Project only covered part of the county, the purpose of combating desertification in the whole Jingtai County, however, was not as concrete as in Zhongwei and Guazhou. Thus, if the concrete purposes in Zhongwei and Guazhou were at the high level, it was at the semi-high level in Jingtai (Table 4.5b). The purposes of combating desertification in Minqin, Dunhuang, and Jinta in the past were even much less concrete than in Jingtai.

Linze had two county-owned stations of desertification control. One was Linze Experimental Station of Desertification Control (also called Linze Station of Vegetation Maintain before it became a subordinate organization of the Linze Forestry Bureau in 1986) which was built in 1976 and cooperated with Chinese Academy of Sciences (CAS) before 2001. Chinese Academy Sciences focused on academic studies, while this station focused on implementation and extension. It became an independent station in 2001 and mainly took charge of desertification control in the north of the county and two state-owned forestry centers (Wuquan and Shahe). The other station was Xiaoquanzi Station which was built in 1988. It mainly took charge of the sandy areas in the south of the county. Linze also had a comprehensive research station (Linze Inland River Basin Comprehensive Research Station) which was first built in 1975 by CAS and was approved to be a desert ecosystem research station of Chinese Ecosystem Research Network (CERN) in 2003. It worked together with the Linze Experimental Station of Desertification Control before 2001 and became an independent station in 2001. Later on, it was subordinate to the Cold and Arid Regions Environmental and Engineering Research Institute, CAS. Because of these, compared with Minqin, Dunhuang, and Jinta, Linze had more concrete purposes than in these three counties, although its purposes were not as concrete as in Zhongwei,

Table 4.5 Differences in mechanisms of organizing scholars' participation

Counties	Zhongwei	Linze	Jingtai	Guazhou	Jinta	Dunhuang	Minqin
a. Percentages of different organizational structures							
Type 1	0.293	0.184	0.310*	0.260	0.262	0.228	0.347*
Type 2	0.154	0.155	0.116	0.087	0.127	0.078	0.159
Type 3	0.214	0.172	0.120	0.147	0.277	0.258	0.204
Type 4	0.361*	0.335*	0.275	0.307*	0.323*	0.375*	0.290
b. Levels of the federal structure and concrete purposes							
Structure	High	Middle	Low	Middle	Middle	High	Low
Purposes	High	Middle	S-high	High	Low	Low	Low
c. Levels of democratic management							
Advantage	0.132	0.167	0.167	0.193	0.258	0.180	0.233
Disadvantage	0.064	0.042	0.058	0.067	0.035	0.030	0.087
Difference	0.068	0.126	0.109	0.127	0.223	0.150	0.146
Levels	Low	Middle	Low	Middle	High	Middle	Middle
Amended levels	Middle	Middle	Low	Middle	High	Middle	Middle
d. Levels of collaborative management							
Levels	High	High	Middle	S-middle	Middle	S-middle	S-middle
Percentages	0.068	0.063	0.058	0.04	0.023	0.018	0.056
Groups	High	High	S-high	Middle	Low	Low	S-high
e. Levels of the federal mechanism of awards and sanctions							
Advantage	0.068	0.172	0.074	0.067	0.1	0.083	0.121
Disadvantage	0.079	0.063	0.066	0.033	0.015	0.028	0.047
Difference	−0.011	0.109	0.008	0.033	0.085	0.055	0.075
Levels	Low	High	Low	Middle	High	Middle	High
Amended levels	High	High	Low	Middle	High	Middle	High
f. Levels of stable scholar-entrepreneurship							
Levels	High	High	High	Middle	Middle	Low	Low

(continued)

Jintai, and Guazhou. Together, I divided the degrees of concrete purposes into these seven counties into four levels: high, semi-high, middle, and low (Table 4.5b).

F9. Democratic and collaborative management and the federal mechanism of awards and sanctions

The interviewees indicated that democratic management (F9.1), collaborative management (F9.2), and a federal mechanism of award and sanctions (F9.3) were important factors influencing the scholars' participation in combating desertification. The evaluation of democratic management was included in ten choices given to evaluate

Table 4.5 (continued)

Counties	Zhongwei	Linze	Jingtai	Guazhou	Jinta	Dunhuang	Minqin
g. Combination between economic development and environmental protection							
Percentages	0.364	0.268	0.233	0.173	0.165	0.315	0.255
Levels	High	Middle	Middle	Low	Low	High	Middle
combination levels	High	S-high	S-high	Middle	Middle	Low	Low
h. Percentages of lacking of valid experiment-extension mechanisms							
Percentages	0.068	0.084	0.074	0.053	0.104	0.120	0.127
Level	High	Middle	High	Middle	Low	Low	Low
i. Percentages of the four kinds of external support							
Financial	0.204*	0.201*	0.213*	0.220*	0.219*	0.218*	0.211*
Technical	0.139	0.146	0.163	0.150	0.092	0.133	0.174
Institutional	0.143	0.071	0.089	0.073	0.012	0.043	0.112
Moral	0.082	0.067	0.050	0.057	0.019	0.028	0.040
j. Levels and non-relevance of stable external support							
Levels and relevance	High	S-high	S-high	Low	Middle	Low	Low

Note Because some people chose multiple choices, the total percentage of the original data in Minqin is over 100%, and the data shown below are revised
* means the highest; S- = Semi-; NR = not relevant

advantages of scholars' organizations in the survey, and the evaluation of the lack of democratic management was included in 13 choices given to evaluate their disadvantages (Table 4.5c). Differences of percentages of the respondents who chose these two choices ("advantage" subtract "disadvantage") were broken down into three groups to evaluate levels of democratic management in the seven counties: high (larger than 0.171), middle (between 0.119 and 0.171), and low (less than 0.119) (also see Table 4.5c). This table indicated that Zhongwei had the lowest level of democratic management. The interview data and archive data, however, showed that democratic management was one of the most important characteristics of combating desertification in Zhongwei, particularly from the 1950s to the 1970s. Because the answers by the respondents in the survey might focus on the current situation and their comparison of current scholars' organizations on those from the 1950s to the 1970s, their answers were biased. Considering this, I amended the level of Zhongwei to the "middle" (Table 4.5c).

The inner collaboration among scholars, farmers, and government officials within the Zhongwei County from the 1950s to the 1970s was mentioned by all the interviewees. After 1980s, firms' participation strengthened its inner collaborative management. Also, external scholars' and different NGOs' participation from all over the world improved its external collaborative management. Thus, its level of collaborative management was labeled "high" (Table 4.5d).

The interviewees in Jingtai indicated that because of the Jingdian Pumping Irrigation Project and the cooperation between its Research Station of Desertification Control (built in 1992) and the Gansu Desert Control Research Institute (GDCRI), the collaboration of scholars' organizations in Jingtai with outside organizations was at the semi-high level. They, however, indicated that the inner collaboration within the county was at the low level. Overall, I characterized its level of collaborative management as middle.

In Linze, the level of inner collaboration, particularly between government and scholars, was very high. The director of the Research Station of Desertification Control was often the vice supervisor of the county. The scholars in Linze also had high-level collaboration with other outside scholars such as from Lanzhou and Beijing. For example, the county-owned Station in Xiaoquanzi was a cooperative station with the Gansu Desert Control Research Institute.

Both of the official interviewees and the scholar interviewees in Minqin indicated that scholars and officials often criticized each other rather than collaborated with each other. Although some of the scholars in Minqin also collaborated with other outside scholars, the total degree of collaborative management in Minqin was at the "semi-middle" level.

The interviewees indicated that on the one hand the inner collaboration in Guazhou was at the low level. Before the 1990s, it did not even have an independent research station of desertification control. On the other hand, they also indicated that although the county-owned station of desertification control had no formal cooperation with some outside organizations such as CAS and GDCRI, because Guazhou was called the world capital of winds, many outside scholars were often interested in studying this county. For example, both Gansu Academy of Sciences and GDCRI built their research bases in Guazhou. The Forestry Bureau also built a Sand and Dust Storm Monitoring Station in Guazhou in 2003. Thus, the collaboration with the outside organizations in Guazhou was at the middle level. Therefore, I characterized the level of collaborative management in Guazhou as "semi-middle."

The Jinta Research Station of Desertification Control was first built in 1959 by CAS, but it was handed over to the county after 1977. Thus, the inner collaboration among the social actors, especially between the scholars' organization and government, was at the high level. After 2005, its Research Station of Desertification Control became a cooperative organization of GDCRI. Thus, its collaboration with outside scholars' organizations was at the middle level. Therefore, I characterize its level of collaborative management as "middle."

On the one hand, although scholars, officials, and farmers often collaborated in Dunhuang, their inner collaboration focused on agricultural planting rather than desertification control. On the other hand, although the Dunhuang Research Station of Desertification Control did not formally cooperate with some outside organizations of desertification control such as CAS and GDCRI, because Dunhuang was a very famous travel site, it often had collaboration with outside organizations and scholars. Several Japanese programs were good examples. Overall, its level of collaborative management was labeled "semi-middle."

Furthermore, the choice of the collaboration with outside scholars' organizations was included in the ten choices given to evaluate the advantages of the scholars' organizations in the survey. The percentages of the respondents in the seven counties were divided into three groups: high (larger than 0.059), semi-high (between 0.051 and 0.059), middle (between 0.034 and 0.051), and low (less than 0.034) (Table 4.5d). This finding was consistent with the finding from the aforementioned interview data.

The evaluation of having the federal mechanism of awards and sanctions was included in the ten choices given to evaluate the advantages of scholars' organizations in the survey, and the evaluation of lacking the federal mechanism of awards and sanctions was included in the 13 choices given to evaluate their disadvantages. Differences of percentages of the respondents who chose these two choices ("advantage" subtract "disadvantage") were broken down into three groups to evaluate the levels of the federal mechanism of awards and sanctions in the seven counties: high (larger than 0.069), middle (between 0.029 and 0.069), and low (less than 0.029) (Table 4.5e). This table showed that Zhongwei had the lowest level of the federal mechanism of awards and sanctions. The interviewees indicated that because of firms' participation in combating desertification the mechanism of awards and sanctions in Zhongwei was marketized. The respondents, however, in the survey often focused on the governmental system. This was also proved by observations. When observers wanted to talk with several people who were fixing sand using straws of wheat, they said they did not have time because they were hired by companies and government and their gains were based on how much work they did. Thus, I amended the level of the federal mechanism of awards and sanctions in Zhongwei to the "high" level (Table 4.5e).

F10. Stable scholar-entrepreneurship

Many interviewees in Zhongwei mentioned the leadership of some famous scholars such as Kezhen Zhu, Shen'e Liu, Ying Liu, Minggang Li, and Junzhao Ji. In Linze, there was a tradition that the head of the Experimental Station could be the vice supervisor of the county. This increased its scholar-entrepreneurship through a formal institutional arrangement. Also, because of the Jingdian Pumping Irrigation Project, scholar-entrepreneurship in Jingtai was strong and stable. Although scholar-entrepreneurship on economic development in Dunhuang was relatively strong and stable, stable scholar-entrepreneurship in the combat of desertification was weak. Finally, because most scholars in Minqin only focused on their own research, its stable scholar-entrepreneurship in the combat of desertification was weak. Thus, I divided the degree of stable scholar-entrepreneurship into the seven counties into three levels: high, middle, and low (Table 4.5f).

F11. Combination between economic development and environmental protection

The interviewees indicated that whether the activities of combating desertification and the improvement of farmers' benefits could be combined was a key factor influencing the effectiveness of combating desertification. Among ten choices given to evaluate advantages of scholars' organizations, more than 17% respondents in the seven counties indicated that these organizations helped people receive more benefits through using knowledge and continually strengthening people' feelings about this (Table 4.5g). These percentages were divided into three levels: high (larger 0.3), middle (between 0.22 and 0.3), and low (less than 0.22) (Table 4.5g). In this table, the levels in Zhongwei and Dunhuang were labeled "high." The interviewees, however, indicated that the difference between Zhongwei and Dunhuang was the activities of combating desertification and the improvement of farmers' benefits were combined very well in Zhongwei, especially through the firms' participation (although there were some tendencies for some firms to begin to pay more attention to economic benefits rather than environmental protection), while more attention was paid to economic development in Dunhuang. In Minqin, the problem was the same as in Dunhuang. Thus, based on the consideration of both the interview data and the survey data, the combination level of economic development and environmental protection was labeled "high" in Zhongwei, "semi-high" in Linze and Jingtai, "middle" in Guazhou and Jinta, and "low" in Dunhuang and Minqin (Table 4.5g).

F12. Experiment-extension governance methods

The interviewees indicated that the experiment-extension governance method played an important role in combating desertification. The choice of no valid experiment-extension mechanism was included in the 13 choices given to evaluate disadvantages of scholars' organizations in the seven counties. The percentages of the respondents were divided into three groups to evaluate the level of experiment-extension governance used in the seven counties: high (less than 0.075), middle (between 0.075 and 0.1), and low (larger than 0.1) (Table 4.5h).

F13. Four Kinds of stable external support

The interviewee indicated that the two most deficient resources to combat desertification were financial and technical supports, particularly the financial support. Twelve choices were given to evaluate the methods to improve the effectiveness of the scholars and the scholars' organization in combating desertification. More than 20% of the respondents indicated that the financial support was the most important one, and more than about 10% indicated that the technical support was the most important one (Table 4.5i). In addition to these two kinds of support, the respondents pointed out that stable institutional support and moral (or spiritual) support were important (Table 4.5i).

The interviewees indicated that if scholars' participation in combating desertification also came from support out of the county, this support should be stable until it had the self-governance capability. The Zhongwei case was a good example. Because of the importance of protecting Yellow River and the Lanzhou–Xinjiang railroad, its external financial, technical, institutional, and moral support was stable. Thus, the level of the stable external support in Zhongwei was high (Table 4.5j). The implementation of the Jingdian Pumping Irrigation Project since the 1970s and current participation of firms from other provinces (such as Shannxi Zhongfu Group Company and Ningxia Meili Paper Industry Co., Ltd.) made the level of the stable

Table 4.6 Comparing different related factors of scholars' participation and its results

Counties factors	Zhongwei	Linze	Jingtai	Guazhou	Jinta	Dunhuang	Minqin
F1							
F1.1	High	High	High	Middle	Middle	High	Low
F1.2	Long	Middle	Middle	Middle	Middle	Middle	Short
F1.3	Middle	Middle	High	High	High	Middle	Low
F2							
F2.1	High	Middle	High	Middle	Middle	Middle	High
F2.2	High	High	High	High	High	High	High
F3	High	High*	High	High	Middle	Low	Low
F4							
F4.1	High	S-high	S-high	S-middle	S-middle	S-middle	High
F4.2	High	S-high	S-high	Middle	Middle	S-middle	Middle
F4.3	S-High	High	Middle	Middle	Middle	High	Low
F5	High	High	Middle	Middle	Middle	High	Low
F6	High	S-High	High	Middle	Low	High	Low
F7	High	Middle	Middle	High	Low	Low	Low
F8							
F8.1	High	Middle	Low	Middle	Middle	High	Low
F8.2	High	Middle	S-high	High	Low	Low	Low
F9							
F9.1	Middle	Middle	Low	Middle	High	Middle	Middle
F9.2	High	High	Middle	Middle	Middle	Middle	S-middle
F9.3	High	High	Low	Middle	High	Middle	High
F10	High	High	High	Middle	Middle	Low	Low
F11	High	S-high	S-high	Middle	Middle	Low	Low
F12	High	Middle	High	Middle	Low	Low	Low
F13	High	S-high	S-high	Low	Middle	Low	Low
Total scores	38.167	32.667	32.083	28.000	23.833	22.667	17.167
Performance	S	SE	SE	SE	SE	U	U

Note S- = Semi; * means this result was revised based on the interview data; NR = Not relevant; S = Successful; Se = Semi-successful; U = Unsuccessful

external support in Jingtai at a "semi-high" level. In Minqin, Guazhou, and Dunhuang, however, the stable external support was at the "low" level. Although there was no strong and stable external support in Linze, people in this county built a relatively stable inner self-supporting system. For example, Linze built two county-owned research stations of desertification control. Thus, I characterized its level and non-relevance of stable external support as "semi-high." Finally, although there was no stable external support for combating desertification in Jingtai, a long history of organizing the county-owned research station of desertification control made Jinta have a relatively stable inner institutional, technical, and even moral support. Thus, I characterized its level and non-relevance of the stable external support as "middle."

4.2.3 Comparing the Seven Counties

To summarize the above findings, I have arranged all of the factors in the seven counties in Table 4.6. This table shows that the higher (or longer) the levels of these factors are the more successful scholars' participation in combating desertification will be.

In order to clarify, I calculated the total scores of these factors after assigning different numbers to different levels. I gave high (long) = 3, middle = 2, and low (short) = 1, semi-high = 2.5, and S-middle = 1.5. Then, I calculated the average scores for F1, F2, F4, F8, and F9 in each county. Furthermore, for each county, I then calculated the total scores from F1 to F13 (Table 4.6). These scores also indicate that higher total scores are correlated with more successful scholars' participation.

References

Dong, K. (2007, April 6). *Gansu Guazhou shachenbao congwei tingxie, bairen huan fengshachen feibing* (*Sandstorms in Guanzhou, Gansu never stops, and 100 people suffers pulmonary disease caused by sand and dust blown by wind*). www.dustdorm.com.cn (*Zhonguo Shachenbao Wang*). Retrieved May 5, 2008, from http://www.duststorm.gov.cn/Article/ShowArticle.asp?ArticleID=7643.

Hu, S. (1928). Zhixue de fangfa yu cailiao. *Xin Yue*, 1(9) (in Chinese).

Hogg, M.A., & Vaughan, G.M. (2002). *Social psychology* (3rd ed.). London: Prentice Hall.

Li, X., Chen, G., Qu, J., Chen, H., & Wang, X. (2003). Chapter 13. In T. Wang (Ed.), *Desert and desertification in China* (pp. 595–639). Shijiazhuang: Hebei Science and Technology Publishing House (in Chinese).

Zhang, Y., Wang, L., Zhang, H., & Li, X. (2003). An analysis on land use changes and their driving factors in Shule River: An example from Anxi County. *Progress in Geology, 22*(3), 53–61.

Chapter 5
Seven Design Principles of Successful Scholar-Based Governance

The Master said, 'I do not open up the truth to one who is not eager to get knowledge, nor help out any one who is not anxious to explain himself. When I have presented one corner of a subject to any one, and he cannot from it learn the other three, I do not repeat my lesson.' (In Chinese: 子曰: "不愤不启, 不悱不发, 举一隅不以三隅反, 则不复也。" (《论语·述而第七》)).

—*Confucian analects* (1900, p. 61)

This chapter analyzes seven design principles for promoting scholars' participation in collective action. Based on reanalyzing the 13 related factors differentiating scholars' participation in combating desertification in the seven counties in the first stage of this study, I first summarize these seven principles. Then, I test their generalizability by analyzing the rest 43 cases in the second stage of this study. Together, based on a comparative study of 50 cases, this chapter indicates that successful scholar-based governance needs to satisfy seven design principles. The better satisfied these principles are, the more successful combating desertification effort is.

5.1 Seven Design Principles and Their Generalizability

Based on reanalyzing 13 related factors differentiating scholars' participation in combating desertification, I summarize a set of seven design principles which characterize successful scholar-based governance (Table 5.1). The first principle includes six subprinciples (Table 5.2). Thus, these seven principles are also consistent with the 13 related factors talked in Chap. 4.

As I already pointed out in Chap. 3, casual generalization as representation and causal generalization as extrapolation are two major causal generalization problems (Shadish et al. 2002). In order to study the casual generalization as extrapolation of the design principles of successful scholar-based governance, the seven design

L. Yang, *Knowledge-Driven Governance*,
https://doi.org/10.1007/978-981-13-2910-4_5

Table 5.1 Seven design principles of successful scholar-participated governance

About characteristics of scholars
P1. Significant participation of steady field-based scholars
Steady field-based scholars must be the key part of scholars and play an important role in governance
About characteristics of scholars' organizations and governance
P.2. Federal organizational structure and concrete, organizational, and stratified purposes Scholars' organizations are mainly bottom-up federal structure and have very concrete, organizational, and stratified purposes
P.3. Democratic and collaborative management with the federal mechanism of appropriate awards and sanctions Practicing democratic and collaborative management, using almost-consensus and face-to-face communication method at the grassroots level, and build a selecting incentive (including both awards and sanctions, but mainly positive) system with differing, graduated, federal types, and multiple ways
P.4. Steady local scholar-entrepreneurship Building steady local scholar-entrepreneurship, no matter whether these scholars are originally local scholars, farmers, and so forth
P.5. Realization of expected benefits Social actors' benefits can be expected, and their expectation can be reinforced with the programs' development. As to the environmental affairs, economic development and environmental protection can be combined
P.6. The experiment-extension method Practicing the method of experimental units and gradual extension governance (abbreviated as experiment-extension governance, and in Chinese it is Shidian-Tuiguang Governance)
For situations where external support is necessary
P.7. Steady external support If external support (financial, technical, institutional, and spiritual) is necessary, it is steady or at least lasts a relatively long term until the system can build its own internal support mechanism, especially for the financial, technical, and institutional supports

principles including the six subprinciples of Principle 1 were used to characterize the 50 cases (including the original seven counties and the 43 cases for testing generalizability) in the second stage of the research. I first analyze 30 cases about combating desertification (Table 5.3). These 30 cases include the original seven counties, the five counties chosen from changing coverage (for detailed case studies, please see Appendix E), the three cases chosen from changing spatial extent (for detailed case studies, please see Appendix F), and the 15 cases chosen from changing cultural context (for detailed case studies, please see Appendix G). I then analyze 20 cases chosen from changing fields with temporal extent (Table 5.4; for detailed case studies, please see Appendix H).

For each case, I indicate which of the seven design principles clearly apply, which apply in a weak form, and which clearly do not apply. I also characterize their achievements as three classes: successful, semi-successful, and non-successful. For combating desertification, when not only the trend of desertification was stopped

Table 5.2 Six subprinciples for Principle 1

P1.1. Relatively independent social identity and high social status Scholars have relatively independent social identity and high social status
P1.2. High enthusiasm and capability of scholars' participation Scholars have high participatory enthusiasm and participatory capability including the capacity to incessantly learn new knowledge and teach other social actors, and leadership and organizing capability
P1.3. Enough, easily understood, and know-how knowledge Scholars have comparative advantages in expert information and easily understood and know-how knowledge over other social actors and also have enough knowledge of local conditions (including local traditions, social norms, etc.), and local social relationships and webs
P1.4. High social capital Scholars have high social capital with other social actors
P1.5. High social responsibility and spirit of practice Scholars have high social responsibility and strong spirit to practice and take concrete actions
P1.6. Respecting other social actors and local knowledge Scholars respect other social actors, especially their ideas and local knowledge

but also much desertification-affected area was restored, the results of desertification control are deemed as "successful"; when desertification became more serious, the results of desertification control are deemed as "unsuccessful"; when the conditions were between the above two types, the results of desertification control are deemed as "semi-successful." For other collective actions, if the purposes of collective action were well realized, its result is deemed "successful"; if they were only partly realized, its result is deemed as "semi-successful"; if they were not realized, its result is deemed as "unsuccessful." The conclusion is the seven principles are all held when multiple coverage, spatial extent, cultural contexts, fields, and temporal extent are addressed. No matter what the scale (larger or small) is, what the hierarchical level (high or low) is, what kinds of countries and cultures are, or what kinds of collective affairs are, the more the seven principles of scholar-based governance are satisfied, the more successful it will be.

The purpose of analyzing their causal generalization as representation is to test whether the seven design principles for successful scholar-based governance in combating desertification as a representative are best described the role and its design principles of scholar-based governance in a whole society. Based on a meta-analysis, I have arrayed the 32 Chinese cases in total studied in this study in Table 5.5. These cases are grouped into three periods: before 1911 (Qing Dynasty and the dynasties before Qing), between 1911 and 1949 (the Republic of China), and from 1949 to present (the People's Republic of China). Their fields and scales are also shown in the table.

Table 5.5 indicates that scholar-based governance is important not only in combating desertification, but also in village governance, rural education, rural development, and economic and political movements in both rural and urban areas. Also, it shows that scholar-based governance has been applied at all levels—village, county,

Table 5.3 Satisfaction of the seven principles of the 30 cases and their results in combating desertification

Cases	Spatial extent	Temporal extent	P1							P2	P3	P4	P5	P6	P7	Results
				P1.1	P1.2	P1.3	P1.4	P1.5	P1.6							
a. *The seven counties*																
1. Zhongwei	County	1949–2007	Y	Y	Y	Y	Y	Y	Y	Y	Y	Y	Y	Y	Y	S
2. Linze	County	1949–2007	W	W	Y	Y	Y	W	W	W	Y	Y	W	W	W	SE
3. Jingtai	County	1949–2007	Y	Y	Y	W	W	Y	W	W	N	Y	W	Y	W	SE
4. Guazhou	County	1949–2007	W	W	Y	W	W	W	Y	W	W	W	W	W	N	SE
5. Jinta	County	1949–2007	W	W	W	W	W	N	N	W	Y	W	W	N	W	SE
6. Dunhuang	County	1949–2007	W	W	N	W	Y	Y	N	W	W	N	N	N	N	U
7. Minqin	County	1949–2007	N	Y	N	W	N	N	N	N	W	N	N	N	N	U
b. *The five cases through changing coverage*																
8. Hotan,	County	1949–2007	Y	Y	Y	Y	Y	Y	Y	Y	Y	Y	Y	Y	Y	S
9. Naiman	County	1949–2007	Y	Y	Y	Y	Y	Y	Y	Y	Y	Y	Y	Y	Y	S
10. Aohan	County	1949–2007	Y	Y	Y	Y	Y	Y	Y	Y	Y	Y	Y	Y	Y	S
11. Ejin Horo	County	1949–2007	Y	Y	Y	Y	Y	Y	Y	W	W	W	Y	Y	Y	SE
12. Yanchi	County	1949–2007	W	W	W	W	W	W	W	W	W	W	W	Y	W	SE
c. *The three cases through changing spatial extent*																
13. Songhe	Village	1949–2007	Y	Y	Y	Y	Y	Y	Y	Y	Y	Y	Y	Y	Y	S
14. Yulin	Prefecture[a]	1949–2007	Y	Y	Y	Y	Y	Y	Y	W	W	W	Y	Y	Y	SE
15. Zhangye	Prefecture	1949–2007	N	N	N	N	N	N	N	N	N	N	N	N	N	U

(continued)

Table 5.3 (continued)

Cases	Spatial extent	Temporal extent	P1							P2	P3	P4	P5	P6	P7	Results
				P1.1	P1.2	P1.3	P1.4	P1.5	P1.6							
d. *The* 15 *cases through changing cultural contexts*																
16. C1	National	1989–2007	Y	Y	Y	Y	Y	Y	Y	Y	Y	Y	Y	Y	Y	S
17. C2	Provincial[b]	1948–2007	Y	Y	Y	Y	Y	Y	Y	Y	Y	Y	Y	Y	Y	S
18. C3	Prefecture	1965–2007	W	Y	W	W	N	Y	N	W	W	W	Y	W	Y	SE
19. C4	Provincial	1981–2007	W	W	W	W	W	W	W	W	W	W	W	W	Y	SE
20. C5	National	1924–2007	W	W	W	W	W	W	W	W	W	W	W	W	W	SE
21. C6	County	1962–2007	W	W	Y	Y	Y	Y	Y	Y	Y	Y	Y	Y	Y	SE
22. C7	Provincial	1932–2007	W	Y	Y	Y	Y	Y	Y	Y	Y	W	Y	Y	W	SE
23. C8	Provincial	1876–1980	N	N	N	N	N	N	N	N	N	N	N	N	N	U
24. C9	Provincial	1900–1980	N	N	N	N	N	N	N	N	N	N	N	N	N	U
25. C10	Prefecture	1900–1982	N	N	N	N	N	N	N	N	N	N	N	N	N	U
26. C11	Provincial	1975–1984	N	N	N	N	N	N	N	N	N	N	N	N	N	U

(continued)

Table 5.3 (continued)

Cases	Spatial extent	Temporal extent	P1							P2	P3	P4	P5	P6	P7	Results
				P1.1	P1.2	P1.3	P1.4	P1.5	P1.6							
27. C12	Provincial	1953–1980	N	N	N	N	N	N	N	N	N	N	N	N	N	U
28. C13	Prefecture	1900–1970s	N	N	N	N	N	N	N	N	N	N	N	N	N	U
29. C14	Prefecture	1900–1980s	N	N	N	N	N	N	N	N	N	N	N	N	N	U
30. C15	Provincial	1869–1980	N	N	N	N	N	N	N	N	N	N	N	N	N	U

Y Yes; *N* No; *W* Weak; *S* Successful; *SE* Semi-successful; *U* unsuccessful
C1 The Landcare Program, Australia; *C2* The Negev Case, Israel; *C3* The Mona Reclamation Experimental Project, Pakistan; *C4* The "LUCDEME" Project, southeast Spain; *C5* Integrated Desert Development and Desertification Control in Turkmenistan; *C6* The Vale Rangeland Rehabilitation Program, Oregon, USA.; *C7* The Dust Bowl Case about Dust Storms in the USA; *C8* The Gascoyne Basin, Australia; *C9* The Region of Coquimbo, Chile; *C10* The Luni Development Block, Rajasthan, India; *C11* The Turan Program, Iran; *C12* The Greater Mussayeb Project, Iraq; *C13* The Eghazer and Azawak Region, Niger; *C14* The Oglat Merteba Region, Tunisia; *C15* The Golodnaya Steppe, Uzbekistan
[a]Prefecture means this case happened in an area which is larger than a county and smaller than a province or state in the country
[b]Provincial means this case happened in a province or state or within several provinces or states

Table 5.4 Cases with changing fields and temporal extent

Case	Field	Scale	Principles time	P1							P2	P3	P4	P5	P6	P7	Results
					P1.1	P1.2	P1.3	P1.4	P1.5	P1.6							
C1	VG	National	221 BC-1905	Y	Y	Y	Y	Y	Y	Y	Y	Y	Y	Y	Y	Y	S
C2	RE	National	1978–2008	Y	W	Y	Y	Y	Y	Y	Y	Y	Y	Y	Y	Y	S
C3[a]	RD	National	1972–1980s	Y	Y	Y	Y	Y	Y	Y	Y	Y	Y	Y	Y	Y	S
C4[a]	RD	Provincial	1970s–2007	Y	Y	Y	Y	Y	Y	Y	Y	Y	Y	Y	Y	Y	S
C5	RU	Village	1927–1930	W	Y	Y	Y	Y	Y	Y	Y	Y	W	N	W	N	SE
C6	RD	County	1929–1937	W	Y	Y	Y	Y	Y	Y	Y	Y	W	W	N	W	SE
C7	RD	County	1931–1937	W	Y	Y	Y	Y	Y	Y	Y	Y	W	W	N	W	SE
C8	RD	National	1921–1949	Y	Y	Y	Y	Y	Y	Y	Y	Y	Y	Y	Y	N	SE
C9	RD	National	1950–1953	W	W	Y	Y	Y	Y	Y	W	W	W	Y	W	Y	SE
C10[a]	IR	Prefecture	1950s–1992	W	Y	Y	Y	Y	Y	Y	W	W	W	Y	Y	W	SE
C11	RD	National	1905–1911	N	N	N	N	N	N	N	N	N	N	N	N	N	U
C12	RD	National	1911–1949	N	N	N	N	N	N	N	N	N	N	N	N	N	U
C13	RD	National	1949–1956	N	N	N	N	N	N	N	N	N	N	N	N	N	U
C14	PM	National	1951–1952	N	N	N	N	N	N	N	N	N	N	N	N	N	U
C15	RD	National	1957–1958	N	N	N	N	N	N	N	N	N	N	N	N	N	U
C16	EM	National	1958–1960	N	N	N	N	N	N	N	N	N	N	N	N	N	U

(continued)

Table 5.4 (continued)

Case	Field	Scale	Principles time	P1							P2	P3	P4	P5	P6	P7	Results
					P1.1	P1.2	P1.3	P1.4	P1.5	P1.6							
C17	RD	National	1958–1983	N	N	N	N	N	N	N	N	N	N	N	N	N	U
C18	PM	National	1963–1966	N	N	N	N	N	N	N	N	N	N	N	N	N	U
C19	PM	National	1966–1976	N	N	N	N	N	N	N	N	N	N	N	N	N	U
C20	PM	National	1955–1978	N	N	N	N	N	N	N	N	N	N	N	N	N	U

Note

C1 The Village Governance System in Ancient China; *C2* The Household Contract Responsibility System and Development in the PRC; *C3* The New Village Movement in South Korea; *C4* People's Planning in Kerala, India; *C5* Tao's Xiaozhuang Normal School Experiment in Nanjing, Jiangsu Province; *C6* Yan's Rural Reconstruction Experiment in Ding County, Henan Province; *C7* Liang's Rural Reconstruction Experiment in Zouping County, Shandong; *C8* The Farmers' Movement and the Farmers' Association Led by the CCP in China; *C9* The Farmers' Association and Land Reform in the PRC; *C10* The Gal Oya Irrigation Project in Sri Lanka; *C11* The Rural Autonomy Movement in Late Qing Dynasty; *C12* The Movement of Local Autonomy in Republic of China; *C13* The Cooperative Movement in the PRC; *C14* The Intellectuals' Ideological Remolding Movement in the PRC; *C15* The Anti-Rightist Movement in the PRC; *C16* The Great Leap Forward Movement in the PRC; *C17* The People's Commune Movement in the PRC; *C18* The Four Cleanup Movement in the PRC; *C19* The Great Culture Revolution in the PRC; *C20* The Rustication Movement of Educated Youth in the PRC

VG village governance, *RE* rural economic development, *RD* rural development, *RU* rural education, *IR* irrigation, *EM* economic movement, *PM* political movement

[a]means that these cases are from non-China countries

The other denotations are the same as in Table 5.3

Table 5.5 A meta-analysis of all the Chinese cases

Fields		Scales	Before 1911	1911–1949	1949–2007
Rural cases	Combating desertification	National			
		Prefecture			2
		County			12
		Village			1
	Village governance	National	1		
		Prefecture			
		County			
		Village			
	Rural education	National			
		Prefecture			
		County			
		Village		1	
	Rural economic development	National			1
		Prefecture			
		County			
		Village			
	Rural development	National	1	2	4
		Prefecture			
		County		2	
		Village			
Rural and urban cases	Economic movement	National			1
		Prefecture			
		County			
		Village			
	Political movement	National			4
		Prefecture			
		County			
		Village			
	Total		2	5	25

prefecture, and national—and in all periods of China—the People's Republic of China, the Republic of China, Qing Dynasty, and the other former dynasties. Although these cases mainly happened in the People's Republic China, the historical analysis of the village governance system in ancient China (see Appendix H) overcomes this problem. This case tells us that scholar-based governance played a very important role in ancient rural China from 221 BC to 1905, especially after the self-governance system was built under the county level in 1071 (the fourth year of Wang Anshi' Reform in the Song Dynasty). Although these cases mainly happened in rural areas (27 cases) and at the county (14 cases) and the national (14 cases) scales, the urban cases (five cases) and the cases at the prefecture (two cases) and the village (two cases) scales have also been covered in this study. Moreover, I can contingently conclude that scholar-based governance plays a very important role in Chinese social governance; the more satisfied the seven principles are, the more successful scholar-based governance in Chinese society will be.

5.2 Describing the Seven Principles in Detail

In this section, I shall discuss each of the seven principles in details.

5.2.1 Principle 1

The scholars in combating desertification may be field-based scholars and non-field-based scholars (Yang 2007a, b). "Field" can refer to a particular subject or a branch of knowledge of special problems such as combating desertification and environmental protection. Also, "field" refers to a particular geographic space or area. Together, I define field-based scholars as those who were born, lived or have been living in the community for many years, know many things about the community, have good relationships with farmers, officials, and other social actors, and have enough field and local knowledge for resolving the related problems such as desertification control and environmental protection. The study shows that the longer scholars stay in the local site, the more successful their participation will be. Also, the study indicated that technical knowledge, local social knowledge, and governance knowledge are the three most important kinds of knowledge in combating desertification, and the more integrated they are, the more successful the desertification control is. Furthermore, this study found that both field-based technical and social scholars are needed. Technical scholars are often familiar with technology and science, but do not have much knowledge about policy and governance problems. On the contrary, social scholars have a large amount of knowledge about policy and governance problem, but often lack understanding of technology and science. Successful scholar-based governance should encourage collaboration between these two groups of scholars.

This study also found six subfactors influencing the local field-based scholars' impact on combating desertification: (1) scholars' relatively independent social identity and high social status, (2) high enthusiasm and capability of scholars' participation, (3) enough, easily understood, and hands-on knowledge, (4) high social capital with other social actors, (5) high social responsibility and spirit of practice, and (6) respect for other social actors and local knowledge. These are deemed as six subprinciples of Principle 1 (Table 5.2).

Independent social identity is a precondition for scholars to effectively participate in collective action. This precondition, however, does not mean that scholars can be completely independent from other social actors. Thus, their independence is relative. High social status is also a necessary condition for successful scholar-based governance. For example, scholars in Zhongwei were highly respected in desertification control, especially in the 1950s, 1960s, and 1970s. Scholars in Minqin only focused on their own academic research and were not respected by officials and farmers in combating desertification.

The interviewees indicated that in addition to lacking of enough practical knowledge, people often complained scholars' low participatory enthusiasm and capacity. They pointed out that scholars must have enough participatory enthusiasm and capacity, if they wanted to play an important role in combating desertification. The case of Zhongwei from the 1950s to the 1970s was a good example. During those eras, many scholars enthusiastically worked day in and day out. Although their living conditions were very poor, and sometimes, they even had no bed to sleep in and lacked food to eat, they did their job very well. The interviewees, however, argued that only high participatory enthusiasm was not enough. They said scholars should also have high participatory capability. In order to build high participatory capability, scholars should have some knowledge about participation and know how to participate. Also, scholars should have enough capability to incessantly understand new problems, learn new knowledge, and teach other social actors. Furthermore, in order to organize themselves or to be scholar-entrepreneurs for other social actors, high leadership and organizing capability were also necessary.

Much modern literature about experts also highlighted the same problem. For example, Hirschhorn (1979) suggested the expert as a facilitator of client learning to assist clients in their own efforts to examine their interests and to plan appropriate courses of action. That is, "professionals must become experts in how clients learn, clarify, and decide" (p. 187). Ettorre (1999), by analyzing the interview data in UK, Finland, the Netherlands, and Greece from 1995 to 1996, argued that experts should be educators or information providers of citizens. Furthermore, Nückles and Stürz (2006), by doing an experiment, suggested that experts should be able to adjust their way of communicating to fit the knowledge of the layperson.

Scholars must have enough comparative advantages in expert information and knowledge over other social actors. In order to communicate with the locals and have concrete and helpful suggestions on specific local problems, they should also have enough knowledge of local conditions. Furthermore, because local social relationships and webs often influence interactions among different social actors and how to resolve collective action problems, knowledge about local social relationships and webs is also necessary. For example, because of lacking scientific and local knowledge, many experts recommended villagers to combat desertification through planting trees, especially white poplars (as in Minqin). The white poplar, however, is not a good species in dry areas because its leaves are so large that they dramatically increase evaporation and deteriorate environments. Also, blindfold activities of planting trees often destroy the original equilibrium of natural conditions.

Trust improves cooperation and reduces conflicts among different social actors. High social capital with other social actors improves the cooperation possibility of scholars with them. Also, it can help build trust or improve trust among the other social actors through a trust chain. For example, if a scholar has a good relationship with both an official and a farmer, through the trust chain among them, the official and the farmer can also have a good relationship through the scholar.

Successful scholar-based governance also depends on scholars' high social responsibility and spirit of practice. For example, all the interviewees in Zhongwei indicated that scholars' high social responsibility and spirit of practice in combating desertification were an important reason of the success of combating desertification from the 1950s to the 1970s. Also, several scholars I met in Zhongwei argued that one of the important reasons which drove them to stay in Zhongwei was their social responsibility.

"Respect" means that people consider others' feelings, ideas, and behaviors when they do something. Respect helps people get along with others and then can improve the possibility of cooperation among different social actors. Many interviewees indicated that scholars must respect other social actors and their ideas to get along with them and improve the possibility to cooperate with each other. Furthermore, because of the complexity of social problems and limited capacity of all individuals, scholars cannot understand everything, even if they have some comparative advantages over other social actors. Thus, other social actors' local knowledge and personal experience or ideas are often very useful. Respect, however, does not mean that scholars must agree with these opinions. Scholars may or may not agree with them, but they should place them into consideration, and these opinions will ultimately influence scholars' own ideas and decisions. For example, the scholars in combating desertification in Zhongwei from the 1950s to the 1970s often highly respected other social actors, and frequently worked, talked, and studied with them.

In summary, I found that all six subprinciples were satisfied in Zhongwei County, especially from the 1950s to the 1970s. In Minqin, although scholars had relatively independent social identity and high social status with relatively enough, easily understood, and hands-on knowledge, they did not satisfy all the other subprinciples. Thus, even if the study of combating desertification made by these scholars was famous in China, they did not contribute too much to practical activities of combating desertification in Minqin.

In the Landcare Program in Australia, landcare groups are local voluntary groups comprised of land users in rural areas that can be deemed as steady field groups. Many leading farmers and conservationists who played a key role in design, implementation, and interpretation of research and development project were included in landcare groups. Through them, research, research extension, and development were made less linear, scientists were put in more direct contact with the community, the boundary between extension and research was dissolved, and practical experience and local knowledge were made greater use (Campbell 1992). They had relatively independent social identity. Because they often earned higher level of cash income, had higher levels of debts, and undertook more land conservation practices than other farmers, they also had relatively higher social status. As far-sighted people, particularly often as activists and conservationists, they are more concerned with the future. As more active seekers of information, they not only had more knowledge advantages than farmers, but also acted as advisers to the farmers (Campbell 1992), legitimate players in research (Chambers et al. 1989), and more aware, informed, skilled, and adaptive resource managers (Curtis and Lacy 1996). Mature landcare groups could gain credibility which formed their social capital. After groups were formed, they grew quickly and developed ongoing relationships with local and state government agencies, local businesses, community groups (including neighboring landcare groups), universities, researchers and consultants, scholars, and other landholders. Also, landcare groups repeatedly had high social responsibility in their own district and of influencing others and could respect other social actors (Campbell 1992).

In the Negev case in Israel, Professor Michael Evenari and his team were a good example of field-based scholars. Although they were professors or researchers, they carefully studied history and many other problems of the Negev and stably took in this program and stayed in the field (Evenari 1989; Evenari et al. 1982). Furthermore, universities such as Ben Gurion University of the Negev at Beer Sheva made a considerable contribution to educate and supply steady field scholars for this region (Schechter and Galai 1980).

5.2.2 *Principle 2*

A federal group is a federation of a number of small groups, each of which may be induced to use its social incentives to get its members to contribute toward the collective goals of the whole group when the federated organization provides some service to them (Olson 1971). The organizational purpose may be a conceptual purpose, which is very abstract and mainly focuses on values. It may also be a concrete purpose, which is relatively material and instrumental and can be used as a tool to pursue a more conceptual purpose. Successful scholar-based governance must have practical and feasible concrete purposes which can be used as tools to realize its conceptual and abstract purposes. For example, the purposes to protect Yellow River and Lanzhou–Xinjiang Railroad in Zhongwei and the Jingdian Pumping Irrigation Project in Jingtai are more concrete than the purposes to combat desertification and protect the environment in Minqin.

The Vale rangeland rehabilitation program in Oregon is also a good example satisfying this principle. For its federal organizational structure, Heady pointed out:

> Planning in the Vale District contributed to the success of the program. Division of the District into three resource or management areas with separate manages spread the workload and responsibilities. The resource areas were further divided into 14 planning units. Because the area managers exercised considerable autonomy in proposing and developing improvement plans, projects, and grazing systems, the mixture of management practices and land treatments differed among the three resource areas. After 1977 the program was reorganized into two resource areas.
>
> Although area managers initiated planning and site selection for improvement projects, the final authority for the coordinated program rested with the district manager. Prior to 1983, district personnel included about 40 permanent employees—more after 1983—and as many as 75 temporary employees during the field and fire season. (Heady 1988, p. 25)

Heady (1988) also listed eight concrete objectives of the Vale Program such as correcting erosion and accompanying downstream sedimentation—and preventing further soil losses, increasing the forage supply for wildlife and livestock, and stabilizing the livestock industry at the present or at increased levels of production.

On the other hand, the Turan program in Iran did not satisfy this principle. Because its management was authoritarian and controlled by government, its organizational structure was centralized (Department of the Environment 1980). The traditional and rural world was also distrusted and ignored. Martin (1982) even argued that the lack of communal organization contributed a lot to environmental destruction in Turan and carefully studied its reasons. In addition, their organizational purpose was not very concrete and often led to contradictory results (Department of the Environment 1980).

5.2.3 *Principle 3*

Democratic institutional arrangements should be applied in scholars' participation governance and almost-consensus and face-to-face communication methods should be used in the decision-making process, especially at the grassroots level. The collaboration among local social actors (such as farmers, scholars, and government officials) and the collaboration between the local social actors and non-local social actors (such as outside scholars, governments, and NGOs) are important to combating desertification. Furthermore, participatory and collaborative research (Fischer 2000) should be encouraged. Both awards and sanctions as selective incentives should be used, though they should be mainly positive (Olson 1971) and graduated (Ostrom 1990) to form a federal system with multiple choices. For example, the Vale Program paid much attention to favorable public opinion, the need for effective publicity, and information by a majority of respondents (Government of the United States of America 1980; Huntsinger and Heady 1988). Considerable autonomy exercised by the area managers also indicted its democratic style (Heady 1988). The cooperation among different resources areas and different managers, between the BLM (the Vale district of the Bureau of Land Management), Congress, community members, program area permittees, and environmentalists, showed the collaborative management of the Vale Program (Heady 1988; Huntsinger and Heady 1988). In this program, monitoring of grazing use was the joint responsibility of the BLM and the permittees. Huntsinger and Heady pointed out:

> The BLM monitors yearly compliance with season of use and stocking levels, but changes are based primarily on measured levels of forage utilization and changes in range condition on permanent plots (BLM 1984). The permittee of the 1980s is given much leeway in day-to-day compliance with grazing plans. The grazing systems designate periods of permitted use, but within these dates, the actual turnout and moving of animals into new pastures is the responsibility of the permittee. Now that good information is available from many years of monitoring range utilization and range condition, the BLM has been able to concentrate on these indicators for needed changes in animal management. (Huntsinger and Heady 1988, p. 63)

In China, the cases of Hotan County, Naiman Banner, Aohan Banner, and Songhe Village are all good examples of meeting Principle 3. For instance, in Naiman, local participation, community-based governance, farmers' voluntary participation, and local and international collaboration (Carle and Ma 2005; Dou and Song 1995; Kim et al. 2006; UNEP 2002; Xu and Yang 1991; Zhang and Tan 1989) demonstrated that democratic and collaborative management was applied effectively. Furthermore, many scholars (e.g., Xu 1995; Wang 2003a) pointed out that the federal mechanism of awards and sanctions was also built.

Decentralized governance, bottom-up and deliberative planning, participatory democracy, and civil society building in People's Planning in Kerala, India (Chettiparamb 2006; Heller et al. 2007; Isaac and Franke 2002) showed that democratic management was applied in this program. The cooperative banks (see Isaac and Franke 2002, pp. 166–167) as well as class coordination (Heller 2000), the synergies civil society, and local government bodies and state government (Törnquist 2000; Véron 2001) showed the internal collaborative management at different levels within India. Extensive network of voluntary organizations and mass movement, which motivated and brought together local representatives, officials in the various line departments, governmental and non-governmental experts relevant to the local planning process (such as KSSP) and the mass citizenry (Fischer 2000, pp. 162–164) also showed this. In addition, the involvement of the Swedish foundations and the like showed its international collaboration.

5.2.4 Principle 4

Scholars can be entrepreneurs of farmers, local communities, firms, and the other social actors (Olson 1971; Yang 2007a). This study found that although both local scholars and outside scholars may be able to help, it is only local scholars who can be most effective. Furthermore, only local scholars can have high scholar-entrepreneurship and finally become real scholar-entrepreneurs themselves. Outsiders, however, cannot be the true leaders. If outsider-scholars temporally become leaders, they should democratically help local people have their local leaders and construct allies with them. For example, Turner et al. 1993) pointed out that in the New Village Movement in South Korea most Saemaul leaders had a residential attachment to their villagers dating back to their grandfathers' or fathers' generation and, thus, were engaged in agriculture. Even if the outsiders paid by the government might perform an important role, they could not be "true leaders." On average, the length of incumbency of the Saemaul leaders was four to five years "with the longest tenures being characteristic of officials of the villages that had developed the most rapidly" (Turner et al. 1993, p. 82).

In the Landcare Program in Australia, Campbell (1992, p. 16) argued that in the phase of maturity, landcare groups as local actors had "settled down, with easy identification of leaders and future leaders, talkers workers, followers, sleepers and hangers-on." The major difference between effective groups and ineffective groups was whether they had good leaders (Campbell 1992; Curtis et al. 2000).

In Zhongwei County in China, at the beginning of the program, many scholars in Zhongwei Experimental Station on combating desertification came from other counties or provinces. Thus, even if they had much scientific knowledge of combating desertification, they could not get enough support from the local farmers to become the real leaders. Then, they tried hard to build good relationships with the local farmers and their leaders by teaching them and helping them find their own scholar-entrepreneurs. Some scholars had become local scholar-entrepreneurs after they had lived in the county for over many years. Furthermore, scholars' leadership was a kind of shared leadership which focuses on clusters of individuals working and growing together rather than on the leader (Denhardt et al. 2002).

5.2.5 Principle 5

Social actors' benefits can be expected, and their expectation can be reinforced with the programs' development. Furthermore, economic development and environmental protection should be combined together. For example, according to my field studies, I found Zhongwei to be a case which combined economic development and desertification control very well through the firms' participation in combating desertification; while on the other hand, Minqin and Dunhuang were two failure cases mainly focusing on economic development which finally led to them destroying the environment.

In the Landcare Program, in addition to soil conservation, conservation farming, farm planning, revegetation, land protection, vermin and weed control, and river management, land users, landcare groups, and other social actors only worked only when the change interests or benefits the people whose change is deemed as desirable, and effective landcare groups must overcome the difficulty—a lack of evidence of immediate benefits of social conservation (Campbell 1992).

In the Negev case in Israel, the use of advanced agricultural techniques and the development of technological innovations decreased the amount of water used per unit of production which reduced production costs and raised the farmers' profit and also freed water for additional production (Schechter and Galai 1980).

5.2.6 Principle 6

The fundamental rule of the experiment-extension method is that social actors first do experiments in relatively small experimental areas or zones and, then, extend them gradually to broader areas after obtaining adequate experience. The more complex and newer the problem is, the more useful this method became. If some new and untested methods are rashly used to a large area, the cost will be very high, even unaffordable, when they fail.

Concretely, there are four ideal types of the experiment-extension method (Fig. 5.1), although they are often mixed in practice. First, the experiment can be extended from a small area to a larger area. For example, an area of five hectare can be expanded to an area of ten hectare, but its hierarchical level is kept the same (Fig. 5.1a).

Second, the experiment can be extended from one unit to many units. For example, one county can be extended to several counties, but the scale and the hierarchical level of the units are kept the same (Fig. 5.1b).

Third, the experiment can be extended from one hierarchical level to another hierarchical level. For example, the county level can be lowered down to the village level (down-extending, or extend to the low level), or the county level can rise to the provincial level (up-extending, or extend to the higher level), and thus the scale and the unit are often changed (Fig. 5.1c).

Fourth, the experiment can be extended from one field to other fields. For example, combating desertification can be changed to irrigation, but the unit, the scale, and the hierarchical level of the experiment are kept the same (Fig. 5.1d).

In the Gal Oya Irrigation Project in Sri Lanka, the ARTI/Cornell team first started developing IOs in a 5000-acre plot area near the head of the system and then gradually expanded it to 25,000 acres (Ostrom 1990), Principle 6 was also satisfied.

In Naiman case in China, the Institute of Desert Research (IDR), Chinese Academy of Sciences undertook a project during the period from 1986 to 1995, which involved three distinct phases: "an experiment conducted in a research station to study desertification mechanisms and to design strategies for their control," "implementation of experimentally proven interventions in the demonstration village," and "encouraging the implementation of successful interventions in other nearby villages" (Duffy and Migongo-Bake 2003, p. 69). The same characteristics of the other projects in this area were also argued by many other researchers (e.g., UNEP 2002; Zhao et al. 1999, 2006).

In the Negev Program in Israel, based on their experiments, the Evenari team reconstructed two of ancient Nabatean farms and "succeeded in obtaining excellent yields of grain, fruits, vegetables, pasture and other crops" (Schechter and Galai 1980, p. 270). The same group of scientists also developed a system of individual micrcocatchments which they then were able to establish a large area for wider scale experimentation. Many other various experiments were also done to resolve the problems such as water desalination, rainfall enhancement, arid-zone irrigation with brackish water, and agricultural technology in the semiarid zone, and then,

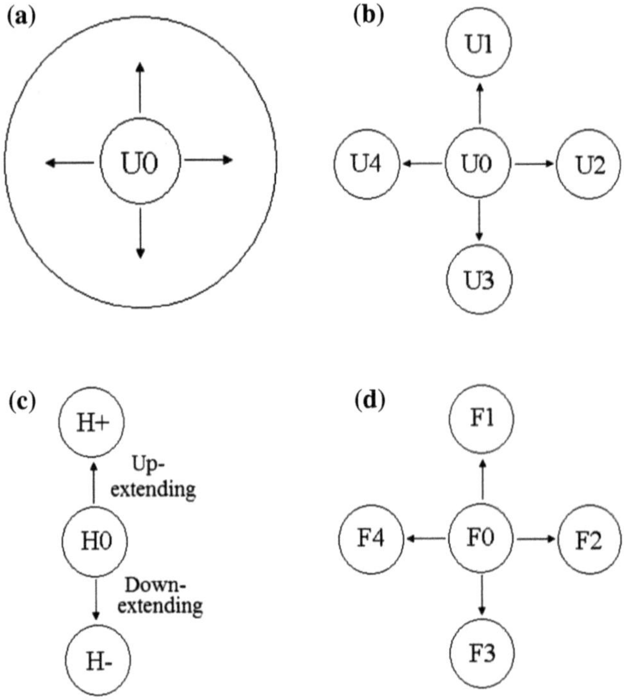

Note: (1) U0, U1, U2, U3, and U4 refer to different Units, and U0 is the original unit.

(2) H0, H+, and H- refer to the original hierarchical level, the higher hierarchical level, and the lower hierarchical level;

(3) F0, F1, F2, F3, F4, and F5 refer to different fields, and F1 is the original field.

Fig. 5.1 Three basic experiment-extension types. **a** Extending to a larger scale. **b** Extending to different units. **c** Extending to different hierarchical levels. **d** Extending to different fields

the successful technologies and methods were extended later (Schechter and Galai 1980).

5.2.7 *Principle 7*

In a self-supported system which can realize its proposed development and desertification control without external helps, scholar-based governance can be successful when the aforementioned six principles apply. For a non-self-supported system, steady external support is necessary. "Steady" here means the support should at least last until its own self-supported mechanism is built. Four kinds of external supports—financial, technical, institutional, and moral—were found important in the Landcare Program in Australia. Campbell (1992, p. (iii) argued: "It is simply unfair and totally unrealistic to expect voluntary groups of people, even if there were one hundred thousand active landcare members, to fix land degradation or develop more sustainable farming systems without a significant external support—financial, technical, institutional and moral."

The analysis of the other cases also indicated this, especially the first three kinds of support. For example, my field study indicated that because of protecting Yellow River and Lanzhou–Xinjiang Railroad in Zhongwei and protecting the Jingdian Pumping Project in Jingtai, these two counties got the significant external support, while the Minqin and Dunhuang cases did not. In the Aohan Banner case in China, except for local scholars, the participation of outside scholars guaranteed the steady external technical and knowledge support for Aohan Banner. External financial support from provincial, central government of China, other counties, the UN, etc., guaranteed its steady external financial support (Chao 2005; McGarry 2005; Michaelowa and Rawat 2007; Wang 2003a, b; Zhan 2002; Zhu 2007). Also, the support from the prefecture, provincial and central government guaranteed its steady external institutional support (McGarry 2005; Wang 2003).

Three famous Experiments in the Republic of China (1911–1949) failed due to external interventions. The basic purpose of Tao's Xiaozhuang Normal School Experiment in Nanjing, Jiangsu (1927–1930), was to develop non-formal education at village level and then to improve the quality of farmers, but his school was closed by Kuomintang's troops after only three years (Keenan 1977). Yan's Rural Reconstruction Experiment in Ding County, Henan (1929–1937), still focused on the rural education, but its scope was broader. Yan even received some external support including governmental support (even President's support) and some international support (especially from the USA). For example, depending on the governmental support, in Hunan Province, Yan even dismissed about 2/3 corrupted officials instead of 5000 scholars and scientists (Hayford 1990). The support, however, was not enough and his experiments could not be extended to many other counties. Also, Liang's Rural Reconstruction Experiment in Zouping County, Shandong (1931–1937) was supported by the provincial government (especially chairman Fuju Han), but it was stopped because of Japanese invasion (Alitto 1986).

5.3 Conclusion and Policy Implications

This chapter found that successful scholar-based governance should meet seven design principles. These seven principles embody essential elements or conditions that help to account for the success of scholar-based governance in combating desertification. From a policy perspective, these seven principles also give us some concrete instructions on how to promote scholars' participation, transform unsuccessful or semi-successful participation to more successful participation, and design new institutions for scholars' participation in combating desertification. I do not claim that these principles cover all the aspects or elements of successful scholar-based governance in combating desertification, though I feel they are essential and core elements.

References

Alitto, G. (1986). *The last Confucian: Liang Shu-Ming and the Chinese dilemma of modernity*(2nd ed.). Berkeley: University of California Press.

Campbell, A. (1992). *Landcare in Australia: Taking the long view in tough times*. Canberra: Third Annual Report, National Soil Conservation Program.

Carle, J. & Ma, Q. (2005). Challenges of translating science into practice: poplars and other species in the Three North Region of China. *Unasylva*, 221(56), 31–37.

Chambers, R., Pacey, A. & Thrupp, L. (Eds). (1989). *Farmer first-farmer innovation and agricultural research*. London: Intermediate Technology Publications.

Chao, L. (2005, February 17). Joint forestry project to see 3000 gectares. *China Daily*. Retrieved May 16, 2008, from http://www.chinadaily.com.cn/english/doc/2005-02/17/content_416948.htm

Chettiparamb, A. (2006). Bottom-up planning and the future of planning education in India. *Journal of Planning Education and Research*, 26(2), 185–194.

Confucian analects. (1900). In *Zhushi jiaozheng huaying sishu* (*Si shu*) (English & Chinese) (Trans. by J. Legge) (pp. 1–218). Shanghai: Shangwu Yinshuguan.

Curtis, A. & Lacy, T. D. (1996). Landcare in Australia: Does it make a difference? Journal of Environmental Management, 46, 119–137.

Curtis, A., Van Nouhuys, M., Robinson, W., & Mackay, J. (2000). Exploring Landcare effectiveness using organizational theory. *Australian Geographer*, 31(3), 349–366.

Denhardt, R. B., Denhardt, J. V., & Aristigueta, M. P. (2002). *Managing human behavior in public & nonprofit organizations*. Thousand Oaks, CA: Sage Publications.

Department of the Environment. (1980). The Turan Programme. In M. R. Biswas, & A. K Biwas (Eds.), *Desertification: Associated case studies prepared for the United Nations Conference on Desertification* (pp. 181–251). Oxford: Pergamon Press.

Dou, Z. Y, & Song, B. M. (1995). Sanbei linglingjiu xiangmu zai Naiman jinzhan shunli. *Journal of Inner Mongolia Forestry*, 1, 11–12 (In Chinese).

Duffy, P., & Migongo-Bake, E. (2003). The application of EIA in the evaluation of sustainable desertification control activity: examples from Namibia and Inner Mongolia. *Impact Assessment and Project Appraisal*, 21(1), 65–73.

Ettorre, E. (1999). Experts as "storytellers" in reproductive genetics: Exploring key issues. *Sociology of Science and Illness*, 21 (5), 539–559.

Evenari, M. (1989). *The awakening desert: The autobiography of an Israeli scientist*. Berlin: Springer-Verlag.

Evenari, M., Shanan, L., & Tadmor, N. (1982). *The Negev: The challenge of a desert*. Cambridge: Harvard University Press.

Fischer, F. (2000). *Citizens, experts, and the environment: The politics of local knowledge*. Durham and London: Duke University Press.

Government of the United States of America. (1980). Vale rangeland rehabilitation programme, Oregon. In J. A. Mabbutt, & C. Flotet (Eds.), *Case studies on desertification* (pp. 269–272). Paris: Unesco.

Hayford, C. W. (1990). *To the people: James Yen and village China*. New York: Columbia University Press.

Heady, H. F. (1988). The Vale Rehabilitation Program. In Heady. H. F. (Ed.), *The Vale Rangeland Rehabilitation Program: An evaluation* (Pacific Northwest Research Station Resource Bulletin

PNW-RB-157) (pp. 20–35). Portland, OR, USA: US Department of the Interior, Bureau of Land Management and US Department of Agriculture, Forest Service.

Heller, P. (2000). Social capital and the developmental state: Industrial workers in Kerala. In G. Parayil (Ed.), *Kerala: The development experience reflections on sustainability and replicability* (pp. 66–87). London: Zed Books.

Heller, P., Harilal, K. N., & Chaudhuri, S. (2007). *Building local democracy: Evaluating the impact of decentralization in Kerala, India. World Development*, 35(4), 626–648.

Hirschhorn, L. (1979). Alternative service and the crisis of the professions. In J. Case, & R. C. R. Taylor (Eds.), *Co-ops, communes, and collectives: Experiments in social change in the 1960s and 1970s* (pp. 153–193). New York: Pantheon.

Huntsinger, L., & Heady, H. (1988). Perceptions of the Vale Program. In H. F. Heady (Ed.), *The Vale Rangeland Rehabilitation Program: An evaluation* (Pacific Northwest Research Station Resource Bulletin PNW-RB-157) (pp. 103–133). Portland, OR, USA: US Department of the Interior, Bureau of Land Management and US Department of Agriculture, Forest Service.

Isaac, T. M. T., & Franke, R. W. (2002). *Local democracy and development: The Kerala people's campaign for decentralized planning*. Lanham: Rowman & Littlefield Publishers, INC.

Keenan, B. (1977). *The Dewey experiment in China: Educational reform and political power in the early Republic*. Cambridge, Massachusetts: Harvard University Press.

Kim, E-S, Park, D. K., Zhao, X. Y., Hong, S. K., Koh, K. S., Suh, M. H., & Kim, Y. S. (2006). Sustainable management of grassland ecosystems for controlling Asian dusts and desertification in Asian continent and a suggestion of Eco-Village study in China. *Ecological Research*, 21(6), 907–911.

Martin, M. A. (1982). Conservation at the local level: Individual perceptions and group mechanisms. In B. Spooner & H. S. Mann (Eds.), *Desertification and development: Dryland ecology in social perspective* (pp. 145–169). New York: Academic Press.

McGarry, D. (2005). *The "current status" of conservation agriculture in China: A report for the People's Republic of China/global environmental facility (OP12)/Asian Development Bank "partnership on land degradation in dryland ecosystems" (first draft)*. In Asian Development Bank. Retrieved May 23, 2008, from http://www.adb.org/Projects/PRC_GEF_Partnership/current-status.pdf

Michaelowa, A., & Rawat, V. R. S. (2007). *CDM afforestation and reforestation baseline methodologies: An analysis of the submission and approval process*. Hamburg Institute of International Economics (HWWI). Retrieved May 17, 2008, from http://www.hwwi.org/uploads/tx_wilpubdb/HWWI_Research_Paper_4–9.pdf

Nückles, M., & Stürz, A. (2006). The assessment tool: A method to support asynchronous communication between computer experts and laypersons. *Computers in Human Behavior*, 22(5), 917–940.

Olson, M. (1971). *The logic of collective action: public goods and the theory of groups*. Cambridge, MA: Harvard University Press.

Ostrom, E. (1990). *Governing the commons the evolution of institutions for collective action*. New York: Cambridge University Press.

Shadish, W. R., Cook, T. D., & Campbell, D. T. (2002). *Experimental and quasi-experimental designs for generalized causal inference*. Boston: Houghton Mifflin.

Törnquist, O. (2000). The new popular politics of development: Kerala's experience. In G. Parayil (Ed.), *Kerala: The development experience reflections on sustainability and replicability* (pp. 116–138). London: Zed Books.

Turner, J. E., Hesli, V. L., Bark, D. S., & Yu, H. (1993). *Villages astir: Community development, tradition, and change in Korea*. London: Praeger Publishers.

UNEP, United Nations Environment Programme. (2002, December). Increasing productivity on fragile soils: Comprehensive desertification control using sustainable moisture conservation, crops and animal husbandry technologies in Naimanqi Banner County, China. In *Success stories in the struggle against desertification: A holistic and integrated approach to environmental conservation and sustainable livelihoods*. Retrieved May 12, 2008, from http://www.unep.org/desertification/successstories/SSS%20Evaluation%20report%20.pdf

Véron, R. (2001). *The "new" Kerala model: Lessons for sustainable development*. *World Development*, 29(4), 601–617.

Wang, G. J. (2003a). Zhili huangmohua de chenggong moshi—Neimenggu Aohan Qi shengtai huangjing jianshe huigu. *Neimenggu Aohan qiwei xuanchuanbu wang*. Retired May 14, 2008, from http://www.aohan.gov.cn/html/cgms.htm (In Chinese).

Wang, S. J. (2003b, February 28). Aohan Qi shuqi zhisha yimian qi. *People's Daily*. Retrid May 16, 2008, from http://iea.cass.cn/jianbao/GRDB003_1714.htm (In Chinese).

Xu, D. W. (1995). Naiman Qi guoyou linchang (pu) gaige jian chengxiao. *Journal of Inner Mongolia Forestry*, 8, 17–18 (In Chinese).

Xu, G. W., & Yang, Z. W. (1991). Linmu suo shalong huangmo bian lüzhou—Naiman Qi zhili shamohua jishi. *Journal of Inner Mongolia Forestry*, 8, 4–5 (In Chinese).

Yang, L. (2007a). Scholar-based governance: A fourth model to collective action. *Chinese Public Administration*, 259 (1), 96–103 (In Chinese with English abstract).

Yang, L. (2007b). Building a multi-collaborative community governance system to resolve the dilemma of collective action—a framework of "product-institutional" analysis (PIA). *Journal of Public Management*, 4(2), 6–23 (In Chinese with English abstract).

Zhan, Y. (2002, April 22). Zhongguo jiaqiang dui shachenbao de zhili. Xinhua News. Retrieved May 22, 2008, from http://news.xinhuanet.com/fortune/2002-04/22/content_368226.htm (In Chinese).

Zhang, B., & Tan, M. Y. (1989). Naiman—Keerqin shadi de lüzhou. *Journal of Inner Mongolia Forestry*, 9, 8–9 (In Chinese).

Zhao, Y. P., Huang, F., & Guo, J. J. (1999). Desertification research based on digital earth. *IEEE*, 646–648.

Zhao, H. L., Zhou, R. L., Zhang, T. H., Zhao, X. Y. (2006). Effects of desertification on soil and crop growth properties in Horqin sandy cropland of Inner Mongolia, north China. *Soil & Tillage Research*, 87(2), 175–185.

Zhu, L. (2007, April 5). Aohan Qi Bao Yongxin jia de zhisha linchang. *Zhongguang Wang* (*The Website of China Radio*). Retrieved May 11, 2008, from http://www.cnr.cn/zhuanti1/scb/qxkb/200704/t20070405_504435799.html (In Chinese).

Chapter 6
Institutional Performance, Institutional Change, and Building a Knowledge-Driven Society

The Master said, 'With one like Ts'ze (also translated as Ci), I can begin to talk about the odes. I told him one point, and he knew its proper sequence.' (In Chinese: 子曰: "赐也, 始可与言《诗》已矣!告诸往而知来者。" (《论语·学而第一》)).

—*Confucian analects* (1900, pp. 8–9)

This chapter focuses on an evaluation of institutional performance, an analysis of institutional change, and an examination of a new model of a humanitarian society. Based on a synthetical analysis of governance results, resilience, and sustainability, I first analyze the institutional performance of the successful and semi-successful cases presented so far in this study. Then, I explore how an institutional arrangement can be more resilient, sustainable, and robust through an analysis of three levels of institutional change and knowledge- or intellect-driven institutional change. Lastly, I propose how to build a knowledge-driven society as a third model of a humanitarian society to compare to Lindblom's intellectually guided society and preference-guided or volition-guided society.

6.1 Institutional Performance

This section analyzes institutional performance of the institutions used in the 50 cases presented in this study. The seven design principles have been summarized and generalized based on the analysis of the 50 cases with successful, semi-successful, and unsuccessful governance results, which evaluate whether their governance purposes such as combating desertification and protecting environments were realized. In Chaps. 4 and 5, however, I did not systematically evaluate the performance of the institutional arrangements. Here, I evaluate institutional performance based on three aspects: (1) the governance result of an institutional arrangement, (2) the degree of

L. Yang, *Knowledge-Driven Governance*,
https://doi.org/10.1007/978-981-13-2910-4_6

resilience of the institutional arrangement, and (3) the degree of sustainability of the institutional arrangement.

Resilience is often deemed as "the ability of systems to cope with shocks and bounce back" (Black and Hughes 2001, p. 16) or "the capacity of a system to tolerate disturbances while retaining its structure and function" (Fiksel, 2003, as cited in Fiksel, 2006). Folke et al. (2002, p. 438) stressed three elements of resilience: (1) "the magnitude of shock that the system can absorb and remain within a given state"; (2) "the degree to which the system is capable of self-organization"; and (3) "the degree to which the system can build capacity for learning and adaptation." Sustainable development, however, is development "that meets the needs of the present without compromising the ability of future generations to meets their own needs" (The U.N.'s World Commission on Environment and Development, 1987, as cited in Sustainable Pittsburgh 2002, p. 5). That is, sustainability focuses on an ethical consideration that "equity must be considered over generations" (Black and Hughes 2001, p. 13).

For both of the finished and the continuing programs, if the institutions showed or have already shown themselves capable of surviving for over 30 years, I presume that they survived or have survived various shocks such as major economic and political changes, droughts, floods, and external intervention and were or are resilient and sustainable. If the institutions of a finished program only survived less than 30 years and no information shows that they survived shocks, I presume that they were not relevant to the topics of shocks, resilience, and sustainability. If the institutions of a finished program only survived less than 30 years and information shows that they did not survive shocks, I characterize them as unresilient and unsustainable. If the institutions of a continuing program only have survived for less than 30 years, I presume that we do not know whether they are resilient and sustainable presently. And then, I break down institutional performance into four categories[1]: (1) When the governance result is "successful" and the institution is "resilient" (or "not relevant") and "sustainable" (or "not relevant"), it is "robust"; (2) when the governance result is "semi-successful" and the institution is "resilient" (or "not relevant" or "unknown") and "sustainable" (or "not relevant" or "unknown"), it is "fragile"; (3) when the governance result is "semi-successful" and the institution is "unresilient" and "unsustainable," it is "failure"; and (4) when the governance result is "successful" and the resilience and the sustainability of the institution are characterized as "unknown," it is characterized as "unknown." Based on these criteria, I have arrayed their institutional performance as well as governance results, status of resilience, and sustainability of the 29 cases with successful and semi-successful governance results in Table 6.1.

Table 6.1 indicates that for all the even cases with the successful governance results, their institutions are "robust," except that we do not know whether the institutions used in the continuing Landcare Program in Australia are robust or not. For the 18 cases with the semi-successful governance results, the institutions used in 15

[1]Here, I do not analyze the cases with unsuccessful governance results, because if their governance results are unsuccessful, their intuitional performance all will be characterized as "failure," and then, it is not necessary to analyze their institutional resilience and sustainability.

Table 6.1 Governance results, shocks, resilience, sustainability, and institutional performance

Cases and sites	Governance results	Years	Shocks	Resilient	Sustainable	Institutional performance
Zhongwei, China	S	59	Yes	Yes	Yes	Robust
Hotan, China	S	59	Yes	Yes	Yes	Robust
Naiman, China	S	59	Yes	Yes	Yes	Robust
Aohan, China	S	59	Yes	Yes	Yes	Robust
Songhe, China	S	59	Yes	Yes	Yes	Robust
The Negev case, Israel	S	60	Yes	Yes	Yes	Robust
The Village Governance System, Ancient China	S	2126	Yes	Yes	Yes	Robust
The Household Contract Responsibility System, China	S	30	Yes	Yes	Yes	Robust
The New Village Movement, South Korea	S	About 18	NR	NR	NR	Robust
People's Planning in Kerala, India	S	About 37	Yes	Yes	Yes	Robust
Linze, China	SE	59	Yes	Yes	Yes	Fragile
Jingtai, China	SE	59	Yes	Yes	Yes	Fragile
Guazhou, China	SE	59	Yes	Yes	Yes	Fragile
Jinta, China	SE	59	Yes	Yes	Yes	Fragile

(continued)

Table 6.1 (continued)

Cases and sites	Governance results	Years	Shocks	Resilient	Sustainable	Institutional performance
Ejin Horo, China	SE	59	Yes	Yes	Yes	Fragile
Yanchi, China	SE	59	Yes	Yes	Yes	Fragile
Yulin, China	SE	59	Yes	Yes	Yes	Fragile
The Mona Reclamation Experimental Project, Pakistan	SE	43	Yes	Yes	Yes	Fragile
The "LUCDEME" Project, Spain	SE	27	Unknown	Unknown	Unknown	Fragile
Integrated Desert Development and Desertification Control, Turkmenistan	SE	84	Yes	Yes	Yes	Fragile
The Vale Rangeland Rehabilitation Program, Oregon, USA	SE	46	Yes	Yes	Yes	Fragile
The Dust Bowl Case, USA	SE	76	Yes	Yes	Yes	Fragile
The Farmers' Movement and Association, China	SE	28	Yes	Yes	NR	Fragile

(continued)

Table 6.1 (continued)

Cases and sites	Governance results	Years	Shocks	Resilient	Sustainable	Institutional performance
The Farmers' Association and Land Reform, China	SE	3	NR	NR	NR	Fragile
The Gal Oya Irrigation Project, Sri Lanka	SE	58	Yes	Yes	Yes	Fragile
Tao's Xiaozhuang Normal School Experiment, China	SE	3	Yes	No	No	Failure
Yan's Rural Reconstruction Experiment, China	SE	8	Yes	No	No	Failure
Liang's Rural Reconstruction Experiment, China	SE	6	Yes	No	No	Failure
The Landcare Program, Australia	S	19	Unknown	Unknown	Unknown	Unknown

Note S—successful; SE—semi-successful; NR—not relevant

cases are "fragile," while used in the other three cases are "failure." The reason why the institutional arrangements in these three cases failed is that they were not resilient and stopped by external interventions (these were shocks). That is, Principle 7 was not satisfied (for detailed analysis, please see Appendix IX). In sum, the seven design principles used to evaluate the governance results also can be used to evaluate the institutional performance, and the more satisfied the seven principles are, the more robust the institutional arrangement is.

Table 6.2 Corresponding relationship among the three levels of institutional arrangements, the seven principles, and the three levels of institutional change

The three levels of institutional arrangements	The seven principles	The three levels of institutional change
The individual level	Principle 1 (including Principles 1.1 through 1.6)	The individual level
The organizational level	Principles 2 through 6	The organizational level
The constitutional level	Principles 3	The constitutional level

6.2 Institutional Change

If robust institutional arrangements of scholar-based governance must satisfy the seven principles, the way to transform the fragile and failure institutional arrangements into robust ones is to change these institutional arrangements to clearly apply to these design principles.

6.2.1 Three Levels of Institutional Change

According to the structural level in a whole institutional system, North (1981, p. 203) divided institutions into three levels—constitutional rules, operating rules, and normative behavioral codes. Kiser and Ostrom (2000, p. 60) also divided institutions into three related but distinct levels: the constitutional level which explains "the design of collective choice mechanisms," the collective choice level which explains "the world of authoritative decision making," and the operative level which explains "the world of action." Also, the seven principles as the rules of the robust institutional arrangements presented in Chap. 4 can be divided into three related but distinct levels. Principle 1 (including Principles 1.1 through 1.6) is about scholars' behaviors, and I call this an individual behavioral rule. Principles 2 through 6 are about scholars' organizations and governance, which I call organizational rules. Principle 7 is related to the rule of the design of the organizations' rules, and it is a constitutional rule. Thus, I also divide institutions into three levels: the individual level, the organizational level, and the constitutional level.

Institutional change can also be divided into different corresponding levels. For example, North and Thomas (1970, p. 10) divided institutional change into two distinct levels: the change of fundamental institutions and the change of secondary institutional arrangements. Here, I consider institutional change from the same three levels of institutions: the individual, the organizational, and the constitutional. I have arrayed the corresponding relationship among the three levels of institutional arrangements, the seven principles, and the three levels of institutional change in Table 6.2.

Individual and organizational rules for robust scholar-based institutional arrangements have been illustrated by Principles 1 through 6. These can be deemed as the ways to transform fragile and failure institutional arrangements into robust at individual and organizational levels. For constitutional rules, Principle 7 indicates that when a robust self-governance system lacks, four types of stable external support (financial, technical, institutional, and moral) are important, but it does not tell us what kinds of constitutional arrangements can be more successful to supply these four types of stable external support. In this study, three are 32 Chinese cases and 18 non-Chinese cases. For the 32 Chinese cases, I assume that their constitutional rules are the same. Let's analyze the 18 non-Chinese cases and see what their differences in constitutional rules are. I have arrayed the 18 non-Chinese cases (15 desertification cases and three non-desertification cases), their regime types,[2] their application of Principle 7, and their institutional performance in Table 6.3.

Table 6.3 indicates that among the 11 cases in democratic countries (including full and flawed democracies), six cases (about 54.55%) clearly applied Principle 7, two cases (about 18.18%) applied Principle 7 in a weak form, and only three (about 27.27%) clearly did not apply. Furthermore, institutional performance in three cases (about 27.27) is robust, in five cases (45.45%) is fragile, and in three cases (27.27%) is failure. Among the seven cases in the undemocratic countries (including hybrid regimes and authoritarian regimes), one case (about 14.29%) clearly applied Principle 7, one weakly (about 14.29%) applied Principle 7 in a weak form, and five (about 71.43%) clearly did not apply Principle 7. Furthermore, institutional performance in two cases (28.57%) is fragile, and in five cases (71.43%) is failure. These findings suggest that democracies have higher possibility to apply Principle 7 and develop robust institutional arrangements. This empirical finding is also consistent with a theoretical argument that democracy serves as the basis of robust, resilient, and sustainable development (Birrer 1999; Burnell and Calvert 1999; Shiva 2005). Thus, at the constitutional level, developing democracy is a way to transform fragile and failure scholar-based institutional arrangements into robust ones.

6.2.2 *Knowledge-Driven Institutional Change and the Paradox of Scholars*

The game theoretical analysis in Chap. 2 and the empirical illustration of the 50 cases indicate that scholars and knowledge play an important role in institutional arrangements, innovation, and change. North (1990) also stressed the importance of entrepreneurs and the development of stock of knowledge to institutional change. He pointed out that "institutions typically change incrementally rather than in discontinuous fashion" (Ibid, p. 6), incremental institutional change often "comes from the perceptions of the entrepreneurs in political and economic organizations that they

---

[2]Democracy index 2006 (Kekic 2007) stressed four regime types: full democracies, flawed democracies, hybrid regimes, and authoritarian regimes.

Table 6.3 Democracy and scholar-participated institutional arrangements

Cases and sites	Regime types	Principle 7	Institutional performance
(1) The Landcare Program, Australia	Full democracy	Yes	Unknown
(2) The Negev Case, Israel	Flawed democracy	Yes	Robust
(3) The New Village Movement, South Korea[a]	Flawed democracy	Yes	Robust
(4) People's Planning in Kerala, India[a]	Flawed democracy	Yes	Robust
(5) The Mona Reclamation Experimental Project, Pakistan	Authoritarian regime	Yes	Fragile
(6) The "LUCDEME" Project, Spain	Full democracy	Yes	Fragile
(7) Integrated Desert Development and Desertification Control, Turkmenistan	Authoritarian regime	Weak	Fragile
(8) The Vale Rangeland Rehabilitation Program, USA	Full democracy	Yes	Fragile
(9) The Dust Bowl Case, USA	Full democracy	Weak	Fragile
(10) The Gal Oya Irrigation Project, Sri Lanka[a]	Flawed democracy	Weak	Fragile
(11) The Gascoyne Basin, Australia	Full democracy	No	Failure
(12) The Region of Coquimbo, Chile	Flawed democracy	No	Failure
(13) The Luni Development Block, India	Flawed democracy	No	Failure
(14) The Turan Program, Iran	Authoritarian regime	No	Failure

(continued)

Table 6.3 (continued)

Cases and sites	Regime types	Principle 7	Institutional performance
(15) The Greater Mussayeb Project, Iraq	Hybrid regime	No	Failure
(16) The Eghazer and Azawak Region, Niger	Authoritarian regime	No	Failure
(17) The Oglat Merteba Region, Tunisia	Authoritarian regime	No	Failure
(18) The Golodnaya Steppe, Uzbekistan	Authoritarian regime	No	Failure

Note [a]About the other fields rather than combating desertification

could do better by altering the existing institutional framework at some margin," and "the perceptions crucially depend on both the information that the entrepreneurs receive and the way they process that information" (Ibid, p. 8). My study found that scholars are often these entrepreneurs, who "may devote their talents or tacit knowledge to ferreting out profitable margins, estimating the likelihood of success, and risking the organization's resources to capture potential gains" (Ibid, p. 87). Because scholars in this study are those who have comparative advantages in knowledge over others, I call this type of institutional change as knowledge- or intellect-driven institutional change.

Various kinds of knowledge need to be integrated in institutional change. North (1990) strengthened differences between communicable knowledge and tacit knowledge in institutional change. Communicable knowledge can be "transmitted from one person to another" (Ibid, p. 74). Tacit knowledge (a term coined by Polanyi 1967), however, is "acquired in part by practice" and can be "only partially communicated;" and "different individuals have different innate abilities for acquiring tacit knowledge" (Ibid). Also, many researchers (e.g., Davis 2005; Lindskog and Tengberg 1994; Reed et al. 2007; Stringer and Reed 2007; Thomas 1997; Thomas and Twyman 2004) indicated that both scientific knowledge and local "non-scientific" knowledge need to be integrated in combating desertification. As stated before, this study also found that scientific knowledge, knowledge about local conditions, and knowledge about local social relationships and webs are three kinds of the most important knowledge in combating desertification. Furthermore, both field-based technical and social scholars should be involved in institutional change.

Institutional arrangements also influence acquisition of knowledge and scholars' actions. North (1990) argued that the payoff—incentives—imbedded in institutional constraints influences members of an organization to acquire different kinds of knowledge, skills, and learning, and structure of monetary rewards and punishments. Also, a society's tolerance of its development influences the incentives to acquire pure knowledge. He emphasized that "it is a two way relationship, that is, the way knowledge developed shaped our perceptions of the world around us and in turn those perceptions shape the search for knowledge" (Ibid, p. 76). He further pointed out that "(1) the institutional framework will shape the direction of the acquisition of knowledge and skills and (2) the direction will be the decisive factor for the long-run development of the society" (Ibid, p. 78).

There is a two-way relationship between scholars and institutional change. On the one hand, scholars can affect institutional change as entrepreneurs; on the other hand, institutional framework and institutional change also influence scholars' actions. Furthermore, this study found that scholars may have both positive and negative impacts on institutional change. On the one side, scholars may promote institutional change due to their perception of knowledge; on the other side, they may also impede institutional change due to their lack of necessary knowledge, stubborn adherence to outworn rules and idea, prejudice, false confidence, etc. The latter is the case of scholars in the late Qing Dynasty (Stavrianos 1998). Ostrom also pointed out:

> The intellectual trap in relying entirely on models to provide the foundation for policy analysis is that scholars then presume that they are omniscient observers above to comprehend the essentials of how complex, dynamic systems work by creating stylized descriptions of some aspects of those systems. With the false confidence of presumed omniscience, scholars feel perfectly comfortable in addressing proposals to government that are conceived in their models as omnicompetent powers able to rectify the imperfections that exist in all field settings. (Ostrom 1990, p. 215)

Thus, I call this is the paradox of scholars. The application of the seven principles and democratic constitutional institutional arrangements, however, is one way to resolve this problem.

This study found that the process of scholar-based and knowledge-driven institutional change can be revolutionary or incremental (or gradual and evolutionary) (Lindblom 1959), leapfrogging or sequential (Ostrom 1990), and imposed or induced (Lin, 1989). Its concrete change forms are determined by the context and many variables shown in the PIA framework. In a long run, however, as Baumgartner and Jones (1993) argued, this institutional change can be deemed as a process characterized by a long period of incremental, gradual, evolutional, and adaptive change punctuated by brief periods of major, revolutionary, leapfrogging, and imposed institutional change. The search for some general equilibrium probably is fruitless. This means that amid seeming stability of the institutional and policymaking process there is continual change and both long period of institutional and policy incrementalism and spasms of change are part of the same equation of issue definition and institutional bias. The case of the village governance system in ancient China is a good example (for detailed analysis, please see Appendix IX).

6.3 Building a Knowledge-Driven Society: A Third Model

6.3.1 Lindblom's Two Models

Based on considering the role of intellect in social organization, in his classical book *Politics and Market*, Lindblom (1977) described two highly sophisticated models of a humanitarian society: an intellectually guided society and a preference-guided or volition-guided society. A representative of Model 1 is the communist systems in the U.S.S.R, while a representative of Model 2 is the market-oriented polyarchies in the USA. Based on Lindblom's analysis, I have summarized the key differences of these two models into five categories: (1) what is people's intellectual competence and how is knowledge consequently produced and protected? (2) How is knowledge owned? (3) How is knowledge tested? (4) What is the relationship among men's needs? (5) How is knowledge applied?

6.3.1.1 Knowledge Production and Protection

Model 1 assumes that there is a match between intellectual capacity and the complexity of the social world. It believes that some wise and informed-enough intellectual leaders can discover correct social organization and produce a comprehensive or synoptic theory to ameliorate social problems and guide social change with a high degree of success. Because of its confidence of the adequacy of intellect and theory, Model 1 also assumes that it is not necessary to protect scientific research, inquiry, public discussion, and dispersed initiatives.

Model 2 assumes a gross mismatch between intellectual capacity and the complexity of the social world and contends that "every one well knows himself to be fallible" (John Stuart Mill, as cited in Lindblom 1977, p. 249). In Model 2, it is impossible for some intellectual elites to comprehend the future well enough to plan it. No synoptic theory as proposed by Model 1 exists. Scholars can produce only scattered partial theories, which are tentative, inadequately tested, and consequently imperfect sources of guidance in practical affair even within their limited domains. Also, no single correct solution exists, discovery is impossible, and social organization has to be chosen based on many people's volitions. Model 2 also recognizes the difficulty of undersupplying analysis and its inadequate competence. Hence inquiry, thought, and science are valued and protected, problem-solving initiatives are encouraged, and incentives to problem solving are widely dispersed by "holding participation in interaction open at all possible points" (Lindblom 1977 p. 260).

6.3.1.2 Knowledge Possession

In Model 1, knowledge is mainly owned by some intellectual elites who are simultaneously some political elites because the society is guided by knowledge rather than volition. For example, "the Party leadership lays claim to a monopoly over the interpretation and application of the sole scientific theory of social development" in the U.S.S.R. (Schwartz, as cited in Lindblom 1977, p. 251).

In Model 2, by contrast, as Hayek (1945, p. 519) said, "the knowledge of the circumstances of which we must make use never exists in concentrated or integrated form, but solely as the dispersed bits of incomplete and frequently contradictory knowledge which all the separate individuals possess."

6.3.1.3 The Test of knowledge and Its Criterion

In Model 1, it is assumed that there always exists a criterion of correctness, and thus, the test of knowledge or an institution or policy is that it is correct. Men's needs, in principle, are often used as the test of institutions and policies. And it assumes that the intellectual leaders sufficiently know what men's real needs are.

Model 2 believes that there is no such a criterion of correctness and people cannot to be competent to know what is correct. Since it is believed that people's needs cannot be known or known with sufficient confidence and thus institutions and policies cannot always be tested based on them, the people's volitions are taken as the best indicator of their needs and wants.

6.3.1.4 The Relationship Among Men's Needs

In Model 1, it is postulated that there is an underlying harmony of men's needs, which can be known to the guiding elite. Model 2, however, assumes that harmony of needs is not only undiscoverable but also nonexistent.

Lindblom further pointed out:

> The two visions, one of harmony and one of conflict, manifest competing views of society that have divided men for at least two millennia: idea society as knit together by consensus, as in Plato, Rousseau, and Hegel; or conflict-ridden society, as in Aristotle, Hobbes and Kant. (Lindblom 1977, pp. 251–252)

6.3.1.5 The Application of Knowledge

In Model 1, the intellectual elites study the problems, discover or find correct decisions by diagnosis or problem analysis, and then guide the other people. Politics is put to an end, and the political system becomes no much more than an administrative organization. Although interactions exist, they only implement elite decisions and problem-solving interactions are suppressed as only a source of disorganization and trouble.

Procedures are not as highly valued in Model 1 as they are in Model 2. Conflict is often suppressed. Men are viewed solely as beneficiaries in such a system for want satisfaction. Society epiphenomena—by-products—are viewed as trouble making or natural with respect to problem solving. Among numerous individual and group differences, only differences among individuals in age, sex, intelligence, strength, and certain special skills are recognized, other individual and group differences tend to disappear, and cultural diversity is not acceptable.

In Model 2, interactions which incorporate with problem-solving politics, market systems, and all other substitutes for analytical problem solving are used as a substitute for analysis. Decisions are made by interactions without diagnostic studies in pursuit of the correct decisions. It, however, does not assume that interactions can always produce perfect solutions to problems. It is only assumed that these solutions will often be superior to the solutions attempted directly by the intellectual elite. Furthermore, interaction in Model 2 never wholly replaces analysis and analysis is adapted to problem-solving interactions.

Procedures shown by some mechanisms such as majority rule, a fair trial, and checks and balances are highly valued in Model 2. Conflict is often positively employed, and many interactions such as separation of powers, tripartite boards, and parliamentary removal of prime minister are designed to create (and then to cope with) conflict. Problem solving itself is valued. Society epiphenomena are often viewed as problem solving, and citizens are to be seen as both doers and achieves. Finally, mutual adjustment and pluralism are conspicuous in Model 2. They embrace individuals, groups, and cultural diversity which "afford broad scope for participants in interactive processes to pursue their own individual or group goals rather than any one or a few overarching national goals" (Lindblom 1977, p. 258).

6.3.2 A Knowledge-Driven Society as a Third Model

The analysis of scholar-based governance, specifically the study of the generalization as representation based on a meta-analysis of the 32 Chinese cases in Chap. 5, indicates that there is an alternative model to Lindblom's an intellectually guided society and a preference-guided or volition-guided society. This new model, as a third model or Model 3 of a humanitarian society, can be called a scholar-based society or a knowledge-driven society in which scholars play a role in social governance as I discussed before in this study. If Model 1 proposes a buoyant, optimistic, or confident distinctive view of man using his intelligence in social organization and Model 2 gives us a more pessimistic or skeptical view of his capacity, Model 3 prefers a view in the middle way.[3] Based on the analysis of the 50 cases in this study, I have set up the five main elements of Model 3 (Table 6.4).

I shall discuss in details of each of the five elements.

[3]In *Introductory Note* of *The Doctrine of the Mean* (*Chong Yung, or Zhong Yong*), Chengzi (Chû Hsî) said: "My master, the Philosopher Ch'ăng (Zicheng in Chinese pinyin) says:—'Being without inclination to either side is called CHUNG; admitting of no change is called is called Yung. By CHUNG is denoted the correct course to be pursed by all under heaven; by YUNG is denoted the fixed principle regulating all under heaven'" (*The Doctrine of the Mean* 1900, p. 246). In Chinese: 子程子曰: "不偏之谓中; 不易之谓庸。中者, 天下之正道。庸者, 天下之定理。" (《中庸》).

Table 6.4 Five main elements of a knowledge-driven society

E1. Dispersed and specialized knowledge production
E2. Dispersed and asymmetric knowledge possession
E3. Knowledge-driven volitions or consensuses as the tests of knowledge
E4. Diverse and heterogeneous needs
E5. Multiple methods of knowledge application

6.3.2.1 Element 1: Dispersed and Specialized Knowledge Production

Everyone who well knows himself knows themselves to be fallible. With people's study, research, analysis, and learning, the complexity of the social world, however, can be at least partially understood. Also, the social problem can be gradually, sequentially, or incrementally resolved—for instance, as showed by the successful and semi-successful cases in this study. There is no a perfect match between intellectual capacity and the complexity of the social world, but it is also not a gross mismatch.

One comprehensive theory of social change cannot be envisioned or designed by some intellectual leaders to serve to guide the society, and scholars only can produce scattered partial theories which are tentative, inadequately tested, and consequently imperfect sources of guidance in practical affairs even within their limited domain. Some better theories, however, can be produced and used to guide practical affairs as shown in the many cases in this study. The reasons of this include knowledge accumulation, specialized production of knowledge, teaching and learning, cooperation and collaboration among scholars, citizens, governments, firms, and other social actors, participatory research, etc. Although all theories are imperfect, tentative, and inadequately tested, some theories may be better or more useful than others. Correct social organization cannot be discovered by some intellectual elite, but social organization is also not chosen based on many people's blind volitions. It must be chosen based on people's volitions, but these volitions can be based on knowledge, intellectual, partial theories, and analysis. Its process is also characterized by a long period of incremental, gradual, evolutional, and adaptive change punctuated by brief periods of major, revolutionary, leapfrogging, and imposed changes.

Everyone is fallible, but everyone also has some intellectual competence to learn, understand, and teach. Knowledge production is really dispersed. This, however, does not mean everyone equally produces the same amount of knowledge. Due to advantages of division of labor or of specialization (Smith 1965; Durkheim 1960; Romer 1987), economies of scale (Christensen and Green 1976; Krugman 1980; Murray and White 1983), economies of scope (Murray and White 1983; Teece 1980), numerous varieties and diversities among different individuals, etc., specialization in knowledge production (Brusoni et al. 2001) is also popular as in any other kinds of production. That is, knowledge production is characterized by a holistic picture of dispersed production accreted by some partial pictures of specialized production.

Knowledge or analysis, as a kind of public goods (D'Aspremont et al. 1998; Stiglitz 1999), is also underprovided (Olson 1971), or "difficult, in short supply, and of an inadequate level of competence" (Lindblom 1977, p. 259) in Lindblom's words. Hence, inquiry, thought, and science are also carefully protected in Model 3. Privatization with competing patent rights, however, is only one method of knowledge production. The difficult of knowledge production as a problem of collection action can be resolved using other methods such as the central authority (e.g., governmental support), self-governance (e.g., governed by scientific communities themselves as knowledge producers), and even scholar-based models. The idea of the tragedy of the anticommons indicates that competing patent rights actually prevent useful and affordable knowledge or products from reaching the marketplace and then leads to less innovation (Heller 1998, 2008; Heller and Eisenberg 1998). Thus, various types of coordination, collaboration, and cooperation such as between scholars, between scholars and other social actors, and between states as shown by the cases in this study are very important in knowledge production. By examining the worldwide trend toward increased privatization of both information and communications technologies, Stiglitz (1999) also called for increased cooperation between states on a regional basis to stop the worldwide trend toward the shrinking role of states and the effects on peoples' access to adequate information goods and services.

6.3.2.2 Element 2: Dispersed and Asymmetric Knowledge Possession

As shown in the aforementioned analysis, knowledge in Model 3 is owned by all the separate individuals and everyone's knowledge is relatively incomplete and frequently contradictory. That is, knowledge possession is dispersed. In contemplation of social or institutional change, no one intellectual elite in Model 3 can monopolize the interpretation and application of the sole scientific theory of social or institutional development. Again, as shown in the aforementioned analysis, this does not mean that knowledge is equally owned by all the separate individuals. First, unless people are living in an ideal world, the asymmetry of knowledge possession is inevitable. Second, even if some institutional arrangements can be designed to distinctly reduce the degree of the asymmetry of knowledge possession, division of labor or specialization, economies of scale and scope, numerous varieties and diversities among different individuals, the limits of human nature, and the complex attributes of knowledge often make it impossible. Thus, knowledge is also asymmetrically owned by separate individuals. Some individuals or organizations often have knowledge more than others—heterogeneity of knowledge possession at the individual or organizational level. Scholars are defined as those who have comparative advantages in knowledge over other social actors such as herders, firms, governments, and the fifth sector in this study. The organizations that have comparative advantages in knowledge over other social organizations can be correspondingly called scholar-organizations.

6.3.2.3 Element 3: Knowledge-Driven Volitions or Consensuses as the Tests of Knowledge

Since it is very hard to fully know men's true physical, psychological, and social needs, it is impossible to fully use the correspondence of an institution or policy to them to evaluate "correctness." It, however, is also not enough, if we only use the volitions of "the people" as a criterion. First, even if using these volitions of "the people" is reasonable, how can we fully know the people's volition? Here, we are facing the same problem as we try to use the correspondence to men's true needs. Sometimes even individuals' preferences or volitions are not very clear. Especially, when it is moved from the individual level to the social level, Arrow's impossibility theorem (1963) argued that it is impossible to formulate a social preference ordering that satisfies the five conditions—unrestricted domain, transitive property, independence of irrelevant alternatives, weak Pareto principle, and non-dictatorship. The Allais Paradox or problem (1953) even highlighted an inconsistency of actual observed choices with the predictions of expected utility theory. Because individuals often pay more attention to the choice with high certainty in the decision-making process (this is called the certain effect), the expected utility theory is not robust enough to capture "bounded rationality" choice (Allais 1953). Second, even if people's volitions can be understood, volitions which people felt and shown may not be true volitions.

This study, however, indicated that in a scholar-based communication, analysis, and discussion, people's volitions can be better understood and their true needs can be included in their true volitions and preferences, if necessary. That is, institutions or policies can be tested based on the people's knowledge-driven volitions. Especially, at the rice-roots level, almost-consensus and face-to-face communication methods can be applied. The theory of deliberative (or discursive) democracy which refers to any democratic system between consensus decision and representative democracy and emphasizes the public deliberation of the citizenry (Bessette 1980, 1994; Cohen 1989; Dryzek 2000; Elster 1998; Habermas 1996) supports this position. Many new institutional innovations such as consensus conference and citizen's conferences also improve the effective partnership between scientific experts and the general public (Guston 1999; Joss 2002).

6.3.2.4 Element 4: Diverse and Heterogeneous Needs

Difference does not mean conflict, and harmony does not mean sameness. Individual needs are diverse and heterogeneous. This, however, does not mean they are always contradictory. Harmony, in Chinese literature, means various and heterogeneous things that can peacefully exist and develop together, peacefully influence each other, and compete with each other (*The Li ki* 1885; *Confucian analects* 1900). That is, harmony must embrace diversity and heterogeneity. Diverse and heterogeneous individual needs may conflict with each other or may only harmoniously exist together. Possible coordination, cooperation, and collaboration as shown in this study even improve the possibility of the latter. Furthermore, based on the rule of majority rather than consensus, public needs or public interests are not only existent but discoverable (for detailed analysis, please see Denhardt and Denhardt 2003, pp. 65–81). This supplies a basis of the idea of knowledge-driven volitions when individual needs are contradictory and were supported by the empirical cases in this study.

6.3.2.5 Element 5: Multiple Methods of Knowledge Application

As in Model 2, interactions in Model 3 also incorporate with "problem-solving politics, market systems, and all other substitutes for analytical problem solving" (Lindblom 1977, p. 254). This study, however, shows that analysis is not only adapted to "problem-solving interactions" in scholar-based governance. If knowledge production is dispersed and specialized, knowledge possession is dispersed and asymmetric, men's needs are diverse and heterogeneous, and institutions and policies are tested based on analysis-based volitions or consensuses, these interactions should also be analysis-based or knowledge-driven interactions or deliberative interactions (Forester 1999) as shown in deliberative democracy. The experiment-extension method used in scholar-based governance supports this position.

Procedures as designed in Model 2 are highly valued in Model 3. In a dynamic and successive process of social interactions, an outcome is always an important starting point of a new procedure as shown by the cases in this study. Furthermore, in a long period, social and institutional change is gradual, sequential, or an incremental process. In some cases, even if the final product is impossible, the intermediate product is desirable. Procedures cannot be used to completely substitute outcomes or products. Based on knowledge-driven or deliberative volitions and interactions, Model 3 tries hard to balance the contradictory values such as procedures, outcomes and equity and efficiency (Yang 2008), and economic development and environmental protection as aforementioned in this study.

Conflict is also positively used, and many interactions such as separation of powers, tripartite boards, and parliamentary removal of prime minister are designed to create conflict in Model 3. Conflict, however, is not similarly highly valued as it is in Model 2. Conflict is deemed as one side of a coin, while cooperation is found to be on the other side. Various and purposive cooperation, coordination, and collaboration as shown in our case analyses in this study are encouraged. Teamwork (Cohen and Levesque 1991; Tambe 1997), coproduction (Ostrom 1996; Whitaker 1980), and the like will totally create a new cooperative culture along with a competitive or conflict culture.

As in Model 2, all individuals are viewed as doers and achievers of democracy in Model 3. Multiple levels of institutions are arranged to systematically resolve various problems. For example, Principles 1 and 1.1 to 1.6 in the seven design provincials illustrate the individual-level rules; Principles 2 to 6 analyze the organizational-level rules. Principle 7 strengthens the importance of the constitutional rules. Some problems are solved as by-products or epiphenomena, while some problems are solved as staples rather than by-products. People can buy something to solve "the problem of resource allocation as a byproduct or epiphenomenon of his own private problem solving" (Lindblom 1977, p. 257), but there are also some people who can simultaneously directly solve some problems (even gradually, incrementally, fallibly) related to the problem of resource allocation as shown by the case of the New Village Movement in South Korea.

Model 3 embraces the individual, group, institutional, and cultural diversity. Diversity, however, does not guarantee pluralism in Model 2 as a robust institutional arrangement. In addition to the collective dilemma in group behaviors (Olson 1971), Lindblom (1977) himself recognized the shortcoming of pluralism or polyarchy. He also discussed its danger easily turning into corporatism when certain groups of elites gain crucial advantages and become too successful and begin to collude with another instead of competition. Even in democracies, when there are controlled volitions, or circularity, masses are often persuaded to ask from elites only what they wish to give them, and thus, a real choices and competition are further limited. Schattschneider (1960) even argued that pluralism in fact is elite pluralism. Interest groups based on self-interests only lead to special interest politics where policy subsystems and politics are often dominated by probusiness and upper-class minority bias. Policy subsystems are policy situations wherein a number of actors control the policy area and exclude the public from the influence in the area. Lowi (1979) also highlighted the problem of policy subsystems and argued that governmental agencies become captive of interest groups and have taken on far too much discretionary power with the immense growth of the liberal state without self-examination and without recognizing problematic consequences of their growth, although they claim that they represent the popular will. Riker (1982) further discussed structurally induced equilibria, where governmental institutions create situations of equilibrium on particular areas, while the nature state of the world is disequilibrium. He pointed out that structurally induced equilibria are maintained by policy subsystems and help create a situation of stability, which exclude the public opinion from policymaking.

How can these problems of pluralism in Model 2 be solved in Model 3? Lowi (1979) calls for the development of a far more legalistic or "juridical" democracy. Juridical democracy has two major components: the juridical ("rule of law operating in institutions") and the democratic ("majority rule democracy" in which the state, through legislative representation, addresses "the citizen in us") (Lowi 1971a, b, 1979). So Grady (1984) also called it "democratic constitutionalism." Baumgartner and Jones (1993) pointed out that policy subsystems can be changed or destroyed when the public becomes mobilized at a high level. To them, new voices and previously excluded interests can be remobilized through dual mobilizations: The first comes during a wave of enthusiasm for a particular policy with a positive policy "image" just as the cycle of "alarmed discovery" and "euphoria" explored by Anthony Downs (1972); the second attacks against the policy status quo through opponents' criticism with negative policy images like the "socialization" of conflict suggested by Schattschneider (1960). In Model 3, all the three levels of institutional arrangements are strengthened as stated above; citizens' participation at all the levels of institutional arrangements are dully mobilized based on deliberative or knowledge-driven volitions; and scholar-based governance and participatory governance of other social actors such as various NGOs and religious groups are encouraged to balance the power of the business interest groups. We can call this generalized pluralism.

6.4 Conclusion

Traditionally, an intellectually guided society and a preference-guided or volition-guided society are often deemed as an either-or choice of the method of social organization. The collapse of the U.S.S.R was often deemed as a symbolic event of the failure of the intellectually guided society and the success of the volition-guided society. Fukuyaman even bravely claimed:

> What we may be witnessing is not just the end of the Cold War, or the passing of a particular period of post-war history, but the end of history as such: that is, the end point of mankind's ideological evolution and the universalization of Western liberal democracy as the final form of human government. (1989, p. 3)

This study, however, proposes a knowledge-driven society as an alternative model to Lindblom's two models. The collapse of the U.S.S.R really shows the failure of the intellectually guided society, but it does not mean the success of the volition-guided society or that the latter should be the only way. As Lindblom (1977, p. 247) already argued, "No society looks defensible." Many problems in the volition-guided society drive us to reform or ameliorate this system. A knowledge-driven society, as a middle way, gives us some hopes. This society, however, is much closer to a volition-guided society than to an intellectually guided society, and it is better for us to deem it as a revised model of a volition-guided society rather than an alternative of the latter. Also, this does not mean I will try to use a knowledge-driven society to substitute a volition-guided society. As the two possible models of the method of social organization, they can harmoniously coexist. Which one is better depends on traditions, culture, level of development, etc. of a society. For example, a knowledge-driven society may be more suitable in China due to its Confucian tradition than in America. This is why this study found many cases of robust institutional arrangements in China which is deemed as an authoritarian regime rather than a democracy (Kekic 2007); while in the other countries, it is found that a democratic society has higher possibilities to develop robust institutional arrangements than an undemocratic society. When China becomes more democratic in the future, it will move closer to a knowledge-driven democracy. As Lindblom (1977) already pointed out, models and actual society, however, should not be confused.

Furthermore, in this study, I open the scholar group to almost all human beings. That is, all human beings can become scholars under certain conditions. Thus, an open society is not only a society in which government is responsive and tolerant, and political mechanisms are transparent and flexible, and then, the state keeps no secrets from itself in the public sense just as this term was originally proposed by philosopher Henri Bergson (1935). Furthermore, it is not only a society which ensures that political leaders can be overthrown without the need for bloodshed in contrast with the closed society where a bloody revolution is needed (Popper 1945). Based on the above analysis, I can conclude that it should also be a society in which everyone should have the right and freedom to learn, to study, and to consume the education and should be given the same probabilities and opportunities to resolve the problems they are facing. The knowledge and the right of getting knowledge and using knowledge to resolve the social problems cannot be hold by only some special groups as Plato claimed.

References

Allais, M. (1953). Le comportement de l'homme rationnel devant le risque: critique des postulats et axiomes de l'école Américaine. *Econometrica,* 21(4), 503–546.

Baumgartner, F. R., & Jones, B. D. (1993). *Agendas and instability in American politics*. Chicago and London: The University of Chicago Press.

Bergson, H. (1935). *The two sources of morality and religion* (Trans. by R. A. Audra & C. Brereton with the assistance of W. H. Carter). New York: Henry Holt and Company.

Bessette, J. (1980). Deliberative democracy: The majority principle in republican government. In R. A. Goldwin & W. A. Schambra (Eds), *How democratic is the Constitution*? (pp. 102–116). Washington, D.C.: AEI Press.

Bessette, J. (1994). *The mild voice of reason: Deliberative democracy & American national government*. Chicago: University of Chicago Press.

Birrer, F. A. J. (1999). Sustainability, democracy and socio-cybernetics. *Kybernetes*, 28(6/7), 810–820.

Black, A., & Hughes, P. (2001). *The identification and analysis of indicators of community strength and outcomes* (Department of Family and Community Services Occasional Paper No. 3). Canberra: Commonwealth of Australia. Retrieved November 11, 2005, from http://www.facs.gov.au/internet/facsinternet.nsf/vIA/occasional_papers/$file/No.3.pdf.

Brusoni, B., Prencipe, A., Pavitt, K. (2001). Knowledge specialization, organization coupling, and the boundries of the firm: Why do firms know more than they make? *Administrative Science Quarterly*, 46(4), 597–621.

Burnell, P., & Calvert, P. (1999). *The resilience of democracy: Persistent practice, durable Idea*. London: Frank Cass.

Christensen, L. R., & Green, W. H. (1976). Economies of scale in US electric power generation source. *The Journal of Political Economy*, 84(4), 655–676.

Cohen, J. (1989). Deliberative democracy and democratic legitimacy. In A. Hamlin & P. Pettit (Eds.), *The Good Polity* (pp. 17–34). Oxford: Blackwell.

Cohen, P. R., & Levesque, H. J. (1991).Teamwork. *Nous*, 25(4), 487–512.

Confucian analects. (1900). In *Zhushi jiaozheng huaying sishu* (*Si shu*) (English & Chinese) (Trans. by J. Legge) (pp. 1–218). Shanghai: Shangwu Yinshuguan.

D'Aspremont, C., Bhattacharya, S., Gérard-Varet, L. (1998), Knowledge as a public good: efficient sharing and incentives for development effort. *Journal of Mathematical Economics*, 30(4), 389–404.

Davis, D. K. (2005). Indigenous knowledge and the desertification debate: Problematising expert knowledge in North Africa. *Geoforum*, 36(4), 509–524.

Denhardt, J. V., & Denhardt, R. B. (2003). *The new public service: Serving, not steering*. New York: M. E. Sharpe.

Doctrine of the mean. (1900). In *Zhushi jiaozheng huaying sishu* (*Si shu*) (English & Chinese) (Trans. by J. Legge) (pp. 246–298). Shanghai: Shang wu yin shu guan.

Downs, A. (1972). Up and down with ecology: The issue attention cycle. *Public Interest*, 28 (summer), 38–50.
Dryzek, J. S. (2000). *Deliberative democracy and beyond: Liberals, critics, contestations*. Oxford: Oxford University Press.
Durkheim, E. (1960). *The division of labor in society* (Trans. by G. Simpson). New York: The Free Press.
Elster, J. (Ed). (1998). *Deliberative democracy*. Cambridge: Cambridge University Press.
Fiksel, J. 2006. Sustainability and resilience—toward a systems approach. *Sustainability: Science, Practice, & Policy (SSPP)*, 2(2), 1–8.
Folke, C., Carpenter, S., Elmqvist, T., Gunderson, L. Holling, C. S., & Walker, B. (2002). Resilience and sustainable development: Building adaptive capacity in a world of transformations. *Ambio*, 31(5), 437–440.
Forester, J. (1999). *The deliberative practitioner: Encouraging participatory planning processes*. Cambridge, Massachusetts and London: The MIT Press.
Grady, R. C. (1984). Juridical democracy & democratic values: An evaluation of Lowi's alternative to interest-group liberalism. *Polity*, 16(3), 404–422.
Guston, D. H. (1999). Evaluating the first U.S. consensus conference: The impact of the 'citizens' panel on telecommunications and the future of democracy. *Science, Technology and Human Values*, 24(4), 451–82.
Habermas, J. (1996). *Between facts and norms: Contributions to a discourse theory of law and democracy* (Trans. by W. Rehg). Cambridge, MA: MIT Press.
Hayek, F. A. (1945). The use of knowledge in society. *The American Economic Review*, 35(4), 519–530.
Heller, M. A. (1998). The tragedy of the anticommons: Property in the transition from Marx to Markets. *Harvard Law Review*, 111(3), 621–88.
Heller, M. A. (2008). *The gridlock economy: How too much ownership wrecks markets, stops innovation, and costs lives*. Philadelphia: Basic Books.
Heller, M. A., & Eisenberg, R. S. (1998). Can patents deter innovation? The anticommons in biomedical research. *Science*, 280 (5364), 698–701.
Joss, S. (2002). Toward the public sphere – Reflections on the development of participatory technology assessment. *Bulletin of Science, Technology & society*, 22(3), 220–231.
Kekic, L. (2007). The economist intelligence unit's index of democracy. Retrieved October 22, 2008, from http://www.economist.com/media/pdf/Democracy_Index_2007_v3.pdf.
Kiser, L. L., & Ostrom, E. (2000). The three world of action: A metatheoretical synthesis of institutional approaches. In M. D. McGinnis (Ed.), *Polycentric games and institutions: Readings from the Workshop in Political Theory and Policy Analysis* (pp. 56–88). Ann Arbor: The University of Michigan Press.
Krugman, P. (1980). Scale economies, production differentiation, and the pattern of trade. *The American Economic Review*, 70(5), 950–959.
Li ki (*The book of rites*) (Trans. by J. Legge). (1885). Oxford: Clarendon Press.
Lin, J. Y. (1989). An economic theory of institutional change: Induced and imposed change. *Cato Journal*, 9(1), 1–33.
Lindblom, C. E. (1959).The science of muddling through. *Public Administration Review*, 19(2), 79–88.
Lindblom, C. E. (1977). *Politics and markets: The world's political-economic systems*. New York: Basic Books, Inc., Publishers.
Lindskog, P., & Tengberg, A. (1994). Land degradation, natural resources and local knowledge in the sahel zone of Burkina Faso. *GeoJournal*, 33(4), 365–375.
Lowi, T. J. (1971a). *The politics of disorder*. New York: Norton.
Lowi, T. J. (1971b). A reply to Mansfield. *Public Policy*, 19(winter), 207–211.
Lowi, T J. (1979). *The end of literalism: The second republic of the United States* (2nd ed.). New York and London: W.W. Norton & Company.

Murray, J. D., & White, R. W. (1983). Economies of scale and economies of scope in multiproduct financial institutions: *A study of British Columbia Credit Unions*. *The Journal of Finance*, 38(3), 887–902.
North, D. C. (1981). *Structure and change in economic history*. New York: W. W. Norton & Company, Inc.
North, D. C., &Thomas, R. P. (1970). An economic theory of the growth of the western world. *The Economic History Review*, 23(1), 1–17.
North, Douglass C. (1990). *Institutions, institutional change and economic performance*. Cambridge: Cambridge University Press.
Olson, M. (1971). *The logic of collective action: public goods and the theory of groups*. Cambridge, MA: Harvard University Press.
Ostrom, E. (1990). *Governing the commons the evolution of institutions for collective action*. New York: Cambridge University Press.
Ostrom, E. (1996). Crossing the great divide: Coproduction, synergy, and development. *World development*, 24(6): 1073–1087.
Polanyi, M. (1967). *The tacit dimension*. Garden City: Doubleday-Anchor.
Popper, K (1945). *The open society and its enemies*. London: Routledge.
Reed, M. S., A. J. Dougill, & M. J. Taylor. (2007). Integrating local and scientific knowledge for adaptation to land desertification: Kalahari rangeland management options. *Land Degradation and Development* (in press).
Riker, W. H. (1982). *Liberalism against populism*. San Francisco, CA: W. H. Freeman.
Romer, P. (1987). Growth based on increasing returns due to specialization. *American Economic Review*, 77(2), 56–62.
Schattschneider, E. E. (1960). *The semisovereign people: A realistic view of democracy in America*. Belmont, CA: Wadsworth/Thompson Learning.
Shiva, V. (2005). *Earth democracy: Justice, sustainability, and peace*. Cambridge, MA: South End Press.
Smith, A. (1965). *An inquiry into the nature and causes of the wealth of nations*. New York: Modern Library.
Stavrianos, L. S. (1998). *A global history: From prehistory to 21st century* (7th ed.). Upper Saddle River, NJ: Prentice Hall Inc.
Stiglitz, J. (1999). Knowledge as a global public good. In I. Kaul, I. Grunberg, & M. Stern (Eds), *Global public goods: International cooperation in the 21st century* (pp. 308–325). New York: Oxford University Press.
Stringer, L. C., & Reed, M. S. (2007). Land degradation assessment in Southern Africa: integrating local and scientific knowledge bases. *Land Degradation & Development*, 18(1), 99–116.
Sustainable Pittsburgh. (2002). *SWPA community indicators handbook*. Pittsburgh: Sustainable Pittsburgh. Retrieved October 10, 2006, from www.sustainablepittsburgh.org.
Tambe, M. (1997). Towards flexible teamwork. *Journal of Artificial Intelligence Research*, 7, 83–124.
Teece, D. J. (1980). Economies of scope and the scope of the enterprise. *Journal of Economic Behavior and Organization*, 1(3), 223–247.
Thomas, D. S. G. (1997). Science and the desertification debate. *Journal of Arid Environments*, 37(4), 599–608.
Thomas, D. S. G., & Twyman, C. (2004). Good or bad rangeland? Hybrid knowledge, science, and local understandings of vegetation dynamics in the Kalahari. *Land Degradation & Development*, 15(3), 215–231.
Whitaker, G. P. (1980). Coproduction: Citizen participation in service delivery. *Public administration review*, 40(3), 240–246.
Yang, L. (2008). *Wanmei quanmian chanpin guanli: Shehui jiegou he guanli de chanpin fenxi (Perfect total product management: A product analysis of social structure and management)*. Beijing: Beijing University Press, Beihang University Press (In Chinese).

Appendix A
The Four Settings of the Harding Game with the Scholar's Information

To Player 1, if Player 2 takes the cooperate strategy, whether he takes the cooperate or defect strategy depends on his payoff of taking the cooperate strategy PF_{1cc} ($=11 + v_{1-dc} - c_1$) and payoff of taking the defect strategy PF_{1dc} ($=11 + v_{1-dc} - c_1$). When $PF_{1cc} > PF_{1dc}$, takes the cooperate strategy; when $PF_{1cc} < PF_{1dc}$, takes the defect; when equal, either one. By the same token, if Player 2 takes the defect strategy, when his payoff of taking the cooperate strategy PF_{1cd} ($=-1 + v_{1-cd} - c_1$) is larger than his payoff of taking the defect strategy PF_{1dd} ($=v_{1-dd} - c_1$), take the cooperate; when $PF_{1cd} < PF_{1dd}$, takes the defect; when equal, either one.

For Player 2, if Player 1 takes the cooperate strategy, whether he takes the cooperate or defect strategies depends on his payoff of taking the cooperate strategy PF_{2cc} ($=10 + v_{2-cc} - c_2$) and payoff of taking the defect strategy PF_{2cd} ($=11 + v_{2-cd} - c_2$). When $PF_{2cc} > PF_{2cd}$, takes the cooperate; when $PF_{2cc} < PF_{2cd}$, takes the defect; when equal, either one. By the same token, if Player 1 takes the defect strategy, when his payoff of taking cooperate strategy PF_{2dc} ($=-1 + v_{2-dc} - c_2$) is larger than his payoff of taking defect strategy PF_{2dd} ($=v_{2-dd} - c_2$), takes the cooperate; when $PF_{2dc} < PF_{2dd}$, takes the defect; if equal, either one.

In sum, the requirements for Players 1 and 2 to simultaneously take the cooperate strategies can be described as follows:

First, there are two strictly necessary conditions: (1) $10 + v_{1-cc} - c_1 > 11 + v_{1-dc} - c_1$; (2) $10 + v_{2-cc} - c_2 > 11 + v_{2-cd} - c_2$. That is $v_{1-cc} - v_{1-dc} > 1$, and $v_{2-cc} - v_{2\text{-}dc} > 1$.

Then, if $v_{1-cd} - v_{1-dd} > 1$, no matter $v_{2-dc} - v_{2-dd} > 1$ $v_{1-cc} - v_{1-dc} > 1$ is larger, less than, or equal to 1, both of the players take the cooperate strategy (see Fig. 3.2, Setting 1).

If $v_{2-dc} - v_{2-dd} > 1$, no matter $v_{2-cd} - v_{2-dd} > 1$ is larger, less than or equal to 1, both of them take the cooperate strategy (see Fig. 3.3, Setting 2).

Second, if $v_{1-cc} - v_{1-dc} = 1$, $v_{2-cc} - v_{2-cd} > 1$, $v_{1-cd} - v_{1-dd} > 1$, and $v_{2-dc} < v_{2-dd}$, the two players also take the cooperate strategy (see Fig. 3.4, Setting 3).

Third, if $v_{1-cc} - v_{1-dc} > 1$, $v_{2-cc} - v_{2-cd} = 1$, $v_{1-cd} - v_{1-dd} < 1$, and $v_{2-dc} > v_{2-dd}$, two players still take the cooperate strategy (see Fig. 3.5, Setting 4).

L. Yang, *Knowledge-Driven Governance*,
https://doi.org/10.1007/978-981-13-2910-4

All these conditions, however, also require two preconditions: (1) $10 + v_{1\text{-}cc} - c_1 \geq 0$; and $10 + v_{2\text{-}cc} - c_2 \geq 0$. That is, (1) $c_1 \leq 10 + v_{1\text{-}cc}$, and (2) $c_2 \leq 10 + v_{2\text{-}cc}$.

Appendix B
The Four Settings of the Central-Authority Game with Incomplete Information but with the Scholar's Information

To Player 1, if Player 2 takes the cooperate strategy, whether he takes the cooperate or defect strategy depends on his payoff of taking cooperate strategy $10 - 2(x + q_c) + v - c$ and payoff of taking the defect strategy $11 - 2(y + q_c) + v - c$. When $10 - 2(x + q_c) + v - c > 11 - 2(y + q_c) + v - c$, that is, $(y + q_c) - (x + q_c)$ (denoted by A) > 0.5 (Setting 1), takes the cooperate strategy. When $A < 0.5$ (Setting 2), takes the defect; when equal (Setting 3), either one. By the same token, for Player 2, whether he takes the cooperate or defect strategy when Player 1 takes the cooperate or defect strategy also depends on whether A is larger than 0.5 (Setting 4) or less than (Setting 5); if equal (Setting 6), either one. These different settings, where two players will take the cooperate strategy, have been shown in Fig. 2.4. In short, the requirement for Players 1 and 2 to take the cooperate strategy simultaneously is $A > 0.5$. Please also remember other five constraints:

(1) $x + y = 1$;
(2) $0 \leq y + q_c \leq 1$;
(3) $0 \leq x + q_c \leq 1$;
(4) $-1 \leq q_d \leq 1$;
(5) $-1 \leq q_c \leq 1$.

The other two requirements are:

$$-2(x + q_c) + v - c > 0;$$
$$-2(y + q_d) + v - c > 0;$$

That is:

$$v - c > 2(x + q_c);$$
$$v - c > 2(y + q_d).$$

L. Yang, *Knowledge-Driven Governance*,
https://doi.org/10.1007/978-981-13-2910-4

Appendix C
The Recruitment Script and the Questionnaire of the Interview

Recruitment Script

I am Lihua yang, a graduate student under the direction of Professor G. Zhiyong Lan in the School of Public Affairs at Arizona State University. I am conducting a research study of scholars' participation in combating desertification in your county.

I am recruiting some interviewees to understand your ideas about scholars' roles in desertification control in your county. It will take approximately one hour.

Your participation is completely voluntary. Furthermore, you must be older than 18 years. You can withdraw from the interview whenever you want. If you have any questions concerning the research study, please call me at (480) 241-6640 or send an email to Lihua.yang@asu.edu.

The Questionnaire of the Interview

The Interviewer:
Date:
Pseudonyms of the Interviewee:
Number:

Dear Sir or Madam:

I am a graduate student under the direction of Professor G. Zhiyong Lan in the School of Public Affairs at Arizona State University.

I am conducting an interview to study scholars' participation in combating desertification. I am asking for your participation. I only want to know your personal ideas, and all the questions I will ask have no correct or incorrect responses. It should not take more than one hour of your time.

Your participation in this study is voluntary. Furthermore, you must be older than 18 years. You can skip any question if you wish. If you choose not to participate or to withdraw from the study at any time, there will be no penalty. Furthermore, there are no foreseeable risks or discomforts to your participation.

L. Yang, *Knowledge-Driven Governance*,
https://doi.org/10.1007/978-981-13-2910-4

As we all know desertification is a very serious problem in our county, your help is very important for us to collaboratively combat this problem.

All records of your answers will be kept in strict and complete confidence and will not be related to you as an individual. Your responses will be anonymous. The results of this study may be used in a dissertation, reports, presentations, or publications but your name will not be known/used. And the result will only be shared in the aggregated form.

If you have any questions concerning the research study, please contact the research team at 480-241-6640 or Lihua.yang@asu.edu. If you have any questions about your rights as a subject/participant in this research, or if you feel you have been placed at risk, you can contact the Chair of the Human Subjects Institutional Review Board, through the ASU Office of Research Integrity and Assurance, at (480) 965-6788.

Sincerely,

Lihua Yang

1. Demographic Questions

    1.1. Gender:
    1.2. Date of birth:
    1.3. Birthplace:
    1.4. Race:
    1.5. Highest level of education:
    1.6. Major:
    1.7. Occupation:
    1.8. Work unit/company
    1.9. Your job/the type of work:
    1.10. Title:
    1.11. Address and contact information

        Address:
        Mailing address:
        Zip code:
        Telephone number:
        E-mail:

2. General Questions of Combating Desertification

    2.1 What do you think the severity of desertification in your county is? Please explain your ideas using examples.
    2.2 What are the major reasons of desertification in your county? Please explain your ideas using examples.

2.3. What are the major problems in combating desertification and dust storms in your county? Please explain your ideas using examples.
2.4. How can the problem of desertification be partly resolved? Please explain your ideas using examples.

3. Participatory Problems of Scholars, Experts, Technicians and Other People Who Are Deemed as the Ones with More Knowledge and Experience

3.1 Do different scholars, experts, and technicians attend in activities of combating desertification and dust storms in your county? How do you know? Please explain your ideas using examples.
3.2 Who are these people? What kinds of backgrounds do they have? Where do they come from, to what kinds of departments do they belong, and what kinds of departments send them to your county? Do they constantly live in your county? Please explain your ideas using examples.
3.3 What do they do? What are their most important tasks and functions? Please explain your ideas using examples.
3.4 Compared with other people, what are their major advantages and disadvantages? Do these advantages and disadvantages are different to different ones coming from different departments, with different backgrounds, and so on? Please explain your ideas using examples.
3.5 Overall, how do you evaluate their functions and importance in the activities of combating desertification and dust storms? If possible, please also evaluate their functions and importance in different decades from 1940 to 2006 (such as in the 1950s, 1960s, 1970s, 1980s, 1990s, and in the twentieth century). Please explain your ideas using examples.
3.6 From the perspective of knowledge, what kinds of knowledge do you think are the most needed in the activities of combating desertification and dust storms in your county? Please explain your ideas using examples.
3.7 Who are the other people who play a very important role in the activities of combating desertification and dust storms and have comparative advantages of knowledge and experience in your county? What are their major functions? Please explain your ideas using examples.
3.8 Overall, how do you evaluate the functions and importance of these people who are not scholars, experts, or technicians but have more knowledge and experience? Please explain your ideas using examples.
3.9 Overall, what are the major problems scholars, experts, technicians, and these other people with more knowledge and experience in activities of combating desertification and dust storms are facing? Please explain your ideas using examples.
3.10 How can these problems be better resolved so these people can play more important roles in activities of combating desertification and dust storms?

4. Interaction and Organizational Problems of Scholars, Experts, Technicians, and Those Other People Who Have More Knowledge and Experience

   4.1 How do scholars, experts, technicians and these other people with more knowledge and experience interact with other people (such as farmers, herders, government officials, employees of other religious, nonprofit organizations, and clans)? What are the special relationships among them? How do you evaluate their relationships with each other? Please explain your ideas using examples.
   4.2 How do scholars, experts, and technicians interact with these other people with more knowledge and experience (such as farmers, herders, governmental officials, employees of other religious, nonprofit organizations, and clans)? What are the special relationships among them? How do you evaluate their relationship with each other? Please explain your ideas using examples.
   4.3 Are the ideas, local knowledge, and experience of the local people respected by scholars, experts, and technicians? How are local "non-scientific" knowledge and scientific knowledge combined? If they conflict with each other, how is the conflict resolved? Please explain your ideas using examples.
   4.4 How are various scholars, experts, and technicians organized in activities of combating desertification and dust storms in your county? Please explain your ideas using examples.
   4.5 Do you know some special organizations of scholars, experts, or technicians in your county (such as the research station of desertification, and research institutes)? What kinds of organizational structure do they have? How are they operated? What are major functions of these organizations in activities of combating desertification? What are relationships among these organizations? Please explain your ideas using examples.
   4.6 Overall, what are the major advantages and disadvantages of these organizations? How can their problems be overcome and let them play more important role in activities of combating desertification? Please explain your ideas using examples.
   4.7 Overall, how do you evaluate the functions of these organizations in combating desertification and dust storms in your county? If possible, please also evaluate their functions and importance in different decades from 1940 to 2006 (such as the 1950s, 1960s, 1970s, 1980s, 1990s, and in the twentieth century). Please explain your ideas using examples.
   4.8 Are the local people with more knowledge and experience also included in these organizations? If so, how are they involved? Please explain your ideas using examples.

Appendix D
The Recruitment Script, the Outline of the Training Content, and the Questionnaire of the Survey

Recruitment Script

I am Lihua Yang, a graduate student under the direction of Professor G. Zhiyong Lan in the School of Public Affairs at Arizona State University. I am conducting a research study to research scholars' participation in combating desertification in your county.

I am recruiting individuals to help your family members, relatives, and neighbors to answer the questionnaire which will take approximately 40 min.

Your participation in this study is voluntary. Furthermore, you must be older than 18 years. Also, the participation of the person you will recruit is voluntary and that person must also be older than 18 years. If you have any questions concerning the research study, please call me at (480) 241-6640 or send an email to Lihua.yang@asu.edu.

Outline of the Training Content for High School Students

1. The Purpose of the Study
   Scholars' participation in combating desertification in your county.
2. The Researcher
   Lihua Yang, a graduate student under the direction of Professor G. Zhiyong Lan in the Department/Division/College of Public Affairs at Arizona State University.
3. Your Participation
   Your participation in this study is voluntary. If you choose not to participate or to withdraw from the study at any time, there will be no penalty.
4. You Tasks
   Understand the purpose of the study and the structure of the questionnaire and then help your family members, relatives, and neighbors answer the questionnaire, which will take approximately 40 min. The return of the questionnaire will be considered your consent to participate.

L. Yang, *Knowledge-Driven Governance*,
https://doi.org/10.1007/978-981-13-2910-4

5. Recruitment

   The participation of the people you will recruit must be voluntary. They can skip any question if they wish. If they choose not to participate or to withdraw from the study at any time, there will be no penalty. You cannot pressure your family members, relatives, and neighbors to participate in this study.

6. Contact Information

   If you have any questions concerning the research study, please call me at (480) 241-6640 or send an email to Lihua.yang@asu.edu.

The Questionnaire Regarding Scholars' Role in Combating Desertification

Date: _______________

Instructions

Dear Sir or Madam:

I am a graduate student under the direction of Professor G. Zhiyong Lan in the School of Public Affairs at Arizona State University.

I am conducting a study to research scholars' participation in combating desertification. I am inviting your participation. Most questions make use of rating scales; please make a check (✓) to the answer that which best describes your opinion. A few questions ask you to answer using your own words. There are no correct or incorrect responses; we are merely interested in your personal point of view. The survey should not take more than 40 min of your time.

Your participation in this study is voluntary. You can skip questions if you wish. If you choose not to participate or to withdraw from the study at any time, there will be no penalty. You must be ordered than 18 years to participate.

As we all know, desertification is a very serious problem in our county; your help is really very important for us to collaboratively combat this problem. There are no foreseeable risks or discomforts as a result of your participation.

All evaluations will be kept in strict and complete confidence and will not be related to you as an individual. Your responses will be anonymous. The results of this study may be used in a dissertation, reports, presentations, or publications but your name will not be known/used. Also, the result will only be shared in the aggregated form.

If you have any questions concerning the research study, please contact the research team at 480-241-6640 or Lihua.yang@asu.edu. If you have any questions about your rights as a subject/participant in this research, or if you feel you have been placed at risk, you can contact the Chair of the Human Subjects Institutional Review Board, through the ASU Office of Research Integrity and Assurance, at (480) 965-6788.

Return of the questionnaire will be considered your consent to participate.

Sincerely,

Lihua Yang

1. Demographic Questions
   The following questions are for demographic purposes only and will not be related to you as an individual.

   1.1. Gender:
   1.2. Date of birth:
   1.3. Birthplace:
   1.4. Race:
   1.5. Highest level of education:
   1.6. Major:
   1.7. Occupation:
   1.8. Work unit/company
   1.9. Your job/the type of work:
   1.10. Title:

Notice:
Before you go to answer the following questions, two things should be noted.

(1) In the following questions, to some people who are not living this county, "your county" means where this survey is done, including Minqin County, Jingtai County, Linze County, Guazhou (Anxi) County, Dunhuang City, and Zhongwei City/County.
(2) The programs organized to combat desertification mean the programs which include building reservoirs, digging trenches, planting trees, bushes and grass, stabilizing sand.

2. General Questions of Combating Desertification

   2.1 What do you think the severity of desertification in your county is? Please select your answer.

   (   ) Very serious
   (   ) Serious
   (   ) Moderately serious
   (   ) Not obvious
   (   ) Greatly Ameliorated

   2.2. Please input numbers in ascending order in the brackets before the major reasons of desertification in your county listed below (1 means the most influential, then 2, 3, etc.).

   (   ) Natural and climatic reasons
   (   ) The competitive use of resources caused by their sparsity
   (   ) Unreasonable use of resources due to the lack of knowledge
   (   ) Unenlightened institutions of resource management and development

2.3. Please input numbers in ascending order in the brackets before the major problems in combating desertification and dust storms in your county listed below (1 means the most significant, then 2, 3, etc.).

( ) Limited financing
( ) Institutional handicaps
( ) The public's unreasonable behavior
( ) A lack of effective knowledge and managerial methods
( ) A shortage of scholars and experts' participation
( ) Insufficient governmental attention
( ) A lack of activities organized by the government
( ) A shortage of participation of religious groups, clans, voluntary organizations, and social intermediary organizations

2.4. Please input numbers in ascending order in the brackets before the methods of dealing with the problems of desertification listed below (1 means the most effective/useful, then 2, 3, etc.).

( ) Increase financial support
( ) Strengthen instructions
( ) Control the unreasonable behaviors of the public
( ) Strengthen the scientificity of the research of related problems, investigations, and managerial methods
( ) Increase the participation of scholars and experts
( ) Strengthen the governmental propaganda and educate the public
( ) Have the government to directly organize the governance activities
( ) Increase the participation of religious groups, clans, voluntary organizations, and social intermediary organizations

3. Participatory Problems of Scholars, Experts, Technicians, and Other People Who Are Deemed as the Ones with More Knowledge and Experience

3.1 Do different scholars, experts, and technicians attend in activities of combating desertification and dust storms in your county? How do you know? Please make a check (✓) to your preferred answer.

(1) Whether participated: yes____; no____.
(2) How you know: personal experience____; heard from other people____; knew form news, TV, newspapers, and the Internet____.

3.2 Who are these people? What kinds of backgrounds do they have? Where do they come from, to what kinds of departments do they belong, and what kinds of departments send them to your county? Do they constantly live in your county? Please make a check (✓) next to your preferred answer.

(1) Who they are (multiple answers): agricultural technicians____; various experts, professors, and researchers____; other kinds of technicians (in fields such as in irrigation and environmental pollution) ____.

(2) Professional background (multiple answers): wind control, sand control, and stabilizing sand____; planting trees____; agricultural planting____; irrigation works____; land development and urban planning____; propaganda and education of science and environmental protection____; propaganda and education of politics, laws, and regulations____; the control of agricultural diseases and insect pests____; the disease control of poultry and livestock____.
(3) Where they come from (multiple answers): the same village____; the same township____; the same county____: the same province____; Beijing____; foreign countries____.
(4) Departments (multiple answers): the county research institute____; the county governmental institute____; the prefecture research institute____; the prefecture governmental institute____; the provincial research institute____; the provincial governmental institute____; the central ore Beijing research institute____; the central governmental institute____; foreign or international institute.
(5) How long they stay (multiple answers): several hours____; several days____; several weeks____; about one month____; over two months____; over a half year____; over tow years____; constantly live here____.

3.3 Please input numbers in ascending order in the brackets before their major work and functions listed below (1 means the most important job, then 2, 3, etc.).

( ) Do their research
( ) Help farmers or herders improve their knowledge level in order to protect environmental resources, control wind, sand, and desertification and improve their economic and living standards.
( ) Directly organize farmers or herders to control wind, sand, and desertification, stabilize sand, plant grass and trees, and protect forestry.
( ) Directly help farmers or herders communicate with governments.
( ) Help farmers or herders communicate and negotiate with companies to protect their benefits.
( ) Directly organize and lead farmers or herders to protect their reasonable rights and pursue their own interests.
( ) Help the government improve their understanding of some problems and to improve their capacity and quality of decision making.
( ) Directly help the government resolve some affairs.
( ) Help firms pursue their own interests.
( ) Help religious groups, clans, voluntary organizations, etc., participate in local activities.
( ) Pursue their private interests.

3.4 Please input numbers in ascending order in the brackets before the major advantages and disadvantages listed below (1 means the most advantageous/disadvantageous, then 2, 3, etc.).

(1) Advantages

( ) Have comparative advantages in scientific knowledge
( ) Understand related laws, regulations, and social governance knowledge
( ) Have a high learning capability
( ) Provide information and knowledge for the public, government officials, and others
( ) Have the capability to educate the public and others
( ) Have strong social responsibility
( ) Easily gain respect from social citizens and government officials
( ) Be humble, and easy to approach and communicate with
( ) Like to participate in practical activities
( ) Respect the local people's ideas and local knowledge
( ) Have good relationships with local people
( ) Have good relationships with the local governments
( ) Have good relationships with prefecture and provincial governments, or important government officials
( ) Have good relationships with entrepreneurs
( ) Have good relationships with religious groups, clans, and voluntary organizations
( ) Like to help various people communicate, understand, and cooperate with each other

(2) Disadvantages

( ) Do not understand the local conditions well
( ) Not easy to approach and communicate with
( ) Relatively stubborn and prejudiced, and do not respect other people's ideas
( ) Cannot impart their knowledge to others
( ) Do not get enough respect from the public and others
( ) Limited capacity of practice
( ) Do not respect the local knowledge of farmers, herders, and others
( ) Do not understand local social relationships
( ) Only help the government
( ) Only help farmers
( ) Only help firms
( ) Only help religious, clans, and various social intermediary organizations
( ) Only pursue their private interests

3.5 Overall, how do you evaluate their functions and importance in the activities of combating desertification? Please make a check (✓) next to your preferred answer.
Very important____; important____; general____; nonessential____; having them makes things worse____; do not know____.

3.6 Please input numbers in ascending order in the brackets before the most needed knowledge in the activities of combating desertification in your county listed below (1 means the most needed, then 2, 3, etc.).

( ) General climatic knowledge
( ) Scientific knowledge about desertification and dust storms
( ) Environmental governance knowledge
( ) Social governance knowledge
( ) Knowledge about laws and regulations
( ) Knowledge about local desertification and dust storms
( ) Knowledge about various local social relationships
( ) Local knowledge and local methods
( ) Irrigation knowledge
( ) Agricultural and forestry knowledge

3.7 Who are the other people who play a very important role in the activities of combating desertification and have comparative advantages of knowledge and experience in your county? What are their major functions? Please answer the questions, respectively.

(1) Who are they? Please input numbers in ascending order in the brackets listed below (1 means the most important, then 2, 3, etc.).

( ) Some experienced and prestigious elders
( ) Some youth having knowledge and experience
( ) Forest rangers
( ) Some students having knowledge and social responsibility
( ) Some middle-school or elementary-school teachers having high social responsibility
( ) Some retired intellectuals
( ) Some retired government officials
( ) Some media reporters with high social responsibility and conscience
( ) Some voluntary and charitarians with knowledge and experience

(2) Functions. Please input numbers in ascending order in the brackets (1 means the most important, then 2, 3, etc.; and skip some if you think they are not relevant)

( ) Help local citizens improve their understanding and working methods

( ) Organize and lead local citizens to carry on various activities of environmental governance
( ) Take charge of the communication among other experts, scholars, and technicians
( ) Take charge of the communication with government officials
( ) Take charge of the communication with firms
( ) Take charge of the communication with religions, clans, and various voluntary and social intermediary organizations

3.8 Overall, how do you evaluate the functions and importance of these people who are not scholars, experts, or technicians but have more knowledge and experience? If possible, evaluate their functions in different eras (please make a check "✓" to your preferred answer).

(1) Overall evaluation
Very important ____; important ____; moderately important ____; dispensable____; negative____; unknown ____.
(2) Evaluation in different eras
In the 1950s:
Very important ____; important ____; moderately important ____; dispensable____; negative____; unknown ____.
In the 1960s:
Very important ____; important ____; moderately important ____; dispensable____; negative____; unknown ____.
In the 1970s:
Very important ____; important ____; moderately important ____; dispensable____; negative____; unknown ____.
In the 1980s:
Very important ____; important ____; moderately important ____; dispensable____; negative____; unknown ____.
In the 1990s:
Very important ____; important ____; moderately important ____; dispensable____; negative____; unknown ____.
In the twentieth century:
Very important ____; important ____; moderately important ____; dispensable____; negative____; unknown ____.

3.9 Overall, what are the major problems scholars, experts, technicians, and these other people with more knowledge and experience in activities of combating desertification and dust storms are facing? (Please input numbers in ascending order in the brackets; 1 means the most important, then 2, 3, etc.).

( ) Society does not realize the importance of their participation
( ) They do not have high enthusiasm of participation
( ) They do not have the capacity to participate

( ) They lack necessary organizational ability and leadership
( ) Citizens do not want to cooperate with them
( ) They do not want to cooperate with citizens
( ) Governments do not want to cooperate with them
( ) They do not want to cooperate with governments
( ) They lack necessary financial support
( ) They lack enough government support for their participation
( ) There are no appropriate organizing and participatory systems and institutional arrangements

3.10 How can these problems be better resolved so these people can play more important roles in activities of combating desertification and dust storms? (Please input numbers in ascending order in the brackets; 1 means the most important, then 2, 3, etc.).

( ) Strengthen propaganda and improve society's understanding of the importance of their participation
( ) Strengthen education, propaganda, and self-cultivation and improve their participatory enthusiasm
( ) Improve their participatory capability through learning
( ) Improve the knowledge level of citizens (including farmers, herders, etc.) and their understanding of the importance of the participation of scholars, experts, technicians, etc., and ameliorate cooperative methods through education and propaganda
( ) Change their own attitude and working methods, and strengthen the cooperation with citizens (farmers, herders, etc.)
( ) Improve the knowledge and managerial levels of government officials and their understanding of the importance of the participation of experts, scholars, technicians, etc., and ameliorate their cooperative methods through education and propaganda
( ) Change their own attitude and working methods, and strengthen the cooperation with government officials
( ) Improve the necessary financial support
( ) Governments should strongly support the participated governance of experts, scholars, general technicians, and other people with knowledge and experience
( ) Construct appropriate organizing and participatory systems and institutional arrangements in order to systematically organize them to participate in combating desertification and dust storms

4. Interaction and Organizational Problems of Scholars, Experts, Technicians, and Those Other People Who Have More Knowledge and Experience

4.1 How do scholars, experts, technicians, and these other people with more knowledge and experience interact with other people (such as farmers, herders, government officials, employees of other religious, nonprofit organizations, and clans)? What are the special relationships among them?

How do you evaluate their relationships with each other? (Please input numbers in ascending order in the brackets; 1 means the most important, then 2, 3, etc.).

(1) The number of two-party communications, talking, working together with the participation of scholars, experts, technicians, etc.

( ) Scholars and officials
( ) Scholars and farmers or herders
( ) Scholars and entrepreneurs
( ) Scholars and religious groups, clans, voluntary organizations, and social intermediary organizations
( ) Scholars and the others having more knowledge and experience

(2) The number of multiple-party communications, talking, and working together with the participation of scholars, experts, technicians, etc.

( ) Scholars, farmers (or herders), and officials
( ) Scholars, farmers, and entrepreneurs
( ) Scholars, farmers, and the other organizations stated above
( ) Scholars, farmers, and the others having more knowledge and experience
( ) Scholars, officials, and firms
( ) Scholars, officials, and the other organizations stated above
( ) Scholars, officials, and the others having more knowledge and experience
( ) Scholars, firms, and the other organizations stated above
( ) Scholars, firms, and the others having more knowledge and experience
( ) Scholars, farmers, officials, firms, and the other organizations stated above
( ) Scholars, officials, firms, the other organizations stated above, and the others having more knowledge and experience
( ) Scholars, farmers, officials, firms, the other organizations stated above, and the others having more knowledge and experience

(3) The degree of the intimacy between scholars and other social actors

( ) Scholars and officials
( ) Scholars and farmers or herders
( ) Scholars and entrepreneurs
( ) Scholars and religious groups, clans, voluntary organizations, and social intermediary organizations
( ) Scholars and the others having more knowledge and experience

4.2 How do scholars, experts, and technicians interact with these other people with more knowledge and experience (such as farmers, herders, government officials, employees of other religious, nonprofit organizations, and clans)? What are the special relationships among them? Please answer the questions, respectively.

(1) Various methods of interaction (Please input numbers in ascending order in the brackets; 1 means the most important, then 2, 3, etc.; and skip some you think they are not relevant)

( ) These people often consult experts, scholars, technicians, etc.
( ) Various experts, scholars, and technicians often consult these people.
( ) These people are often invited to attend activities such as meetings and education and training programs organized by experts, scholars, and technicians, etc.
( ) Experts, scholars, and technicians often invite these people to attend local activities and affairs
( ) Experts, scholars, and technicians often understand local conditions through these people
( ) These people pursue more knowledge and information through experts, scholars, and technicians
( ) They get to know and communicate with each other through various activities organized by governments

(2) Number of the times and importance of communication

( ) They have frequent communications and interactions
( ) They have some communications and interactions, but not frequently
( ) They only have a few communications and interactions
( ) They basically do not have communications and interactions

(3) The nature of relationships

( ) They highly trust and respect each other
( ) Only these people trust scholars, experts, technicians, etc., but scholars, experts, technicians, etc., do not trust and respect them
( ) Only scholars, experts, technicians, etc., trust and respect these people, but these people do not trust and respect them
( ) They do not trust and respect each other
( ) They always oppose and conflict with each other

4.3 Are the ideas, local knowledge, and experience of the local people respected by scholars, experts, and technicians? How are local "non-scientific" knowledge and scientific knowledge combined? If they conflict with each other, how is the conflict resolved? Please answer the questions, respectively.

(1) The degree of respecting local knowledge, ideas, and experiences of local citizens (please check your preferred answers)

( ) Highly respected
( ) Moderately respected
( ) Only get a little respect
( ) They only superficially respect them and actually do not care about them
( ) They never care about these
( ) They often despise these local knowledge, ideas, and experiences

(2) The methods of combining the knowledge of scholars, experts, and technicians, etc. (Please input numbers in ascending order in the brackets; 1 means the most important, then 2, 3, etc.; and skip some you think they are not relevant)

( ) Communication and understanding through voluntary activities of scholars, experts, technicians, etc., with local citizens
( ) Voluntary reports by local citizens to scholars, experts, and technicians
( ) These kinds of knowledge are often actively proposed in meetings and activities in which all these people participate
( ) These kinds of knowledge are often considered when the scientific knowledge of experts and the like cannot resolve the problem

(3) Contradiction and conflict resolution

( ) Synthetically considered and negotiatorily resolved based on facts
( ) Mainly depend on local knowledge
( ) Mainly depend on scientific knowledge
( ) Two parties often do not compromise each other, and eventually cannot reach a consensus

4.4 How are various scholars, experts, and technicians organized in activities of combating desertification in your county? (Please input numbers in ascending order in the brackets; 1 means the most important, then 2, 3, etc.; and skip some you think they are not relevant)

( ) These people only do their research in relevant institutes, and then their research results are included in various governance policymaking
( ) These people often voluntarily take part in concrete activities driven by their own research interests and social responsibility
( ) These people are involved in activities organized by government organizations
( ) These people are invited by citizens to attend governance activities

( ) These people are invited by firms to attend governance activities
( ) These people are invited by religions, clans, voluntary organizations, and various social intermediary organizations to attend governance activities

4.5 Do you know some special organizations of scholars, experts, or technicians in your county (such as the research station of desertification, and research institutes)? What kinds of organizational structure do they have? How are they operated? What are major functions of these organizations in activities of combating desertification? What are relationships among these organizations? Please answer the questions, respectively.

(1) The level of understanding (please check your preferred answer)

( ) Know them very well
( ) Heard of all these organizations, but do not know who they are
( ) Only know some of these organizations, but have heard of the others
( ) Only heard some of these organizations and do not know the others
( ) Never heard of these organizations

(2) Functions and activities (please select and rank)

( ) Organize scientific research activities combined with local reality
( ) Supply policy advice to governmental institutes about resolving environmental problems, such as combating desertification and dust storms
( ) Directly participate in decision making and policymaking of environmental governance, such as combating desertification and dust storms
( ) Collect and summarize information and experience about local environmental governance
( ) Directly organize local people to carry on some activities about environmental governance
( ) Directly participate in various concrete activities and programs about environmental governance, such as combating desertification and dust storms
( ) Supply suggestions about environmental governance to various firms' development
( ) Cooperatively participate in various environmental governance activities with religions, clans, voluntary organizations, and different social intermediary organizations

(3) The relationship among organizations (Please select and rank)

( ) Controlled by vertical leadership
( ) They are not subordinate to each other and are no communication and association among them
( ) They are not subordinate to each other, but there are some communication and association with different degrees
( ) There are diverse relationships: although they are not subordinate to each other, they build good institutions of cooperation and communication among them; or, although they are subordinate to each other, the lower-level organizations are often self-governed, and their relationships with the upper-level organizations are often cooperative rather than strictly controlled by the superiors

4.6 Overall, what are the major advantages and disadvantages of these organizations? How can their problems be overcome and let them play more important role in activities of combating desertification? Please answer these questions, respectively.

(1) Advantages (please select and rank)

( ) They are local organizations which press close to the local reality and concrete problems
( ) They have close relationships with the local people and organizations, and then their research results can be easily applied
( ) They can help people believe that they can gain more advantages, and in fact they really help people feel that they obtain more advantages and gradually strengthen this feeling
( ) They respect citizen's ideas, thoughts, and democratic rights. They often try very hard to find a consensus resolution through negotiation and communication, and they do not use strict coercive and doctorial methods
( ) They build an effective mechanism of awards and sanctions which include various concrete methods which are related to each other
( ) They have relationships with external research institutes and can gain more information and knowledge
( ) They directly communicate with local people
( ) They have good relationships with local government officials
( ) They have good relationships with local firms
( ) They have good relationships with religious groups, clans, voluntary organizations, and intermediary social originations

(2) Disadvantages (please select and rank)

( ) They have no good arrangements of organizational structure
( ) They lack persons with ability
( ) They lack enough financial support

( ) Their research and experimental results are too academic and unpractical, cannot improve people's real income, and cannot be extended to other places
( ) Although their research and experiment results are practical and they would improve people's real income if they were applied, there are no appropriate mechanisms of the experiment-extension method
( ) Their participated activities of environmental governance are limited
( ) They are often dictatorial and do not respect the local's ideas, thoughts, and democratic rights; their adopted methods are often ones which are not discussed by citizens
( ) They have no effective mechanisms of awards and sanctions
( ) They have complicated relationships with local users and cannot carry out some activities
( ) They cannot be respected by other departments and organizations because of their limited position and capacity
( ) They cannot gain stable external support
( ) They have no institutional guarantee about their status and functions

(3) How to improve

( ) Strengthen reform and design better organizational structure which is more suited to concrete conditions
( ) Having more persons with ability through improving treatment
( ) Improve financial support
( ) Improve research
( ) Strengthen the relationship between research and social needs and develop more research which can really improve people's interests
( ) Build an effective experiment-and-gradual-extension mechanism
( ) Actively participate in various concrete local activities about combating desertification and dust storms
( ) Respect local people and try hard to find a consensus solution through negotiation and discussion
( ) Build effective mechanism of award and sanctions
( ) Strengthen their cultural and moral building and neatly resolve their relationships with various organizations
( ) Improve the understanding of these organizations through propaganda
( ) Give more support and guarantees for their legal organizational status and activities from both the national and local institutional levels

4.7 Overall, how do you evaluate the functions of these organizations in combating desertification in your county? If possible, please also evaluate their functions and importance in different decades. Please check your preferred answers.

(1) Total evaluation
Very important ____; important ____; moderately important ____; dispensable____; negative____; unknown ____.

(2) Evaluation in different eras
In the 1950s:
Very important ____; important ____; moderately important ____; dispensable____; negative____; unknown ____.
In the 1960s:
Very important ____; important ____; moderately important ____; dispensable____; negative____; unknown ____.
In the 1970s:
Very important ____; important ____; moderately important ____; dispensable____; negative____; unknown ____.
In the 1980s:
Very important ____; important ____; general ____; nonessential ____; having them make things worse ____; do not know ____.
In the 1990s:
Very important ____; important ____; moderately important ____; dispensable____; negative____; unknown ____.
In the twentieth century:
Very important ____; important ____; moderately important ____; dispensable____; negative____; unknown ____.

4.8 Are the local people with more knowledge and experience also included in these organizations? If so, how are they involved? Please answer these questions, respectively.

(1) The involvement in these organizations (Please check your preferred answer)

( ) Many of them are involved in these organizations
( ) Only a few of them are involved in these organizations
( ) No any of them are involved in these organizations
( ) I do not know

(2) Organizational forms (Please input numbers in ascending order in the brackets; 1 means the most important, then 2, 3, etc.; and skip some if you think they are not relevant)

( ) They formally participate in these organizations and become members
( ) They are invited by these organizations to do some work or are deemed as affiliated member of these organizations
( ) They are deemed as reliable people to communicate and consult and are often invited to attend many programs
( ) Although they are reliable, they are often invited to attend related activities or deemed as the object to communicate and consult

Appendix E
The Approval Document by the Institutional Review Board

ARIZONA STATE UNIVERSITY
RESEARCH AND ECONOMIC AFFAIRS

Office of Research Integrity and Assurance

To:	Gerald Lan UCENT
From:	Mark Roosa, Chair Soc Beh IRB
Date:	12/03/2008
Committee Action:	**Exemption Granted**
IRB Action Date:	12/03/2008
IRB Protocol #:	0811003478
Study Title:	Scholar-Participated Governance: Combating Desertification and Other Dilemmas of Collective Action

The above-referenced protocol is considered exempt after review by the Institutional Review Board pursuant to Federal regulations, 45 CFR Part 46.101(b)(2) .

This part of the federal regulations requires that the information be recorded by investigators in such a manner that subjects cannot be identified, directly or through identifiers linked to the subjects. It is necessary that the information obtained not be such that if disclosed outside the research, it could reasonably place the subjects at risk of criminal or civil liability, or be damaging to the subjects' financial standing, employability, or reputation.

You should retain a copy of this letter for your records.

L. Yang, *Knowledge-Driven Governance*,
https://doi.org/10.1007/978-981-13-2910-4

Appendix F
Case Studies by Changing Coverage

Case 1: Hotan (Hetian) County, Xinjiang Autonomous Region (Successful)
Hotan Oasis is located at the southwestern margin of the Taklimakan Desert in Xinjiang. Its annual average precipitation is only 34.8 mm, while its evaporation is 2564 mm. From the 1950s, 1960s, and 1970s to 2007, its desertification was successfully combated, through adopting many methods such as undertaking new water conservancy projects, building complete sand-protecting plantation and windbreak forest system around the oasis, and stabilizing moving dunes and leveling individual dunes. In 1989, it was elected to the rank of Global 500 Roll by the United Nations Environment Program (UNEP) (Zhou 2000). Because of significant scholars' participation, its governance can be deemed as a successful case of scholar-based governance. Now let's analyze how the seven principles had been satisfied.

Stable field-based scholars significantly took part in combating desertification in Hotan County. These scholars often had a relatively independent social identity and high social status, high enthusiasm and participatory capability, enough expert knowledge, local knowledge and the knowledge about local social webs, high social capital, high social responsibility, and the spirit of practice. Furthermore, they also respected other social actors. For instance, the stories of Kailong Gong, Fen Long, and Guiwu Wu were good examples (for detailed analyses, please see Bezlova 2006; Yang and Zhang 2005). Professor Mingting Liu was another good example. He helped the local people plant Desertliving Cistanche (Wang 2005). So Vice President of the Chinese Academy of Forestry in China, Ci (2004), argued that desertification prevention plans should be drawn up scientifically and experts' participation should be highly considered. That is, *Principle 1 (Principles 1.1–1.6) in Hotan County was satisfied.*

In order to build various organizations to combat desertification, a federal organization structure was guaranteed in Hotan. Furthermore, because this region is a minority region, more democratic and diverse organizations were also built. This strengthened its federal structure. As stated above, its program of combating desertification had very concrete purposes, which were developing new water

L. Yang, *Knowledge-Driven Governance*,
https://doi.org/10.1007/978-981-13-2910-4

conservancy projects, building complete sand-protecting plantation and windbreak forest system around the oasis, and stabilizing moving dunes and leveling individual dunes. Thus *Principle 2 was satisfied.* Citizens' participation (Cheng 2007; Xinjiang Uygur Autonomous Region government 2007), local army's participation (Li and Zhang 1991), NGOs' participation (e.g.,the Poverty Alleviation Association of Hotan Prefecture), other provinces'support (e.g., Zhejiang), international cooperation (such as Japan and UNEP), etc., also showed its democratic and collaborative managements. The federal mechanism of awards and sanctions was also built (Zuo 2002). Thus, *Principle 3 was satisfied.* Because stable scholars' participation also created stable local scholar-entrepreneurship as shown by the aforementioned stories of Kailong Gong and Fen Long, Guiwu Wu, and Professor Mingting Liu, *Principle 4 was satisfied.*

Hotan combined desertification control and economic and social development very well. For example, the policies of stopping using firewood, turning to methane gas, and planting Desertliving Cistanche (Cheng 2007; Wang 2005) not only protected environments but also improved farmers' income. Thus, *Principle 5 was satisfied.* Because the aforementioned methods were also gradually developed, tested, and extended, *Principle 6 was satisfied.* Finally, outside scholars' participation, governmental support, and support from various NGOs, UNEP, and other organizations also guaranteed its stable external financial, technical, and institutional supports (e.g., see Xinhua News Agency 2007). Particularly, because this region is in a minority area, government financial support was often stronger than in other areas. Thus, *Principle 7 was satisfied.*

Case 2: Naiman Banner (Qi), Inner Mongolia (Successful)

Naiman Banner is a semiarid county located on the eastern edge of the Horqin Desert, Inner Mongolia. Its annual precipitation is 362 mm and mainly falls between June and August. Its potential evapotranspiration, however, is 1935 mm and the mean annual wind speed is 3.5 m (Duffy and Migongo-Bake 2003). Because of dramatic land use change and over utilization of vegetation and water, desertification in this area expanded at a growing rate from the middle of the 1950s to the 1980s. This process, however, was gradually reversed from the middle of the 1990s to the 2007 due to reafforestation and vegetation restorations. In 1997, it was evaluated as a potential success story on desertification control by UNEP (Duffy and Migongo-Bake 2003). Since 2000, grazing had also been excluded. Now let's analyze how the seven principles were satisfied.

Naiman Research Station of Desertification (NRSD) is located at Naiman County (120° 42′E, 42° 55′N, 358 m a.s.l., about 180 km Southwest of Tongliao City) and is one of 35 field stations of Chinese Ecosystems Research Network. Since 1973, researchers from former Desert Research Institute of Chinese Academy of Sciences had carried out research in desertification in this area. The research station was inaugurated in 1985 and was subordinate to the Cold and Arid Regions Environment and Engineering Research Institute (CAEERI), CAS. Until 2007, NRSD was staffed by 15 research workers whose majors covered 10 diverse research fields. It had five laboratories and eleven field experiment plots (Chinese National Ecosystem Observation & Research Network 2006). Their achievement received a "Saving the Dry lands" certificate awarded by UNEP in 1998. These showed that scholars played a very important role in desertification control in Naiman and they also had enough expert knowledge (UNEP 2002). Many scholars stayed in Naiman over 10–20 years and could be seen as stable local field scholars. Furthermore, they had high enthusiasm and capability to combat desertification, often taught the local farmers, learned local knowledge, had good relationships with local people, had high social responsibility and spirit of practice, and liked to work with local people and respect them (Chen 2008). The stories of local self-taught scholar Ligao Bao (Li 2008) and other local field-based scholars and leaders of forestry centers and reservoirs such as Shan Li (Sun and Buhe 2000; Wang 2003d), Zhe Li (Xu and Wang 2007), Yuanbin Liu (Sun and Liu 1997), Zhong Liu (Sun et al. 1998), and Jingshuang Zhao (Buhe et al. 2001) showed these. Thus, *Principle 1 (Principle 1.1–1.6) was satisfied in Naiman.* Various laboratories, field experiment plots, forestry centers, etc., formed a federal organization of desertification control and made its purposes of combating desertification more concrete (UNEP 2002). Support from Tongliao City, provincial governments, Lan Zhou, UN, etc., also strengthened its federal structure (Carle and Ma 2005). Thus, *Principle 2 was satisfied.*

Local participation, community-based governance, farmers' voluntary participation, local and international collaboration (Carle and Ma 2005; Dou and Song 1995; Kim et al. 2006; UNEP 2002; Xu and Yang 1991; Zhang and Tan 1989) demonstrated that democratic and collaborative management was applied very well in this area. Furthermore, the federal mechanism of awards and sanctions was built (e.g., please see Xu 1995; Wang 2003d). Thus, *Principle 3 was satisfied.*

The aforementioned stories of local self-taught scholar Ligao Bao (Li 2008) and other local field-based scholars and leaders of forestry centers and reservoirs such as Shan Li (Sun and Buhe 2000; Wang 2003d) and Zhe Li (Xu and Wang 2007) showed that stable local scholar-entrepreneurship was also built in Naiman during the process of desertification control. Thus, *Principle 4 was satisfied.*

Also, the activities of desertification in Naiman combined economic development and environmental protection very well (Shi et al. 2005, 2006; UNEP 2002; Wang 2003e). Thus, *Principle 5 was satisfied.*

The experiment-extension method was also applied. For example, the Institute of Desert Research (IDR), Chinese Academy of Sciences, which has its headquarters at Lanzhou, Gansu Province, undertook a project during the period from 1986 to 1995, and it involved three distinct phases: "an experiment conducted in a research station to study desertification mechanisms and to design strategies for their control," "implementation of experimentally proven interventions in the demonstration village," and "encouraging the implementation of successful interventions in other nearby villages" (Duffy and Migongo-Bake 2003, p. 69). The similar characteristics were also argued by other researchers (UNEP 2002; Zhao et al. 1999, 2006). Thus, *Principle 6 was satisfied.*

The participation of many external scholars from Lanzhou, Beijing, and other counties supplied stable external technical support for its desertification control. Government and international financial support (e.g., from Japan) and credit service

(Shi et al. 2005) guaranteed it stable external financial support. Furthermore, in addition to the institutional support at the village, township, and county levels (UNEP, 2002; the story of Qinghe Bao, the vice head of Naiman Banner, was a good example, please see Sheng and Yang 1997), the provincial and central governmental support guaranteed its stable external institutional support (Chen 2008; Dou and Song 1995; Liu and Zhu 1998; Zhang and Tan 1989). Thus, *Principle 7 was satisfied.*

Case 3: Aohan Banner (Qi), Inner Mongolia (Successful)

Aohan Banner, an autonomous county, is a part of Chifeng City in Inner Mongolia. Unit 2007, its total area was 8300 km^2, the population was 593 thousand, and it included seven towns and 22 townships. After many years of desertification control, its vegetation cover was improved from 1% in the early period of the new China to 40% in 2003. In 2002, it was elected to the rank of Global 500 Roll by the United Nations Environment Program (UNEP) (Wang 2003d). Now let's analyze how Principles 1–7 were satisfied.

Stable local field-based scholars played an important role in desertification control in Aohan County. During the first 30 years of New China, because people did not realize the importance of science and scientists in desertification control, the activities of desertification control were not successful. Then they realized that it is necessary to combat desertification relying on science and knowledge (Wang 2003a, b). After over 20 years, they achieved a lot. Furthermore, these scholars had high enthusiasm and capability to combat desertification, liked to teach the local farmers and learn local knowledge, had good relationships with local people, high social responsibility, and spirit of practice, and liked to work with local people and respect them. The story of Yongxin Bao was a good example (Zhu 2007). The stories of Rui Li, Haichao Ma, Jiali Sun, Xianming Zheng, and Jiusen Yu were also good examples (Wang 2003a, c). Thus, *Principle 1* (*Principle 1.1–1.6*) *was satisfied.*

Various methods of desertification control were applied in different areas of Aohan Banner. Various projects were also implemented in Aohan, and over ten projects were awarded Autonomous Regions or National Awards of Science and Technology Advancement (Wang 2003a, c). Because of these, federal organization of desertification control was built and concrete organizational purposes were set up. Thus, *Principle 2 was satisfied.*

The methods of democratic and collaborative management (including international collaboration such as with Japan, Germany, Italy, the UN, and Asian Development Bank) and citizens' participation were applied (Chao 2005; McGarry 2005; Michaelowa and Rawat 2007; Wang 2003a, c; Zhan 2002; Zhu 2007). Also, the necessary system of awards and sanctions was built (Zhan 2002; Zhi 2001). Thus, *Principle 3 was satisfied.* The aforementioned stories of the scholars such as Yongxin Bao and Rui Li showed that Aohan Banner also built stable local scholar-entrepreneurship in the process of desertification control. Thus, *Principle 4 was satisfied.*

The activities of desertification control in Aohan County also combined environmental protection and economic development very well (Wang 2003a; Wu

2006). The story of Yongxin Bao (Tang 2007; Zhu 2007) was a good example. Thus, *Principle 5 was satisfied.* Furthermore, the experiment-extension method was applied, and this was called model demonstration (*dianxing yingdao*) in Aohan (Wang 2003a, c). That is, *Principle 6 was satisfied.*

The participation of external scholars guaranteed the stable external technical and knowledge support for Aohan Banner. External financial support from provincial, central government of China, other counties, the UN, etc., guaranteed its stable external financial support (Chao 2005; McGarry 2005; Michaelowa and Rawat 2007; Wang 2003a, c; Zhan 2002; Zhu 2007). Furthermore, the support from the prefecture, provincial, and central governments guaranteed its stable external institutional support (McGarry 2005; Wang 2003a). So *Principle 7 was satisfied.*

Case 4: Yanchi, Ningxia Hui Autonomous Region (Semi-successful)

Yanchi County is located in the middle-eastern part of Ningxia where animal husbandry is the dominant economic activity. After combating desertification many years, the desertificated area was reduced from 5.39 million *mu* to over 1 million mu (a Chinese unit of area, 1/15 of a hectare) (Zhuang 2006). Thus, until 2007, it could be deemed as a semi-successful case of combating desertification. Now let's analyze how the seven principles of successful scholar-based governance had been satisfied.

Many scholars had taken part in the activities of combating desertification in Yanchi County. The Yanchi Research Station in a desert-steppe zone did a lot of pioneering work on desert research and desertification control. Applied technology, even local knowledge, was developed to combat desertification (Mitchell et al. 1998). For example, as a famous desertification-control scholar and heroine, Chunlan Bai was originally a farmer. She first came to Yanchi in 1980. In order to combat desertification, she and her husband tried very hard to learn from other scholars in research station, test many methods, and teach other farmers to work together with them. Her story showed how technology and scholars played a very important role in combating desertification in Yanchi. Furthermore, her story showed that scholars in Yanchi had high enthusiasm and capability of participation, liked to teach farmers, have high social responsibility and spirit of practice, and friendly work with farmers and respect them (for detailed stories, please see People's Daily 2007; *Wuzhong Dangjian Wang* 2005). The importance of the scholars' roles in Yanchi County in total, however, was less than in Hotan (Hetian) County, Naiman Banner, and Aohan Banner. The number of stories of scholars such as Chunlan Bai found in this county was less than the number in these three counties. Thus, *Principle 1* (*including Principle 1.1–1.6*) *was weakly satisfied* in Yanchi.

The cooperation among local self-taught experts and others scholars, the participation of research stations in combating desertification, the scholars in government agencies, various forestry centers, etc., together ambiguously formed a federal organization of combating desertification. In Yanchi, various methods were applied to combat desertification in different areas. In northern sandy area, the major methods for combating desertification were building fenced belts and aerial seeding. In middle area, the major way was to build Agricultural Guard Forest

System and develop fodder forest on grassland. In southern area, however, the major method was developing non-wood forest (Zhou and Wang 2006). These methods guaranteed that people could have relatively concrete purposes when they tried to combat desertification. Thus, *Principle 2 was weakly satisfied.*

The stories of Bai Chunlan, citizen participation (for the story of Maoxian, see Chen 2000), cooperation among local self-taught scholars and other experts (for instance, the story of Bai Chunlan and other scholars), international cooperation (such as with Japan, Germany, and the UN), the participation of private companies (for detailed analysis, please see Ningxia Shangwu Ting 2005), etc., showed that the management in Yanchi was democratic and collaborative. The evidence of the federal mechanism of awards and sanctions, however, could not be found. Thus, *Principle 3 was weakly satisfied.* Also, the story of Chunlan Bai showed that there was relatively stable local scholar-entrepreneurship. This scholar-entrepreneurship, however, was not very strong. Except for the stories of Bai, no other similar stories were found. Thus, *Principle 4 was weakly satisfied.*

Although the activities of combating desertification were combined with economic development and environmental protection in Yanchi, this combination was not very well (the stories of Bai Chunlan and Maoxian were good examples). Thus, *Principle 5 was weakly satisfied.* The experiment-extension method, however, was applied very well (CCICCD 1996; Ho 2000). Thus, *Principle 6 was highly satisfied.*

Finally, central governments, international investments, external scholars' participation, etc., guaranteed its stable external financial and technical support. Although some scholars, Wang (2008), for one, indicated that support from governments at different levels also let it have relatively stable institutional support, no much evidence of this support had been found. Thus, *Principle 7 was weakly satisfied.*

Case 5: Ejin Horo (Yijinhuoluo) Banner (Qi), Inner Mongolia (Semi-successful)

Ejin Horo Banner, also known as Ejin Horo Qi, is a part of Ordos City. It is an autonomous county situated at the southeastern part of the Inner Mongolia Autonomous Region and the northeastern fringe of Mu Us Desert. Until 2007, its total area was 5600 km^2. After many years of combating desertification, Bai et al. (2006) found that from 1990 to 2000 its total value of ecosystem services decreased 0.21 billion with local economic growth. Based on a spatial analysis, they also found that the ecological value distribution was unbalanced, and it increased in a larger area while it decreased in a local area. Zhang et al. (2002) also found that desertification threat in this area still existed although desertification had been controlled somewhat. Thus, I deem Ejin Horo Banner as a semi-successful case of combating desertification from 1949 to 2007. Now let's analyze how the seven principles were satisfied.

Stable local field-based scholars played a very important role in combating desertification in Ejin Horo Banner. Network for Desertification Information and Data Exchange argued:

> Since the late 1970s, for developing the technical methods to control quicksand and wind erosion under the local natural and socioeconomic conditions, several special scientific research stations, most of which are the field experimental stations, have been established in the Yijinhuoluo County. On the basis of the existing traditional technology, knowledge, know-how and practical experiences for combating desertification and degraded lands, the stations scattered over this region have developed a series of effective techniques for building windbreaks, stabilized mobile dunes and fixing shifting sands. Meanwhile, the stations are throwing themselves into demonstrating the new techniques in the real situations and extending them to the local people and institutions. The great success has been achieved in combating desertification and in improving the local ecological environments by these field stations. (Network for Desertification Information and Data Exchange 2008)

It also pointed out:

> Located at the Mo Us Sandland, the Xinjie Experimental Station for Desert Control has a technique-demonstrating area of 16,186.7 ha. By long-term experimental research and practices, the station brought forward very successful technical guidelines for quicksand control. For example, to plant trees, shrubs and grass in the low-lying lands at first, then in dune slopes; and, to establish the sand biological barriers before and behind dunes, to give up middle of dunes to wind force, etc. Besides, the Xinjie Experimental Station for Desert Control has greatly contributed to the arboriculturation of the drought-resistance species. Totally, the station has sorted out more then ten of tree species suitable to the local conditions form 200 of the crossbred tree species introduced for whole region extension. Until 1989, the Xinjie Experimental Station had the total demonstrating afforestation of more than 15 thousands ha, and the breeding nursery per year had reached to 38 ha. Additionally, hundreds of the locals had been trained in the station annually. (Ibid)

It further argued:

> Besides the local experimental stations, the Experimental Station for Pasture Ecology of Erdos Plateau set up by the Institute of Botanic Sciences, the Chinese Academy of Sciences, working together with the local technical institutions, also significantly contributes to the research on ecology of arid region, the control of land degradation and the introduction of botanic species for land rehabilitation etc. (Ibid)

For instance, it demonstrated that several scientific expeditions from the Chinese Academy of Sciences and other institutions carried out their investigations in this region during last several decades. Two most important examples were "the Comprehensive Expedition of the Chinese Academy of Sciences for the Inner Mongolia and Ningxia Autonomous Zones of the 1950s" and "the Comprehensive Expedition of the Chinese Academy of Sciences for the Loess Plateau of the 1980s" (Ibid).

The scholars in these aforementioned stations and institutions also had high enthusiasm and capability to combat desertification, liked to teach the local farmers, learn local knowledge, have good relationships with local people, had high social responsibility and spirit of practice, and liked to work with local people and respect them. The story of Yushan Wang, a former leader of Xinjie Experimental Station for Desert Control, was a good example (for detailed stories, please see Inner Mongolia Daily 1990; Ren 1990). Thus, *Principle 1 (Principles 1.1–1.6) was satisfied.*

Different local experimental stations and the Ordos Sand Grassland Ecosystem Research Station formed in 1991 and supervised by Institute of Botany, CAS (Institute of Botany, Chinese Academy of Science 2003) formed a federal structure of scholars' organization in Ejin Horo Banner. This period, however, was not very long. From 1991 to 2006, there were only 16 years. Furthermore, in addition to desertification control, the more concrete purposes of combating desertification had not been set up. Thus, *Principle 2 was weakly satisfied.*

Citizens' participation, governmental support (Bai 1982; Inner Mongolia Daily 1990; Wang and Li 1982), the participation of international organizations such as Asia Bank and the UN (Li 1999), etc., showed that collaborative management existed in Ejin Horo. Evidence of bottom-up democratic management, however, was not been found. Also, the necessary system of awards and sanctions was built (Ai 1995; Wu et al. 2005). Therefore, *Principle 3 was weakly satisfied.*

The analysis of Principles 1 (including Principle 1.1–1.6) showed that stable local scholar-entrepreneurship was relatively built in Ejin Horo. Yushan Wang's story was a good example. The stories about other self-taught scholars as farmers or entrepreneurs, however, were not found. That is, *Principle 4 was weakly satisfied.*

The activities of desertification control in Ejin Horo Banner combined economic development and environmental protection very well (Bai et al. 2006; Wang and Li 1982; Yang et al. 2005; Yu and Zhang 1984; Zhang et al. 2002). Thus, *Principle 5 was satisfied.*

As shown by the cited paragraphs from Network for Desertification Information and Data Exchange (2008), the experiment-extension method was also applied in Ejin Horo Banner. Thus, *Principle 6 was satisfied.*

Also, governmental financial and institutional support (Ai 1995; Wu et al. 2005) and external scholars' participation guaranteed its stable external financial, technical, and institutional supports. Thus, *Principle 7 was satisfied.*

Appendix G
Case Studies by Changing Spatial Extent

Case 1: Songhe Village, Minqin County (Successful)
Songhe Village is in Xuebai Township, Minqin County. Until 2007, it included 13 production teams and about 418 families. Its population was 1767, and the total area was 95 km^2 (Shilongwei of Wuwei City 2008). After the activities of desertification control for about 50 years, it created a "Songhe Model" for desertification control (Lü 2007). Let's analyze how the seven principles were satisfied.

Stable local field scholars played a very important role in Songhe Village. In 1974, the Comprehensive Desert Control Experimental Station in Minqin sent five science researchers and technicians to Songhe Village to teach people how to combat desertification. After 1985, the Forestry Department of Wuwei City arranged three science researchers and technicians to help local people develop economy and control desertification (People's Daily 2000a). A farmer, Yuzhu Shi, as a local self-taught scholar, was also a good example. His and other scholars' stories showed that these scholars had high enthusiasm and capability to combat desertification, liked to teach the local farmers, learn local knowledge, and have good relationships with local people, had high social responsibility and spirit of practice, and liked to work with local people and respect them (Bu 2005; Liu et al. 2005; Nie 2005; People's Daily 2007; Tebie Baodaozu 2007; Zhou and Liang 2003). Especially, Shi often learned from other science researchers and technicians (Nie 2005). Thus, *Principle 1 (including Principle 1.1–1.6) was satisfied* in Songhe Village.

As a small village, families' and farmers' participation and support from the township government, the county government, the experimental station and others social groups together formed a federal organization in combating desertification. They also had a very concrete purpose of desertification control—to protect their village from stopping sand-moving. The small-scale guaranteed that they could relatively easily implement their policies and purse this purpose. Thus, *Principle 2 was satisfied.* According to the observation and interviews by the author in 2007, the methods of citizen's participation, democratic management, necessary mechanisms of awards and sanctions were applied in Songhe Village. Furthermore,

L. Yang, *Knowledge-Driven Governance*,
https://doi.org/10.1007/978-981-13-2910-4

various inner and external social groups' participation also showed its collaborative management. Thus, *Principle 3 was satisfied.* The aforementioned stories of Shuzhu Shi and other scholars demonstrated that stable local scholar-entrepreneurship was built in Songhe village. Especially, Shi had led other farmers to combat desertification for over 50 years. Thus, *Principle 4 was satisfied.*

Desertification control in Songhe Village combined economic development and environmental protection together. For example, from 1985, farmers in Songhe Village began to develop ecological nonwood forestry and rural courtyard economy (People's Daily 2000b) which significantly improved farmers' income (Nie 2005; Zhou and Liang 2003). Thus, *Principle 5 was satisfied.* The experiment-extension method was also applied in Songhe Village. Shuzhu Shi's story showed this. That is, *Principle 6 was satisfied.*

Science researchers and technicians from Minqin Experimental Station of Comprehensive Desert Control, Wuwei City, and other organizations at the provincial level provided enough stable technical support for desertification control in Songhe Village. Support from the township governments, the county governments, the provincial governments, and even the central government (e.g., Chinese premier Jiabao Wen) guaranteed its stable institutional support. Furthermore, according to the author's interviews in Songhe Village, external investment from the governments and the other organizations also guaranteed its external financial support. Also, because its scale was relatively small, its dependence on external financial support was not very large. Thus, *Principle 7 was satisfied.*

Case 2: Yulin Prefecture, Shaanxi (Semi-successful)

Yulin Prefecture is a region facing serious threats of desertification in northern part of Shaanxi Province, the south fringe of Mu Us desert. Its annual precipitation is 414.6 mm. Under the prefecture, there are one city, eleven counties, and 262 townships. Forest Division of Yulin Prefecture, Shaanxi Province (1998) pointed out that in Yulin Prefecture "the total area of the sand land is 24,400 km^2, which accounts for 56.7% of the total land area of the prefecture and covers 95 townships with 1,010,000 residents of seven counties (city), including Yulin, Shenmu, Fugu, Hengshan, Jingbian, Dingbian, and Jiaxian. The Yulin sand area is one of the main construction areas for 'three norths' protection forest belt system." Up to 1997, the rate of forest cover of the region increased from 1.8% in the early 1950s to 30.5%. Forest area reached 1.3 million hectares. Originally, the total moving sands in this region were 573.3 thousand hectares. Local people stabilized 400 thousand hectares of moving sands through afforestation. Furthermore, the days of sandstorm also decreased from 66 to 24 days per year (Ma 1999). That is, a remarkable success had been achieved in combating desertification in Yulin district in the course of the last 40–50 years. Using remotely sensed data and geographic information systems (GIS), Zhang et al. (2008), however, found that although land desertification had to some extent been controlled in the areas by planting grasses and trees during the past 40 years from 1960 to 2006, land desertification in the area had been developing rapidly. That is, the issue of land desertification in this region was still serious. Thus, I deem this region as a semi-successful case of combating desertification from 1949 to 2007. Now let's analyze how the seven principles were satisfied.

According to its Web site, the Institute of Sand-Control of Shaanxi (ISCS) was established in Yulin City in 1960, based on Yulin Investigating Station of Comprehensive Desert Control of Chinese Academy of Science. Its chief task was to do research on "comprehensive sand control with biological technical measures and control models for government policymaking and for local farmer combating against desertification, sand disaster and development of the resources in sandy areas" (Institute of Sand-Control of Shaanxi 2008). Its Web site also said:

> It has 143 members of staff, 110 of whom are scientists and technicians in desert control, forestry, water and soil conservation, geography, pedology, horticulture, forest protection, agriculture and so on. There are 36 technical personnel with senior technical titles and 33 with technical titles of middle rank. Two scientists have prominent contributions in the state-level and the provincial level, and four scientists are topnotch. There are 5 research fellows who enjoy the government special subsidiaries among them. It has 2500 hectares of experimental sand, in which 300 hectares of Sand Botanical Garden is built… In recently years, 15.4 hectares of experimental land have been built in it, and have been developed into the demonstration land for desert control. Rare Psammophyte Protection Base accounts for 178 hectares.
>
> Our institute has attained a large number of scientific research achievements… At present, 54 items of research result have carried out the approval and the appraisal, in which 5 items reached advanced level in the world and 17 items reached advanced level at home. Forty items have been awarded and 24 on prefecture level. At the same time, more than ten contract projects on sand control and afforestation have been completed… Batches of research achievements… have been introduced not only in Yulin desert region, but also to Inner Mongolia, Gansu, Ningxia. (Ibid, 2008)

This institute deems Yulin desert region as a "big battlefield." It has formed complete set of desert control approaches in this region, including "plane sowing by air seeding," "closing hillsides to facilitate afforestation and artificial transplantation." Also, it has found a batch of effective ways of development in this desert region. The Institute of Sand-Control of Shaanxi (2008) also pointed out:

> Forest and grass area has amounted to 993,000 hectares in the Yulin desert region, of which the vegetation coverage rate has increased from 1.8% to 35%. 400,000 hectares of wind-drift sand has been fixed. 100,000 hectares of farmland has formed network of forest belts. 87,000 hectares of farmland has been newly built. 153,000 hectares of meadow has been restored and improved. Preliminary control rate has reached 68.4% in the desert region. Average grain owned per person in the desert region is over 500 kilograms, and average annual cash income is nearly 2000 yuan per person…the generalization area for each research achievement adds up to 470,000 hectares only in the Yulin desert region, and the generalization area in the adjacent provinces amounts to more than 200,000 hectares. The institute has provided 18,000,000 seedling trees, different kind of seed 250,000 kilograms. The generalization benefit of all achievements is 1700,000,000 Yuan. (Ibid)

From above statements, it can be concluded that local stable-field scholars played a very important role in combating desertification in Yulin. The Forest Division of Yulin Prefecture in Shaanxi Province also pointed out:

> The achievement is the 40-year hardship and sweat embodiment of the scientist, technicians and leaders of all previous communist party committees and governments at different levels… The scientists and technicians of a few generations publicize advanced and practical techniques. Leaders and technicians work on spot wherever there are sand control projects. Many leaders and technicians got ill due to over-working, and some over 60 years old were still working in the forefronts of sand control. A series of scientific management responsibility systems of sand harnessing for leaders has been formed in the process of the long-term practices, and leaders at different levels have their own targets contract.
>
> …After the national conference on sand control was held in Lanzhou, over 1000 scientists and technicians from the levels of prefecture, counties and townships organized by the prefecture, have worked out the plan of the integrated control and development for Yulin sand area on the basis of investigation, which has been brought into the line with the 10-year state plan. (Forest Division of Yulin Prefecture, Shaanxi Province 1998)

Thus, I conclude that *Principle 1(including Principles 1.1–1.6) was satisfied.*

Yulin Prefecture built the Institute of Sand-Control of Shaanxi, which included a management committee and different scientific research departments such as various laboratories and wilderness inspection stations, farms, parks, and botanical gardens (Institute of Sand-Control of Shaanxi ISCS 2008). The federal organizations of heresiarch stations in the whole area, however, could not be built. Furthermore, the Forest Division of Yulin Prefecture, Shaanxi Province argued:

> The scientific and technical organization subordinated by the departments of agriculture, forestry, water conservancy and animal husbandry at all levels, contract some counties or townships and offer technical services for them. Correspondingly, the systems of check, acceptance, reward and punishment were established. All local governments in the sand area pay much attention to the management, at same time they supply farmers with comprehensive services such as planning, designing, fund collection, seedlings and so on. (Forest Division of Yulin Prefecture, Shaanxi Province 1998)

Public participation was also emphasized. For example, the Forest Division of Yulin Prefecture, Shaanxi Province, argued:

> In the 1950s, co-operation and job responsibilities were executed to plant trees and bring the sands under control in the area. In the 1980s, government has implemented the policy that the man who contracts and manages sand land will get whole profit from the sand land. At the moment, 100,000 households contract 330,000 ha of sands. Mr. Zhang Jianwang, a farmer of Jingbian County who has serious illness, has turned 820 ha of barren sands into fertile land. Mr. Shiguangyin, a farmer of Dingbian County, organized 95 households and established the Barren Sand Control Corporation, which collected 100,000 Yuan and has brought 50,000 mu of sand under control. In the past 40 years the people in the sand area had worked hard and never stopped harnessing, and even some people give their lives to the sand control. According to statistics, the people in the area have invested 900 million RMB Yuan of cash and labor. (Ibid)

Collaborative management with external scholars and international scholars was also practiced. For example, the Institute of Sand-Control of Shaanxi argued that it has cooperated with many scientific research institutions and academies including The Northwest University of Agriculture and Forestry, Beijing Normal University, Northwest Institute of Soil and Water Conservation of Ministry of Water Resources, Institute of Earth Environment of Chinese Academy of Science, Institute of Meteorological Sciences of State Meteorological Administration, and Chinese Research Institute of Forestry Science. Also, it has established academic exchange and cooperation with many universities and scientific research institutions in Japan, USA, France, Israel, Belgium, Australia, and so on (Institute of Sand-Control of Shaanxi 2008).

The collaboration between scholars and governments has also been shown by the above citation. In summary, although cooperative organizations were built, public participation was emphasized (including woman and army participation), collaborative manage was applied, and the systems of check, acceptance, reward, and punishment were established, it could not build a systematic federal organization to combat desertification, because its management was mainly the top-down hierarchy method rather than a more democratic bottom-up method. And because its area was very large, if the federal-type organization could not be built, its purposes of combating desertification in different counties were not concrete and stratified correspondingly. Thus, *Principles 2 and 3 were only weakly satisfied.*

Scholars in the Institute of Sand-Control of Shaanxi could be deemed as stable local field scholars, and they could partly lead local people and help government combat desertification. Some local self-taught scholars with high scholar-entrepreneurship such as Bai Chunlan in Yanchi County and Wang Youde in Lingwu County in Ningxia County, however, could not be found. Thus, *Principle 4 was weakly satisfied.*

As the aforementioned paraphrases from the Web site of the Institute of Sand-Control of Shaanxi (2008) show us that the activities of combating desertification in Yulin combined economic development and environmental protection together. Thus, *Principle 5 was satisfied.* The experiment-extension method was applied. For instance, the Forest Division of Yulin Prefecture, Shaanxi Province (1998) pointed out, "Each county in the sand area has its own demonstration area of sand control to extend experience from point to area and speed up the control and development of the sand area." That is, *Principle 6 was satisfied.*

The above analysis of scholars' participation and external and international collaboration shows that its external technical support was stable and obvious. The Forest Division of Yulin Prefecture, Shaanxi Province (1998) also pointed out, "From 40 years ago on, especially since the implementation of 'three norths' protection forest project, a foundation has been laid for the control and development of Yulin sand area." Furthermore, it also argued, "All local governments in the sand area pay much attention to the management, at same time they supply farmers with comprehensive services such as planning, designing, fund collection, seedlings" Then, I can conclude that the necessary institutional support was guaranteed. Thus, *Principle 7 was satisfied.*

Case 3: Zhangye Prefecture, Gansu (Unsuccessful)

Zhangye Prefecture is located in the Middle of Hexi Corridor. It includes Ganzhou District, Shandan County, Minle County, Linze County, Gaotai County, and Sunan County. Until 2005, its total area is 62.886 million mu, and its total population is 1.2733 million (Forestry of Gansu 2005). Historically, Zhangye was called "golden

Zhangye" or "the breadbasket" of western China, but now desertification becomes very obvious. Although during last several decades, different levels of desertification control had been completed (e.g., see Fan and Zhou 2001) and the land lost to desertification in Zhangye City (Ganzhou District) rapidly decreased from 1949 to 1995 (the increased area of desertification was −32,705 km^2, 71,143 km^2 in Linze County, 11,450 km^2 in Gaotai County, 10,987 km^2 in Shandan) due to the wide development of fixed sand dune for irrigation agriculture accompanying population increase (Feng et al. 2001), the desertification problem in this whole region was still very serious. The Gobi Desert area near Zhangye City was even blamed for the frequent occurrence of dust and sand storms in remote areas such as the Beijing and HuaBei regions. Wang et al. (2005) found that the total desertificated area in Zhangye Prefecture was once increased from the foundation of People's Republic of China to 1939 and increased 59,265 hm from 1993 to 1999. From the late 1950s until 2000, underground water in Zhangye Prefecture fell by nearly 100 m (People's Daily 2000c). Luo et al. (2005) also argued, "Although the development and reversion of sandy desertification coexist, the sandy desertification in this area seems serious and is attributable to the irrational use of water and land" (p. 90). Thus, desertification control in Zhangye Prefecture was unsuccessful from 1949 to 2007. Now let's analyze how Principles 1–7 were satisfied.

In Zhangye Prefecture, various experimental stations of desertification control such as Zhangye Saline Soil Grassland and Animal Production Experimental Station, and Xiaoquanzi Desertification-Control Station and Wulidun Desertification-Control Station in Linze County were built. Research stations in other counties were not built or not very famous, although their desertification was also serious. Water conservation in Zhangye Prefecture was relatively successful (Liu et al. 2005). Therefore, although scholars played some roles in combating desertification in Zhangye Prefecture, their participation was limited, especially for local stable field scholars' participation. Dorr (2004) even argued that dead trees and roadside ditches outside of Zhangye were evidence of a failed experiment. Thus, *Principle 1(including Principles 1.1–1.6) was not satisfied.*

The activities of desertification control in Zhangye Prefecture were not systematically organized. Except for in Linze County and Suzhou District, necessary organizations in others areas in Zhangye were not built, and good organizations or management in Suzhou and Linze were not extended to other areas. Furthermore, except for water conservation, in desertification control, concrete organizational purposes were not set up in the whole prefecture. Thus, *Principles 2 and 6 were not satisfied.*

Although citizens' participation, international participation (such as ODA loans), government support, firms and NGOs' involvements showed the same characteristics of democratic and collaborative management and some mechanisms of awards and sanctions were also applied, *Principle 3 was not satisfied* from a perspective of the whole prefecture. For example, Zhang pointed out:

> In early 2002, the Ministry of Water Resources (MWR) of China initiated an experimental project to establish a water-saving society in Zhangye city in the Heihe river basin in

> north-west China – this project was the first of its kind in China. The aim of the project was to establish a new water use rights (WUR) system with tradable water quotas and to reallocate water resources reasonably and efficiently through market-based instruments. This report presents the research done on the system and water markets. It has been found that that the system is hard to implement well and that WUR trading is not popular. The barriers to implementing a WUR system are social and administrative in nature. Local farmers cannot be forced to limit their water use because they cannot endure losses caused by water shortage. Local water agencies have no incentive to restrain local farmers from using excessive water. On the other hand, WUR trading faces management, legal, administrative, and fiscal barriers. There are management risks for farmers in switching to low water-intensive crops. It is also difficult for water buyers to buy rights to land and water use from farmers with small parcels. Farmers are discouraged from selling water to the government whom they fear will reduce their water quotas, and divert irrigation water to other sectors. (Zhang 2005, p. 1)

Based on the analysis of Principle 1 including Principles 1.1–1.6, it can also be concluded that local stable field scholar-entrepreneurship could not be built in Zhangye Prefecture. Thus, *Principle 4 was not satisfied.*

Although the activities of desertification control in Zhangye Prefecture tried to combine environmental protection and economic development and in fact grain yields had increased rapidly to 700 kg capita with increasing desertification control and the popularization of interplanting techniques (Fan and Zhou 2001), their efforts were not successful. For example, Li and Qian (2006) pointed out that Zhangye had failed to take advantage of the opportunities to go through the transition from traditional and agricultural based economy to a more diversified economic structure through assessing the Great Western Development Strategy (GWDS) (xibu da kaifa) in Zhangye. Zhang (2005) also argued that an experimental project of establishing a water-saving society in the Heihe River in Zhangye, initiated by the Ministry of Water Resources (MWR) of China, encountered various barriers such as the barriers to implementing a water use right (WUR) system, difficulty forcing local farmers to limit their water use because they could not endure losses caused by water shortage, less incentives to local water agencies to restrain local farmers from using excessive water, and other management, legal, administrative, and fiscal barriers. Thus, *Principle 5 was not highly satisfied.*

The lack of external scholars' participation means that stable external technical support in Zhangye could not be guaranteed. Also, external financial support was not stable and enough. For example, Wei and Fang (2006) argued that foreign direct investment (FDI) remained very limited in Zhangye City and it was only US $70,000 in 2001. Because the economy in Gansu Province was not very good, it can be concluded that the investment from provincial government was also limited. Much evidence of central governmental financial support was not found. Before it was elected as a prefecture area of demonstration on desertification control (Qi 2006), its institutional support was also limited. Thus, *Principle 7 was not satisfied.*

Appendix H
Case Studies by Changing Cultural Contexts

Case 1: The Landcare Program, Australia (Successful)
Australia also has a desertification problem. According to Christie (1986), about 42% of the five million km^2 of arid and semiarid lands, some two million km^2, in Australia is estimated to be either in a severe (26%) or very severe (16%) state of desertification. Two national programs, Drought Alert and Landcare, have been developed to control land degradation (Ludwig and Tongway 1995). In a broad sense, "landcare" embraces what used to be known as "soil conservation," "conservation farming," "farm planning," "revegetation," and "land protection"; to a lesser extent, it embraces vermin and weed control and river management (Campbell 1992). It tries to build a partnership between the community, government and business to protect and repair our environment. Landcare groups are a critical element in the landcare movement, and now thousands of volunteer community landcare groups including bushcare urban landcare, rivercare, coastcare, and sustainable agriculture groups are tackling land degradation in every corner of Australia (LAL 2008). Since this program initiated in 1989, many achievements have been gotten. This program has been widely deemed as a relatively successful effort to combating land degradation and a model of community action contributing to more sustainable management of natural resources (Campbell 1994; Curtis 1998). Victorian landcare groups are seen as the key element of an emerging Australian success story (Curtis and Lacy 1996). Now let's analyze how the seven principles of successful scholar-based governance were satisfied from 1989 to 2007.

(1) *Stable, local and community-based landcare groups as stable field scholars were the key part of scholars and played an import role in Landcare Programs.*

As stated above, landcare groups were a critical element in the landcare movement. As local voluntary groups of people, landcare groups were comprised of land users in rural areas. Thus, they could be deemed as stable field groups. Many leading farmers and conservationists were included in landcare groups, who played a key role in design, implementation, and interpretation of research and development project. Through them, the extension, research, and development was made

L. Yang, *Knowledge-Driven Governance*,
https://doi.org/10.1007/978-981-13-2910-4

less linear, scientists were put in more direct contact with the community, the boundary between extension and research was dissolved, and practical experience and local knowledge were made grater use (Campbell 1992). Furthermore, Campbell pointed out:

> The farmers who are in landcare groups or receive advice from landcare groups, on average are younger, earn higher levels of cash income, have higher levels of debt, are more active seekers of information from a wider range of sources, they are more concerned about the future, more positive about and receptive to government and importantly, they undertake more land conservation practices than other farmers. (Ibid, p. iv)

The landcare group effectiveness was also hugely influenced by knowledge, and knowledge was seen as a key manipulable factor in voluntary change (Röling 1991, as cited in Campbell 1992, p. 14). Many landcare groups were deemed as legitimate players in research (Chambers et al. 1989). Campbell (1992) argued that through participating jointly in research and development projects with scientists and extension staff from state agencies, CSIRO and universities, landcare groups can also be potentially involved in generating information as well as sharing it. Curtis and Lacy (1996, p. 119) also pointed out that "the assumptions underlying landcare are that limited funding of group activity will produce more aware, informed, skilled, and adaptive resource managers with a stronger stewardship ethic, will increase the adoption of sustainable practices, and will assist the move to more sustainable resource use." The importance of education, awareness, and research and extension as factors influencing the effectiveness of landcare groups was also highly stressed by Campbell (1992). All these tell us that successful landcare groups were scholar-like groups.

(1.1) *Landcare groups had a relatively independent social identity and a high social status.*

As local voluntary groups, landcare groups were independent social actors. Furthermore, because the members of landcare groups often earned higher level of case income, had higher levels of debts, and undertook more land conservation practices than other farmers, they had relatively high social status (Campbell 1992).

(1.2) *Members of landcare groups had high enthusiasm and capability of scholars' participation.*

Members of landcare groups were far-sighted people, who were often activists and conservationists and more concerned about the future. Also, they gained intellectual stimulation and excited new knowledge and the satisfaction of doing something constructive in their district and of influencing others. Landcare groups improved land users understanding of land degradation problems and their solution. Furthermore, through extension activities, they shared information about better land management practices between group members and to non-group members (Ibid). That is, these members had not only the enthusiasm and capability to take part in landcare but also the enthusiasm, capability, and leadership to teach other local land users.

(1.3) *Landcare groups had relatively enough knowledge.*

First, as local scholars, they had more scientific knowledge about landcare than other local social actors. Second, they learned scientific knowledge from other researchers and scientists as local legitimate players in research and translators of scientists' knowledge. Third, as local members, they had more knowledge about local conditions and local social relationships and webs than other outside scholars. So Campbell (1992) argued that they developed local inventories of natural resources (such as remnant vegetation and seed sources) and documented local knowledge about land and its management.

(1.4) *Landcare groups had high social capital.*

Mature landcare groups could gain credibility which formed their social capital. After groups were formed, they often grew quickly and developed ongoing relationships with local and sate government agencies, local businesses, community groups (including neighboring landcare groups), universities, researchers and consultants, scholars and other landholders. Effective groups often enjoyed constructive relationships with land conservation agencies. Also, they had credibility in their local community, which was reinforced by a group identity and recognition of members (Campbell 1992).

(1.5) *Landcare groups had high social responsibility and spirit of practice.*

As local voluntary groups, landcare groups had high social responsibility in their own district and of influencing others. Landcare groups not only played a key role in design, implementation and interpretation of research and development projects but also were active participants in programs introduced to alleviate structural unemployment. For example, through a year of environmental service, school leavers or university graduates received training and practical experience in land management, nature conservation, mapping and inventory skills, waste management and rehabilitation techniques. These show that landcare groups had strong spirit of practice, and in fact, as already pointed out, that they undertook more land conservation practices than other farmers and did things on the ground (Ibid).

(1.6) *Landcare groups respected other social actors.*

Through building good relations with other social actors as stated above, landcare groups showed their respects to other social actors. Especially, as local voluntary groups, they understood rural farmers and respect them. Farmers, however, said that urban people should understand them more (Ibid, 1992).

(2) *Landcare groups had federal organizational structure and concrete organizational purpose.*

Campbell argued:

> A farmer, local activist or department person (or any combination of these), see a land management issue, fell that a landcare group is the way to go, talk it over with

> friends/neighbours/extension staff and call a meeting. The meeting elects a steering committee, which investigates local problems, interests, resources and assistance available, then calls another meeting to form a group and elect a committee (although sometimes this happens at the first meeting), which may comprise the entire group, or be an executive subset of the group. (Campbell 1992, p. 15)

Furthermore, landcare groups had very complex relationships with other social actors including other landcare groups. Thus, they formed a federal web.

The broad goal of most landcare groups was to tackle land degradation and develop more sustainable systems of land management. A mature landcare group, however, had a clear understanding of its role and goals and was well known within the local community. So Campbell (1992) clearly pointed out that effective groups had a clearly defined and understood problem or goals and knew where they were going and had a clear, achievable plan of how to get there. Lack of clear goals or an achievable of attack, however, was an important negative characteristic of ineffective landcare groups. This was also tested by a survey study (Curtis et al. 2000).

(3) *Landcare groups applied democratic and collaborative management and a federal mechanism of awards and sanctions.*

The effectiveness of landcare groups depended on non-coercive policy instruments (as education, information and persuasion), anticipation, participation, "upward" communication, voluntary behaviors, equity (especially between men and women), democratic management, bottom-up decision making and the collaboration among land users, group members, different groups, and different other social agencies (Campbell 1992). Especially, landcare groups played a facilitation role and a coordination role. In summary, the Landcare Program applied democratic and collaborative management. Especially, it tried hard to build partnerships between communities and government (Alexander 1995; Campbell 1994) through promoting and supporting a national-wide of community landcare groups (Lockie and Vanclay 1997, p. 1). Also, it encouraged even empowers community-based groups of stakeholders to work together to mange resources in a sustainable manner (Chamala 1995).

Landcare groups encouraged respected but disinterested farmers to get involved. Some regulations had also been developed. For example, when analyzing what forms of support would be most effective, Campbell (1992) listed three regulations: (1) "*Have laws passed that anyone that degrades the land is fined.*" (2) "*Stronger enforcement by states of land users' responsibility to the land. The Ag Dept is loathe to use its regulatory functions.*" (3) "*A central plan that covers the State. Maybe most people would think it is an encroachment on their rights*" (p. 110; italic in original). Furthermore, he pointed out that both the Commonwealth government and local governments should do a lot to encourage land conservation among landusers (Ibid; for the analysis in detail, please see pp. 114–117). Various groups around Australia were interested in land resource monitoring and in developing indices of sustainability. Even landcare groups' involvement in monitoring was considered. All these tell us that Australian Landcare Program built a federal mechanism of awards and sanctions.

(4) *Landcare groups built stable local scholar-entrepreneurship.*

As local actors, landcare groups had four phases of group development—formation ("communication—talking up the group approach"), establishment ("setting direction, discrete projects, external funding"), consolidation ("programs rather than projects, goals, strategies, sharing the work"), and maturity ("autonomy, self-reliance, internal resources, outward and forward looking") (Campbell 1992, p. 17). They, however, were relatively stable scholar-entrepreneurs in landcare. First, when in the phase of maturity, the group had settled down, with "easy identification of leaders and future leaders, talkers, workers, followers, sleepers and hangers-on" (Ibid, p. 16). Second, members of landcare groups are far-sighted people as mentioned above. Third, leadership was also important to effective landcare groups. Campbell (1992) argued that effective groups "have **good leaders**, with vision, who delegate, share responsibilities and workload, **involve members** in planning, making decisions and actives" (p. 22; emphases in original), while ineffective groups "**lack leadership**, lack sufficient committed people, or **lack access** to responsive technical, financial or administrative support" (p. 23; emphases in original). All these tell us that effective landcare groups built stable local scholar-entrepreneurship.

(5) *Landcare groups tried hard to protect and increase landusers' benefits.*

Land users, landcare groups, and other social actors worked hard to find economically profitable solutions. Röling (1991) pointed out, "they will work only when the change interests or benefits the people whose change is deemed as desirable" (cited from Campbell 1992, p. 14). Effective landcare groups must overcome the difficulty—"*a lack of evidence of immediate benefits of social conservation*" (Keith and Roberts 1990, as cited in Campbell 1992, p. 27; italic in original).

(6) *Landcare groups applied an experiment-extension method.*

Campbell pointed out:

> Land conservation agencies in Australia seem to be rushing like lemmings to re-structure, re-organise, amalgamate, re-name, re-locate. First (and still) Victoria, then Tasmania, New South Wales and Queensland and now Western Australia and South Australia have moved to rationalize land conservation agencies. (Campbell 1992, p. 34)

For using knowledge and technologies, a research-extension method was also used. Presently, thousands of volunteer community landcare groups including bushcare, urban landcare, rivercare, coastcare, and sustainable agriculture groups were tackling land degradation in every corner of Australia. All these tell us that landcare groups used an experiment-extension method.

(7) *Landcare received stable external support, especially financial, knowledge and institutional support.*

Because of landcare groups' efforts, many universities, research institutes, researchers, etc., directly took part in local Landcare Programs. Government and other social actors also supplied financial, technical, institutional, and moral support. Thus, Campbell argued:

> It is simply unfair and totally unrealistic to expect voluntary groups of people, even if there were one hundred thousand active landcare members, to fix land degradation or develop more sustainable farming systems without a significant external support—financial, technical, institutional and moral. (Ibid, p. iii)

He also pointed out that one of regular activity of the mature landcare groups was "actively drawing from a wide range of support-government and non-government-local, regional, state and national." Especially, he argued that "Government must provide a significant level of resources to support what has to be a profound change in Australian agriculture" (Ibid, p. 45). First, government should educate landusers and support scientists and researchers meaningful involvements in landcare. Second, financial support was very important, because "*money-this is more important than defense*" (Campbell 1992, p. 114; italic in original). Thus, landcare was deemed as "an example of state-sponsored rural development in a developed nation" (Curtis 1998, p. 563). Third, governmental or institutional support was also necessary (for detailed analyses, please see Campbell 1992, pp. 31–37). Fourth, farmer and their efforts also should be understood and respected, and then moral support was also necessary (for detailed analyses, please see Ibid, pp. 37–42).

Case 2: The Negev Case Israel (Successful)

The Negev represents 62% of the entire area of Israel with the pre-1967 boundaries and encompasses approximately 12,000 km^2. Since the establishment of the state of Israel, the implementation of an intensive program of research and development as well as several resettlement projects has transformed large sections of the region into important productive areas. This has contributed to the progress and well-being of the entire nation of Israel (Schechter and Galai 1980). It actives of combating desertification still continues. As a successful case of combating desertification with scholars' participation, this case study focuses on the analysis whether their efforts from 1948 to 2007 were still consistent with the fundamental principles of successful scholar-based governance.

(1) *Stable field scholars significantly took part in the Negev program.*

Stable field scholars were the major contributors of the Negev program. For example, because water was a major factor in recovery of arid area and the oblivious direction for this recovery was "the supply of additional water and more efficient use of already available supplies" (Scheckter and Galai 1980, p. 265), stable field scholars contributed a lot to resolve this problem and the most modern civilization methods were utilized to develop a highly intensive farming system and combat desertification. Professor Michael Evenari and his team were a good example (see Evenari 1989). Although they were professors or researchers, they were stable field scholars who carefully studied history and many other problems of the Negev and stably took part in this program. Scheckter and Galai (1980) pointed out that the use of advanced agricultural techniques and the development of technological innovation was the most important one among the factors which contributed the high productivity of Israeli agriculture. Furthermore, universities contributed a lot to educate and supply stable field scholars for this region. Schechter and Galai argued:

> Ben Gurion University of the Negev at Beer Sheva has made a considerable contribution to the region. Founded in 1968, it now has an academic staff of 900, and 4500 students. Granting of tenure to staff is conditional on residence in the Negev. Before its establishment, all the students from the Southern District studied either at Jerusalem, the Haifa Technion or Tel Aviv University. Now 50% of the District's students attend their local university; 75% of Beer Sheva students are studying in Beer Sheva… An indicator of the University's impact on the life of the region is the fact that 26% of the students are of Asian and African origin, a percentage twice as high as in any other Israeli institution of higher learning. (Scheckter and Galai 1980, p. 298)

In summary, as Schechter and Galai (1980, p. 307) summarized:

> Applied research and development have played a vital role in development of the Negev. Considerable effort has gone into the adaptation of known technology to Israeli conditions, and into local innovation. An efficient extension service brings the results of research to the production system. Experimental stations and research institutions, especially of an applied nature, have much to contribute, from the technological, educational and sociological viewpoints. (Ibid, p. 307)

(1.1) *These scholars had a relatively independent social identity and a high social status.*

In this program, the scholars often had an independent social identity and a high social status. This also could be seen from their biographies (e.g., see Evenari 1989).

(1.2) *These scholars had high enthusiasm and capability to take part in the Negev program.*

The experience of professor Evenari and his team (Please see Evenari 1989, pp. 137–147) demonstrated scholars' high enthusiasm and their capability to take part in the Negev program. Considering their dream to build a new state, we could not doubt their enthusiasm to combat desertification. Furthermore, they tried hard to teach other people and make their technology or methods relatively simple to operate even by inexperienced farmers (Schechter and Galai 1980).

(1.3) *These stable field scholars had enough knowledge.*

Enough knowledge had been developed for combating desertification. First, enough expert information and knowledge was developed. For example, in addition to technology of water use, agricultural technology in the semiarid zone was developed (Ibid). The book *The Negev: The Challenge of a Desert* (Evenari et al.

1982) summarized more than two decades of scientific study in the Negev desert by the authors. Second, enough knowledge about local conditions, local traditions, social norms, etc., and local wisdom or knowledge was developed. For example, Professor Michael Evenari and his team tried hard to answer the following questions:

> Where did the desert farmers get their water? Was the Negev a desert in ancient times as today, or did it have a more humid climate with more rainfall? Was agriculture restricted to a certain type of soil? Did they employ different farming methods? Were the Nabateans the only group to carry out farming activities in the desert? (Evenari 1989, p. 139).

Third, as Israel scholars, these people had adequate knowledge about their own social relationship and webs. Fourth, education helped them improve knowledge of all the citizens. For example, Schechter and Galai pointed out:

> Along with the establishment of population centers, cultural, educational and health facilities were set up. The dearth of primary school teachers was made up by volunteers who fulfilled their army service requirements by teaching in the settlements and cities of the Negev. High schools and vocational schools were established and oriented to meet the industrial and professional requirements of the area. A modern hospital and health care service was established to serve the entire southern area of Israel. (Schechter and Galai 1980, p. 290)

(1.4) *These scholars had high social capital.*

As all Israeli who came together to build a new county, the self-identity and trust among them were high. Also, these scholars had high social capital not only among themselves but also with governmental officials, local farmers, and so on. Furthermore, development of the Negev was a part of the settlement policy of Israel. When new people came together to do something for the same purpose, the social capital among themselves could be built quickly.

(1.5) *These scholars had high social responsibility and spirit of practice.*

This also can be seen from the story of Professor Evenari and his team (see Evenari 1989; Schechter and Galai 1980). Evenari argued:

> These expeditions to the Negev were difficult since most of the roads did not exist at that time. We could only travel over rough terrain in jeeps and had to carry guns in case we encountered Arab marauders. On army orders, we had to leave the Negev before dark and stay in Beer Sheva overnight. The logistics fell upon Liesel. She had to see that there was enough food and prepare meals fro the research team and volunteers, up to ten people in all. Her "desert salad," into which she put anything that was edible, became well known. One guest said jokingly that she may have even put his sandals in one of her concoctions.
>
> During our Negev excursion we were sometimes in physical danger as well. Once we were surveying Wadi Abiad, near Shivta, where there were extensive traces of ancient agriculture. Before starting out on this trip we had to inform the military authorities as they often carried out training exercises in this area and ask for permission to work there. While surveying the area, shells started exploding all round us. We later found out that the army command had forgotten to inform the battery exercising there of our presence in the area.

> In retrospect, it seems remarkable how we worked together harmoniously for so long, although there were problems.... (Evenari 1989, p. 138)

(1.6) *These scholars respected other social actors and local knowledge.*

For example, Professor Evenari at Avdat did an experiment to test the technique method of utilizing runoff water of ancient desert agriculture (Schechter and Galai 1980). A case study submitted by the Environmental Protection Service, Government of Israel argued:

> Some early attempts to construct Bedouin villages were unsuccessful because they failed to take account of social differences among the Bedouins. It was mainly the younger wage earners who occupied the houses provided. Recently policy has been to provide lots on which the Bedouins can build, allowing for those extended families still engaged in the traditional economy, giving a better accommodation of tribal and social groupings, and leading to greater involvement of the Bedouins themselves in the development of the settlement. It is also recognized that the establishment of settlements must be linked with the provision of local employment. Experience indicates that adjustment from nomadic to settled life must be gradual, that the design of settlements must take account of social structures, that the local population must be involved in planning, and that the establishment of settlement must be linked with the provision of employment. (Government of Israel 1980, p. 265)

While some people criticized that "too much confidence was placed on foreign technology and inadequate effort given to local research and development" (Government of Israel 1980, p. 265), in the Negev, local knowledge (here means the research in Israel) in fact was respected as stated above. Furthermore, the fact that Israeli realized this problem in 1980 also meant that they began to change this trend at that time.

(2) *The Negev program had a federal organizational structure and concrete organizational purposes.*

Federal organizations from individual farmers, various regions and municipalities, and the Government had been built. For example, the Israel "Water law" legislated in 1959 further empowered "the Government to set up an administrative system to allocate water quotas to various regions and municipalities and to individual farmers" and constituted "the legal basis for the inter-regional transport of water the arid Negev" (Government of Israel 1980, p. 266).

The Negev Program had concrete organizational purposes. Schechter and Galai (1980) argued that the obvious direction for recovery of arid areas was to supply additional water and more efficiently use already available supplies. This task was even undertaken as a concrete effort through combining technological means with social and economic methods. Their research institutes also had concrete purposes. For instance, when analyzing settlement and development of the Negev, Schechter and Galai (1980) pointed out that there were two scientific research institutes (they formed the nucleus for the establishment of the Ben-Gurion University at a later date), which were founded in the South in the nineteen fifties and attracted a large number of scientists and engineers. Especially, one of them was oriented to solve problems of the arid zone.

(3) *The Negev Program applied democratic and collaborative management and the federal mechanism of awards and sanctions.*

Democratic managements among the scholars themselves were found in Evenari's autobiography (Evenari 1989). Also, the involvement of the local people in planning of settlement showed the democracy of its management (Environmental Protection Service 1980).

The collaboration between the scholars and government, the scholars and farmers, and government and farmers was obvious in the Negev Program (Schechter and Galai 1980). Even collaboration among different scholars was also important. So Evenari argued:

> When we began our work, the meteorology, climatology, soil conditions and geology of the Negev were very poorly documented. We were fortunate in that several specialists in these fields also began their investigation at around the same time as we did. Their data helped a great deal. (Evenari 1989, p. 139)

Furthermore, laws were passed to manage the water use in Israel. In order to encouraged industry to move south, the Israeli Government even undertook special incentives "for establishment of industrial enterprises and for tourism" (Schechter and Galai 1980, p. 307).

(4) *The Negev program built stable local scholar-entrepreneurship.*

The aforementioned building of Ben-Gurion University and its function guaranteed the stable local scholar-entrepreneurship in the Negev area.

(5) *The Negev program created expectable benefits and reinforced benefits.*

For example, the use of advanced agricultural techniques and the development of technological innovations decreased the amount of water used per unit of production. This not only reduced production costs and raised the farmer's profit but also saved water for additional production. Furthermore, injecting fertilizers directly into the irrigation water resulted both in their effective utilization and reduced costs. As to water desalination, there were also several desalination plants operating in the Negev. Especially, the manufacture of desalination units for export became a very lucrative industry (Ibid).

(6) *The Negev Program applied an experiment-extension method.*

For example, based on the experiments, the Evenari team reconstructed two of ancient Nabatean farms and succeeded in obtaining excellent yields of grain, fruits, vegetables, pasture, and other crops. The same group of scientists developed a system of individual micrcocatchments, and then a large area was established for wider scale experimentation. Many other various experiments were also done to resolve the problems such as water desalination, rainfall enhancement, arid-zone irrigation with brackish water, and agricultural technology in the semiarid zone, and then the successful technologies and methods were extended later (Ibid).

(7) *The Negev Program had stable external support.*

The major supports were financial, technical, and institutional supports. When talked about management of cultivated semiarid ecosystems, Schechter and Galai pointed out:

> In Israel, prices are usually guaranteed and government-supported drought compensation removes part of the economic hazard in the wetter part of the semi-arid zone. The main management objective in the field is thus to produce economically the maximum amount of marketable products under conditions of climatic uncertainty. The agro-technical practices available to the farmer can be applied so as to improve the moisture regime, to increase soil fertility, to reduce competition from weeds and to control plant disease. (Ibid, p. 280)

A case study submitted by the Environmental Protection Service, Government of Israel (1980, p. 265) criticized that "excessive centralization of governmental control over development in the Negev has tended to deter investment from overseas governments." Also, there was other external support. For example, Schechter and Galai (1980) argued that the availability of external inputs (such as fuel, machinery, fertilizer, plant protection chemicals, and sometimes water) guaranteed a wide range of management options open to farmers in Israel's arid and semiarid zone.

Case 3: The Mona Reclamation Experimental Project, Pakistan (Semi-successful)

Pakistan has a long history of irrigation. The Mona Reclamation Experimental Project (MREP) was located in the North central part of the Chai Doab in Punjab, and its total area was about 44,500 ha, and the cultivated area was about 41,300 ha (Government of Pakistan 1980). Before the rehabilitation project, in the irrigation period (1901–65), although crop yields increased, they were still lower than those of many other countries. Because of the extension of irrigation, the existing flora prospered. Salinity, however, was increased. The case study supplied by the government of Pakistan argued:

> With the passage of time, unfortunately, the ill-effects of irrigation decreased the productivity of the land, and consequently the income of the people decline. According to the 1965 survey, most of the people in the study area were earning less than 5000 Rs. per family per year, which barely provided a satisfactory livelihood. (Government of Pakistan 1980, p. 238)

The rehabilitation project was implemented from 1965 to 1976. Although there were still many problems, its achievement was also evident. The case study by the Government of Pakistan pointed out:

> Land utilization has improved, cropping intensity has gone up, the adoption of improved innovations has been hastened and net returns to the farmers have increased, thus making a substantial contribution to the welfare of the area and making it a demonstration model for other developing rural areas of the county. (Ibid, p. 246)

Furthermore, the very first three-strainer well which was installed at Phullarwan Research Farm of Mona Reclamation Experimental Project (MREP) in 1974 was still "in operation with minor modifications in configuration and is used to supplement the irrigation requirement of the farm" presently (Saeed et al. 2003, p. 168). There were also many scholars who took part in this project, and thus this case can be deemed as a semi-successful case of scholar-based governance. Let's analyze how the different principles had been satisfied between 1965 and 2007.

(1) *Many scholars took part in this project but there were no many local stable field scholars.*

Many scholars took part in this project. For example, as the case study by the Government of Pakistan argued:

> Before initiating the programme of research, an intensive socio-economic survey of the area, in co-operation with the Department of Agricultural Economics and Rural Sociology of the University of Agriculture, Lyallpur, was carried out in 1965. An evaluation of the hydrological factors and of ground-water was also carried out and the installation of observation pipes, water-stage recorders and rain gauges was completed. Since 1973, research studies in the field of water management have been carried out in co-operation with advisers from Colorado State University, U.S.A.
>
> The research work in the project started during 1966–67 with the development of laboratory facilities for soil, water and plant analyses… (Government of Pakistan 1980, p. 239)

Furthermore, the research program of the Mona Reclamation Experimental Project mainly focused on applied research and investigation. This also made these scholars pay more attention to the field and to have more connections with local places. Much evidence about the activities of local stable field scholars, however, was not found. This is also proved by the project's unsuccessful agricultural extension which will be reanalyzed later when I talk about the experiment-extension method.

(1.1–1.6) *There was more knowledge about the expertise, but the knowledge about the local conditions, traditions, social norms, etc., was not enough.*

Many aforementioned studies provided a lot of knowledge about the necessary expertise and applied research. Advisory services on the choice of crops for particular soils were good examples. However, because of lacking local scholars, the knowledge about the local conditions, traditions, social norms, etc., was not enough, and the output of research was not successfully transmitted to the farmers. Furthermore, although the level of people's education in the area was continually improved during both the irrigation period (1901–65) and the period of the rehabilitation project (1965–76), it never improved at a rate sufficient to meet the requirement of development. The literacy figures were still low. Thus, *Principle 1 and 1.3 were only partly and weakly satisfied.*

The scholars had a lot of enthusiasm, spirit of practice and might also have had relatively independent social identity and high social status. But, because of lacking local stable scholars, their ability of participation was limited, and their social capital was also low. The local knowledge and local people were reasonably respected, but their responsibilities were not fully carried. Brian Spooner argued:

> Traditionally, the farmers on each watercourse worked out their own system of distribution (known as koccha warabandl). These traditional systems have the advantage of flexibility in that it is easy for individuals to enter into arrangements to swap shares in order to meet specific crop requirements outside the cycle. However, although their ability to make such arrangements is not restricted by the system, it may be restricted by their personal relations. In fact, their personal relations, deriving from the range of roles and statuses in the general social structure, may lead to a degree of injustice in the operation of the system that renders it intolerable for the poorer and less influential parties to a particular watercourse. (Spooner 1984, p. 37)

Thus, *Principle 1.2 was weakly satisfied, Principles 1.1 and 1.5 might be satisfied, but Principle 1.4 and 1.6 were not satisfied.*

(2) *It had a federal organizational structure and concrete organizational purposes.*

A federal-style method was applied in its organizational structure. For detailed analysis, please see the analysis of administrative structure provided by the Government of Pakistan (1980, p. 250). The main purpose of the Mona Reclamation Experimental Project was "to collect operation research information adaptable to areas being developed under Salinity Control and Reclamation Projects" (Government of Pakistan 1980, p. 217) or "to derive such information from operational research as to make it possible to identify the problems of SCARP (Salinity Control and Reclamation Projects, added by the author) areas and develop the most adaptable techniques to solve these problems" (Ashraf 1981, p. 68). Ashraf also carefully listed the specific objectives of the Water Management Research Programme at Mona (Ibid, p. 69). In the period of the rehabilitation project, its purposes were even reported by the Government of Pakistan (1980) according to its principal areas (for detailed analysis, please see pp. 230–240).

(3) *Democratic and collaborative management was partly applied but the analysis about the federal mechanism of awards and sanctions was not found.*

Much evidence about collaborative management was found. For example, the above citation talking about scholars' participation in this project clearly shows collaboration among different scholars and research institutes even with some universities in America. Spooner also studied the cooperative behaviors among herders. He said:

> How can such a situation come about? In this case, the answer is relatively clear, and provides an excellent illustration of the more general problem of technology in the larger sense - the organization of sets of tasks over time. The present population is of three different origins. The composition of the original population of the area before the development of the irrigation system appears to have been the main determining factor in the present structure. These people were semi-nomadic herders with subsidiary interests in dry farming and, perhaps, a little irrigation. From what is presently known of traditional modes of pastoralism in the area, it is immediately evident that the biradari structure correlates usefully with the requirements for cooperation, movement, and exploitation in this type of traditional pastoral system, because it provides for flexibility and cohesion on the scale of the camp group which must change frequently. As the development of irrigation led to increased carrying capacity and the labour requirements, settlers were brought into

> supplement the population, mostly from similar backgrounds. A third component of the present population consists of refugees that have come into the area from India since Partition. These come from more diverse backgrounds, but appear not to have modified the structure of the society significantly. This interpretation not only explains the present problems in cooperative arrangements- it draws attention to the fact that sociocultural systems do not necessarily adapt to changing ecological or technological conditions as quickly or predictably as they are often assumed to. (Spooner 1984, pp. 37–38)

The collaboration among different government agencies was reported by the analysis of administrative structure by the Government of Pakistan (1980, p. 250). Furthermore, this project established a great deal of credibility with the farming community and most research studies were conducted on the farmers' fields with their cooperation (Ashraf 1981). Although direct analysis about democratic analysis was not found, these analyses about collaborative management show some characteristics of democratic management. The analysis about the federal mechanism of awards and sanctions, however, was not found.

(4) *No strong stable local scholar-entrepreneurship was found.*

Although evident scholars' participation in this project was obvious, not much local scholars' participation was found (Government of Pakistan 1980; Spooner 1984). Thus, it can be concluded that even if the scholars' participation was relatively stable, strong stable local scholar-entrepreneurship was not built.

(5) *The principle of expectable and reinforced benefits was satisfied.*

For example, the case study by the Government of Pakistan argued:

> With the installation of tube-wells, the problems of waterlogging and soil salinization have been arrested. The additional water provided by the tube-wells encouraged crops which require more water. The area under wheat, rice, maize, sugar-cane and orchards has accordingly increased, while the area under cotton, bajra (millet), etc., decreased. The land-utilization pattern has improved. In 1965–66 the area sown more than once per year was 14 per cent of the total farm area, and this rose to about 30 per cent in 1970–71. The cropping intensities also increased from about 100 per cent to nearly 130 per cent. The rotation of crops has become pragmatic. Wheat no longer follows year after year, but it rotated with crops such as sugar-cane, cotton, rice, etc. The value of livestock increased from 10 million Rs. in 1965–66 to about 17 million Rs. in 1975–1976, and the values of crops have doubled in the same period. (Government of Pakistan 1980, p. 243)

Furthermore, both the net farm income and the per capita income had gone up. For detailed analysis, please see the case study by the Government of Pakistan (1980, pp. 245–246). Some scholars also claimed that conservation agriculture as a technological package contributed to environmental conservation along with enhancing production on a sustainable basis (Hobbs et al. 1997). Particularly, after 1996–1997, zero tillage as an innovation not only offered conservation of water and energy resources but also resulted in better crop yields in Punjab (IIMI 1999; OFWM 2000).

(6) *The experiment-extension method was applied but not very successful.*

Transfer of knowledge to farmers and encouragement of its adaptation was proposed at the time of planning of the rehabilitation project. Before embarking on large-scale project development, some pilot projects were set up. Agricultural extension was deemed as the bridge and the direct point of contact between provincial departments of agriculture, research organizations, and farmers. Many effective techniques such as "seventh-day schools," "periodic farmers' gatherings," "model farms and field demonstrations," "libraries," "agricultural competition," "farm-guide clubs," and "seed-bank schemes" were adopted. The available evidence, however, indicated that this contact generally has been of limited effectiveness (Government of Pakistan 1980). The Government of Pakistan also pointed out:

> This was in part due to general recruitment problems and administrative arrangements. Added to these was the problem of transport, which prevented extension workers from carrying their responsibilities in the whole of the areas assigned to them. Many farmers who were keen to accept new ideas and advice could seldom get an opportunity of meeting extension workers. (Ibid, p. 251)

(7) *It got stable and enough external support.*

First, the central government supported this project and planned and financed the major development project. Thus, it obtained adequate financial and institutional support. Spooner (1984, p. 39) even argued, "In an integrated system as large as that of Indus River Basin, investment can only be organized by government." Second, many external scholars' participation guaranteed enough external technical support for this project (Government of Pakistan 1980).

Case 4: The "LUCDEME" Project, South-East Spain (Semi-successful)

The LUCDEME Project (Lucha contra la desertificatión en el Mediterráneo = Desertification control in the Mediterranean) was established in 1981 by the National Institute for the Conservation of Nature (Ministry of Agriculture of Spain) in the southeastern area of Spain (30,000 km^2). This area covered the most eroded land and desertificated areas of this country (Perez-Soba and Barrientos 1986), and desertification still continues. The scope of the project includes the greater part of the Province of Murcia, the whole of Almeria, and the southern part of the Province of Granada. It administrative area is about 32, 622 km^2, and the ecological area is about 22,597 km^2. One aspect of its background is: people realized that desertification "too often through lack of knowledge, the lack valid alternatives, or through attempts to extract yields incompatible with the conservation of resources, human degrade the natural environment where they settle" (Ibid, p. 138). Until now, it has gotten some achievements. Oñate et al. (2001) pointed out that the economic and environmental evaluations of the results of the project were generally positive. Through combined use of GIS and environmental indicators, however, Paz et al. (2006) found that "around 29% of the area corresponding to the Valencian Community is affected by high to very high physical and soil degradation, 36% by high to very high biological degradation, and 6% by high to very high chemical degradation" (p. 151). Thus, I deem it as a semi-successful case of scholar-based governance in combating desertification from 1981 to 2007.

The stable involvement of a lot of scholars and their significant studies and investigations as discussed by Perez-Soba and Barrientos (1986, p. 144) showed that scholars played an important role in combating desertification in this project and relatively enough knowledge had been developed. In fact, the LUCDEME project was initiated in order to establish a program to study and monitor erosion in Spain (Carrera et al. 1981). The attention to local scholars and local knowledge, however, was not enough. Thus, *Principles 1 and 1.3 are only weakly satisfied.* Then *Principles 1.1, 1.2, and 1.4 to 1.6 are also only weakly satisfied.*

The involvement of different organizations in the project showed its federal-style organizational structure combining global, continental, regional, national, and even local organizations. However, there was not enough attention paid to grassroots organizations. This project was framed within the contribution to the plan of Nairobi, which was sponsored by the United Nations. Its purposes were relatively concrete. As Perez-Soba and Barrientos (1986) pointed out that it was to achieve "control of desertification in the Mediterranean area" (p. 139). Because desertification is normally a slow process, the final solution should also be long term. The possibilities of short-term action, however, were also considered and based on restoration and improvement of land use. Perez-Soba and Barrientos (1986, pp. 141–142) carefully analyzed the project's concrete and stratified objectives including its aim, medium-term objective, and immediate objectives. For example, its medium-term objectives included analysis of various resources and factors involved in the processes of desertification, determination of systems and techniques which can be applied to combat desertification, education, training and extension on the project's content, and anti-desertification combat activities. Thus, *Principle 2 was relatively satisfied.*

Participating organizations also showed its democratic and collaborative management. Although this project was supervised and dominated by the National Institute for Nature Conservation (ICONA) (Thornes 1995), it collaborated with the University and the Research Centers (Carrera et al. 1981). Various organizations such as Ministry of Agriculture, "universities, technical schools, institutes of scientific investigation together with government technical bodies concerned with agriculture, hydrology geology" (Perez-Soba and Barrientos 1986, p. 143), and meteorology were involved in this project. The universities and research centers which participated in this project included "University of Murcia, University of Granada, University of Sevilla, University of Almería, University of Valencia, the Zaidín Experimental Station (Granada) and the Desertification Research Centre-CIDE (these last two centers belonging to the Spanish Research Council)" (Ibáñez et al. 2003, p. 352). It also had international collaboration. The Spanish government was interested in promoting cooperation among the Mediterranean nations on desertification (Perez-Soba and Barrientos 1986). In 1994, Almeria even hosted an International Conference on Social Aspects of the Desertification Convention (Thornes 1995). Its involvement of local people and local scholars at the rice-roots levels, however, was not enough. Thus, *Principle 3 was only weakly satisfied.*

The stable involvement of scholars and research institute guaranteed relatively stable scholar-entrepreneurship, but its neglects on local scholars and grassroots research institutions explained its relative lack of local stable entrepreneurship. Thus, *Principle 4 was only weakly satisfied.*

The combination among ecological, economic, demographic, and social aspects rather than to simply pursue an increase in income levels showed its effort of balancing environmental protection and economic development. Its lack of attention to grassroots organizations, however, reduced its effectiveness. Thus, *Principle 5 was only weakly satisfied.*

One of its medium objectives—education, training, and extension on the project's content—and one of its immediate objectives—extension programs—showed its application of the experiment-extension method. Its less attention on the grassroots organizations, however, reduced its effectiveness. Thus, *Principle 6 was only weakly satisfied.*

The supervision by ICONA guaranteed its external financial and institutional support, while the collaboration with scholars and research institutes from other countries and the significant involvement of scholars in Spain guaranteed its external technological support to the activities of combating desertification in these three provinces. Thus, *Principle 7 was satisfied.*

Case 5: Integrated Desert Development and Desertification Control in Turkmenistan (Semi-successful)

Turkmenistan was a constituent republic of the Soviet Union until 1991 and called the Turkmen Soviet Socialist Republic. After 1991, it becomes an independent country. The sandy desert of karakum and a number of other desert territories in Turkmenistan cover "90.8%, mountains 5.1%, and oases approximately 4.1% of the Republic's territory" (Babaev et al. 1980, p. 481). In addition to the natural factors, anthropogenic factors directly aggravate its processes of land degradation (Durikov and Annaklycheva 2005). This case study focuses on its efforts to combat desertification both in the Soviet period and the independent period 1924–2007. During the Soviet period, Kharin argued:

> Scientific investigations on ecological problems were conducted in the Academies of Science of these republics, in Department of Forestry and Agriculture, in universities and in state reserves. It was a harmonious (from the bureaucratic point of view!) system which stayed under the control of Moscow. A special Committee on Coordination of Scientific Works was functioning in the USSR Academy of Sciences. Training technical specialists was organized through many universities and institutions. Many leading Russian scientists participated in these activities.
>
> But this high potential was not used in full measure because of ideological considerations. Leaders of the USSR couldn't tell the truth about the failure in realization of great project on nature transformation. Severe censorship and corruption ingratiated the situation. Ecological concerns, in the minds of ordinary people, were absent. (Kharin 2002, p. 157)

That is, although a lot of scholars participated in combating desertification in the Soviet period, their functions were limited. Especially, there were no many local scholars and a lot of local knowledge. After 1991, however, Kharin argued:

> The hand of Moscow" was not functioning any more. The system created over 70 years was fully destroyed. The activity of the Academies of Sciences of these republics was paralyzed because of the lack of money, low salary of scientists and corruption. Russian scientists left the territory of Central Asia en mass. Attempt to continue training of scientific cadres through local VAKs (High Certifying Commissions) failed. As a result the countries of Central Asia made a step back in their technical and cultural development. (Kharin 2002, p. 157)

Cadres were trained in local language, and local specialists were educated. Furthermore, self-governmental institutions and citizens' participation made local people more knowledgeable and trained some local scholars. Then, more local knowledge, local traditions, local social webs were considered. Thus, when both of the Soviet period and the Independent period are considered, *Principles 1 and 1.3 were weakly satisfied*. According to the aforementioned analyses and other studies (e.g., Babaev et al. 1980; Kharin 2002), it can be concluded that *Principles 1.1, 1.2, and 1.4–1.6 were weakly satisfied*. For example, the scientific tile showed that scholars had relatively high social status and were respected. Also, the participation of citizens, NGOs, and religious organizations showed that other social actors were respected.

The Soviet system was a federal structure, but it was organized mainly by a top-down method rather than a bottom-up method. For example, Babaev et al. pointed out:

> To combat desertification in Turkmenistan recommendations in the form of long-term projections or specific annual programes of assimilating achievements of theory and practice are widely used. These pogrammes are backed by appropriate allocation both from the state budget and from the budgets of collative and state farms. (Babaev et al. 1980, p. 519)

After 1991, subregional institutions such as Subregional Council for Control of Implementation of Programs, Subregional Scientific Council, Subregional Monitoring Centers (SMC), and National Centers of Desertification Monitoring also showed this. Also, during the Soviet system, the relatively concrete organizational purposes were set up (see Babaev et al. 1980). Thus, *Principle 2 was weakly satisfied.*

During the Soviet period, the activities of combating desertification were often conducted collectively and using a top-down method. Thus, they were not very democratic and many conflicts with local people arose by widening the areas of reserves and national parks (Kharin 2002). After 1991, because of increasing participation of local people and decentralization, the degree of democracy was improved. The cooperation among different government departments and various scholar organizations showed limited collaborative management during the Soviet period. Particularly, citizens' participation and the involvement of NGOs, religious organizations, and other actors significantly increased its level of collaboration. Particularly, after joining the UN Convention to Combat Desertification (UNCCD) in 1996, Turkmenistan closely cooperated with a number of international agencies such as the Secretariat of the UNCCD, Global Mechanism of the UNCCD, German Agency for Technical Cooperation (GTZ), Asian Development Bank, TACIS, UNDP, USAID, and others (Durikov and Annaklycheva 2005). The international cooperation with Peru, the Republic of Mali, Iran, Israel, China, and Germany was also developed (Babaev et al. 1999). Its limited inter-agency collaboration, however, harmed effective implementation of the Convention to Combat Desertification (CCD) (Saigal 2003). Also, the federal mechanism of awards and sanctions was not systematically built. Thus, *Principle 3 was only weakly satisfied.*

From the analysis stated above, it can also be concluded that there was a little stale scholar-entrepreneurship, especially after 1991. Thus, *Principle 4 was weakly satisfied.*

In the Soviet period, the use of technology partly increased the economy of Turkmen Soviet Socilist Republic (Please see Babaev et al. 1980, pp. 501–518). Thus, Bavaev et al. suggested:

> Successful and rational development of deserts necessitates a balancing of exploitation of desert natural resources and their protection and restoration. This is an integrated task and it can only be settled by applying ecosystem monitoring and a range of other actions. (Ibid, p. 522)

Kharin, however, argued:

> In all countries of the region, including Turkmenistan, people are preoccupied with economic survival. The great number of refugees testifies to bad governance and instability. Legal and administrative reforms are conducted in all countries of Central Asia. Several good laws and regulations have been adopted, but the governments can't realize them because of economic disruption, corruption of the administrative bodies and the passive role of the population. According to the old communist traditions, many laws are purely declarative. Economic criteria don't play an important role, and a modern infrastructure is not created. (Kharin 2002, p. 153)

O'Hara (1997) even claimed that although the expansion of Turkmenistan's irrigation network, particularly since the late of the 1950s, had significantly increased its agricultural output, especially of cotton, more and more land was becoming saline in Turkmenistan. Furthermore, detailed information about this had only recently been readily available to western researchers, while the problems of waterlogging and salinization of its irrigation systems had been reported in the past. In summary, *Principle 5 was only weakly satisfied.*

The stage-by-stage construction of the karakum canal "made it possible to introduce the necessary adjustments, specifically concerning seepage losses, the determination of the canal's bed stable dimensions in the easily washed-out rock, the water losses control at oases, the construction of different types of drainage systems and an array of other questions" (Babaev et al. 1980, p. 500). Some experiments in the spirit of "nature remarking" were also conducted in Turkmenistan (Kharin 2002). All these showed that although the experiment-extension method was not evidently announced, it was also partly used. Thus, *Principle 6 was weakly satisfied.*

External scholars' participation and strong government during the Soviet period guaranteed the relative adequate technical, financial, and institutional support for combating desertification during the Soviet period. After 1991, the support from Moscow was suddenly stopped, but the support within Turkistan was not supplied as stated above. Saigal (2003) proposed two types of constraints why donor agencies stopped or scale-downed their assistance to Turkmenistan: "(i) the lack of absorptive capacity; and (ii) the Government's gradual approach to some crucial areas of policy reforms, which has led donors to defer aid commitments tied to creating an enabling policy environment for effective development outcomes" (p. x). Furthermore, he argued that the fairly extensive range of environmental legislation of Turkmenistan lacked clarity and enforceability (Saigal 2003). Thus, *Principle 7 was only weakly satisfied.*

Case 6: The Vale Rangeland Rehabilitation Program, Oregon, USA (Semi-successful)

The Vale District of the Bureau of Land Management (BLM) was located in southeastern Oregon. It occupied 2.6 million ha (6.5 million acres). The Vale Rangeland Rehabilitation Program was mainly implemented from 1962 to 1972. After the first evaluation of this program done by Heady and Bartolome in 1977, the Vale BLM District was enlarged to include most of Baker County, Oregon (Heady 1988a). Heady argued:

> The analysis of the Vale Program in 1977 showed that rehabilitating 10% or less of an area can permit rapid improvement of the remaining 90% through proper management. Thus, the Vale Program dealt with improvement of the whole district, not just the areas plowed, seeded, and sprayed. This is an example of improving range and forage conditions by intensively treating part of the land and managing all of it. (Heady 1988b, p. 27)

Thus, this program is often deemed as a relatively successful case of combating desertification and still continues (BLM 2008). This program was organized by the BLM. Because the BLM was a local independent and professional organization (Huntsinger and Heady 1988), this program also satisfied some principles of successful scholar-based governance. Now let's analyze how the seven principles were satisfied from 1962 to 2007.

(1) *Local stable field scholars played a relatively significant role in the program.*

Officially the BLM was a governmental agency in Malheur County; but it was also a local independent and professional organization (Huntsinger and Heady 1988). Its members were professional range managers (Heady 1988b). The members of the BLM and its former organization local advisory boards did their work like scholars. For example, they defined grazing capacity and made management decisions based on their professional judgment about grazing capacity, seasonal grazing, and proper use (Bartolome and Heady 1988). Other local scholars also took part in this program and played an important role. For instance, Heady (1988a) pointed out that scholars of range research at the Squaw Butte Experiment Station near Burns, Oregon, and at other places in the sagebrush/grass type also contributed to the Vale Program. Furthermore, experts contributed to the proposal and the individual projects from the beginning in the subjects such as wildlife, recreational facilities, watershed values, and included mention of riparian conditions (Heady 1988b).

(1.1) *Local scholars had a relatively independent social identity and a high social status.*

The BLM was a local independent and professional organization (Huntsinger and Heady 1988). The members of the BLM were both local scholars and manager, who generally had relatively high social status.

(1.2) *Local scholars had high enthusiasm and capacity to participate in program.*

Employees in the BLM and other scholars enthusiastically took part in the program. Heady pointed out that one of the important factors which were in place to favor initiation of a range rehabilitation program was that "local people, politicians, and the BLM were anxious to put a land management program in place that would end disagreements over adjudication of grazing permits and management of wasting resources" (1988, p. 17). The BLM directly managed the program. The program educated both the BLM members and community members, and this was applauded by many social actors such as permittees, BLM, and local community members (Ibid). Huntsinger and Heady wrote:

> A permittee described the educational value of the program as follows: "Most Vale ranchers are resource-oriented, they would not tear up an area. I think the Vale program contributed to that education, to good rangeland management, and that is still an active educational process. For example, the Grazing Advisory Board suggested allocating excess forage on a temporary basis and monitoring to make sure allocations are not just based on production in a good year. The board is resource-conservative, and this is the only place I've seen this." (Huntsinger and Heady 1988, p. 132)

(1.3) *Local scholars had enough knowledge.*

This could be easily understood according to the above analysis. Especially, they could supply some knowledge, data and information which could be applicable to large managerial units, should be accurate and pertinent, and should be usable by non-research personnel (Government of the United States of America 1980).

(1.4) *The BLM members built high social capital.*

Huntsinger and Heady argued:

> The Vale Program changed the relationship of the BLM with the ranching community from one of suspicion to one with foundation in trust. Regardless of whether or not ranchers and BLM employees agree or disagree on particular questions, there is now a mutual belief that compromise is attainable. In the words of one permittee, "Today the BLM people are different. BLM and rancher relations years ago were completely adversarial; prior to the program they wouldn't sit at a table together. Now people have come together, both sides have been educated." A senior BLM employee commented, "We gain ranchers' trust by association. Sometimes they do hammer on us in public, and outsiders see this, but at other times we share coffee and cookies. (Huntsinger and Heady 1988, p. 131)

This paragraph tells us that the BLM members built high social capital.

(1.5) *The BLM members had high social responsibility and spirit of practice.*

As the direct organizers and managers, the BLM employees had high social responsibility and spirit of practice (Huntsinger and Heady 1988).

(1.6) *The BLM members respected other social actors.*

This can be known from the above analyses.

(2) *The Vale Program had a federal organizational structure and concrete organizational purposes.*

Heady said:

> Division of the District into three resource or management areas with separate manages spread the workload and responsibilities. The resource areas were further divided into 14 planning units. Because the area managers exercised considerable autonomy in proposing and developing improvement plans, projects, and grazing systems, the mixture of management practices and land treatments differed among the three resource areas. After 1977 the program was reorganized into two resource areas.
>
> Although area managers initiated planning and site selection for improvement projects, the final authority for the coordinated program rested with the district manager. Prior to 1983, district personnel included about 40 permanent employees—more after 1983—and as many as 75 temporary employees during the field and fire season. (Heady 1988b, p. 25)

Heady also listed eight objectives of the Vale Program:

1. To correct erosion and accompanying downstream sedimentation—and prevent further soil losses.
2. To increase the forage supply for wildlife and livestock.
3. To stabilize the livestock industry at the present or an increased level of production.
4. To facilitate fire control by replacing high hazard cheatgrass and sagebrush with low hazard perennial grasses and improving detection and suppression facilities.
5. To prevent the encroachment and spread of noxious and poisonous weeds.
6. To accomplish necessary land tenure adjustments.
7. To safeguard public lands from improper recreational use.
8. To provide for the development of access roads and service roads in the vast areas of untapped recreational potential. (Heady 1988b, p. 21)

These tell us that the Vale Program had a federal organizational structure and concrete organizational purposes.

(3) *The Vale Program applied democratic and collaborative management and had a federal mechanism of awards and sanctions.*

The Vale Program paid more attention to favorable public opinion and the need for effective publicity (Government of the United States of America 1980). Also, considerable autonomy exercised by the area managers indicted its democratic style (Heady 1988b). Huntsinger and Heady pointed out:

> The BLM simply cannot please every constituency with every decision. A proportion of every group agreed with BLM on every issue, and BLM was relied upon as a good source of information by a majority of respondents in each group, but it is clear that BLM has not been co-opted by any one group. (Huntsinger and Heady 1988, p. 133)

The cooperation among different resources areas and different managers and between the BLM and Congress, the community members, program area permittees, and environmentalist showed the collaborative management of the Vale Program (Ibid). Heady said:

> Beginning in 1964, a number of Congressional acts and court decisions directly affected land management on public lands administered by the BLM and other Federal agencies. These actions were in response to a heightened public awareness of and concern for the environment, natural resource conservation, and recreational opportunities on public lands. Their impacts on management activities in the program area are exerted through altering a static or declining budget… The following are some of the regulations altering resource management in the study area. (Heady 1988b, p. 33)

About the monitoring system, Huntsinger and Heady pointed out:

> Under the AMP's, monitoring of grazing use becomes the joint responsibility of the BLM and the permittees. The BLM monitors yearly compliance with season of use and stocking levels, but changes are based primarily on measured levels of forage utilization and changes in range condition on permanent plots (BLM 1984). The permittee of the 1980s is given much leeway in day-to-day compliance with grazing plans. The grazing systems designate periods of permitted use, but within these dates, the actual turnout and moving of animals into new pastures is the responsibility of the permittee. Now that good information is available from many years of monitoring range utilization and range condition, the BLM has been able to concentrate on these indicators for needed changes in animal management. (Huntsinger and Heady 1988, p. 63)

That is, a federal mechanism of awards and sections was built.

(4) *The BLM members had relatively stable local scholar-entrepreneurship.*

As both local managers and local scholars, the BLM members showed their strong scholar-entrepreneurship (for detailed analysis, please see Huntsinger and Heady 1988; Heady 1988a, b, c).

(5) *The Vale Program's benefits were expectable and were reinforced.*

Allen pointed out:

> The Vale Program produced many benefits in addition to rangeland rehabilitation. Water developments, for example, improved habitat for many species of wildlife and provided new recreational opportunities. …Although *some* of these changes, particularly in wildlife populations, cannot be directly related to the Vale Program, improvements in range condition and water distribution have played an important role. (Allen 1988, p. 65)

Bartlett, Bartolome, and Quigley also argued:

> Economic evaluation of the Vale Program involves both direct and indirect benefits and costs. Direct benefits include increased grazing of livestock, increases in some wildlife populations, decreased erosion, increased range condition and plant diversity, and greater recreational opportunities. Direct costs include treatment, construction, maintenance, and administration costs.
>
> A major investment program like the Vale Program has direct and indirect economic impacts on the local economy. By avoiding grazing reductions through range rehabilitation, the livestock sector avoided economic losses which in turn helped other sectors of the local economy avoid losses. The large public expenditure for the program increased the level of activity in the local economy. Finally, increases or decreases in other uses of the program area had impacts on the local economy. (Bartlett et al. 1988, p. 86)

These benefits were not only perceptible and large but also were reinforced. In some cases, even if it was unforeseen at first glance, it could be seen after the program was implemented. Furthermore, Heady pointed out that some unforeseen benefits might occur. For example, patches of brush and irregular borders left by plowing (because the machinery could not operate on steep slopes and rocky areas) created a mosaic of vegetation, which provided desirable wildlife cover and edge habitats (Heady 1988b).

(6) *The Vale Program used the experiment-extension method.*

The Vale Program was first implemented in 2.6-million-ha area. Since the first evaluation in 1977, untreated areas had been included in the program, and then it included almost most of Baker County, Oregon (Government of the United States of America 1980; Heady 1988b). This means that the experiment-extension method was used. Also, Heady (1988b) argued that methods developed during the 1950s at the Squaw Butte Experiment Station near Burns, Oregon, and limited experience in range seeding on the Vale ELM District before the start of the program heavily supported seeding practices of this program.

(7) *The Vale Program received the stable external financial and knowledge support.*

First, this program obtained funding from Congress (Heady 1988b). Its support was both financial and institutional. Second, the range research at the Squaw Butte Experiment Station and at other places guaranteed its external knowledge support (Heady 1988a). So Heady (1988a) claimed that one of the important factors which were in place to favor initiation of a range rehabilitation program was that "information on what to do and how to do it was available" (p. 17).

Case 7: The Dust Bowl Case in United State (Semi-successful)

Although the boundaries of the Dust Bowl are never precise, it is usually applied to cover a 97 million acre section of the southern half of the Great Plains having an average annual precipitation of less than 20 inches. In other words, that portion of the area lies with southeastern Colorado and northeastern New Mexico, western Kansas, and the panhandles of Texas and Oklahoma (Hurt 1981, p. 2; Stephens 1937, p. 750). During the 1930s (1931–1939), called the Dust Bowl era (Baumhardt

2003), the people of the Dust Bowl experienced the most awesome "black blizzard," and the USA experienced one of the most devastating droughts of the past century (Schubert et al. 2004). The reasons of dust storms were complex, but the major ones were the composition of the soil, the severity of the climate, and the settlement of men. The major producers of dust storms in the Dust Bowl could be divided into two parts: the nature and human beings (Schubert et al. 2004). Here I focus on three groups of human beings—farmers, scholars, and governments. Farmers' improper cultivating and grazing behaviors were the major cause of dust storms in the Dust Bowl. The lack of scholars' study was also the significant cause of dust storms. The policy failure was the third one.

Let's see how the Dust Bowl problem came? During the 1880s, a large-scale agricultural settlement of the Great Plains began, especially when the Plains tribes removed to the reservations. By 1910, almost all of land of southern Plains was claimed mainly with wheat farming. As wheat acreage was increased and farm sizes were enlarged, the pastures declined, and cattle overgrazed the remaining grass. At that time, although wind erosion had happened, it was almost neglected. The settlers needed money through planting wheat, and most of them really could not afford to practice soil conservation, and some even did not think it was necessary. Some people might want to do but they did not know-how, although there were some farmers as the minority who used plows to roughen the soil to slow the movement of the wind and planted drought-resistant grain sorghums. Even the use of new technology, like the large machines such as headers, steam engines, threshing machines, the sulky or riding plow, the lister plow, endgate seeders, grain drills, rotary-drop corn planters, check-row corn planters, and tractors stimulated the expansion of reclamation and wind erosion and desertification, along with the high yields, stable prices in market, and even adequate precipitation. Furthermore, the suitcase farmers made the situation worse, because most of them in the region gave little attention to maintain or increase soil humus. The mechanical difficulties, the lack of labor and equipment, and the increased costs often made them not to steadily work into the ground. When all the claimed land was grown, the wind erosion was not very serious. Once the land was abandoned because of the drought and other reasons, the wind erosion was exacerbated and dust storms were formed when other conditions were satisfied, like the incessant wind. This was the fact from 1932 to 1940 in the Dust Bowl. Under such a situation, when the farmers found that they had no confidence to combat desertification and dust storms, they would turn to religious support. For example, during the 1930s, some ones who stayed in the Dust Bowl began to fight dust storms religiously and psychologically as well as physically (Hurt 1981, p. 55). Even the service of the church, however, could not be afforded. For example, one Texan demanded wrote, "We are rather proud of our sandstroms," "because they tell us the West Texas is on the move. There dust storms are intrusive aliens from Kansas and Nebraska" (Hurt 1981, p. 58).

Because of the severity of the dust storms, the problems in the Dust Bowl were gradually noticed by more and more people, especially some scholars. These scholars included some journalists or reporters of media who propagated the problems to the public and government, some conservationists, some researchers who supplied the scientific support and necessary technology, and so on. For example, because of the severity of the twin problems of drought and social blowing, some conservationists, soil scientists, and southern Plains residents began to classify dust storms (Hurt 1981, p. 31). The first group of these scholars was those peoples who grew from the natives or lived in the Dust Bowlsuch as John L. McCarty, the editor of the *Muleshoe* (Texas) Journal (Hurt 1981, pp. 59–60). Then other scholars followed. They were some photographers (such as Arthur Rothstein and Dorothea Lange), some producers of films (such as Rexford Tugwell and Pare Lorentz), some of their critics (such as T. E. Johnson and Eugene Worley), some painters (such as Alexander Hogue), some newspapers, and so on. Their work showed the problems of dust storms to the public, scholars, and government. Based on these developments, the farmers themselves began to rethink the problems, deeply understood the problems of dust storms and their behaviors and found the new method to resolve the encountered problems. For example, by tying a logging chain to a tree, some Dust Bowlers supposedly learned to test wind velocity (Hurt 1981, p. 57). Also, many experts, including some natural scientists, like the researchers in the Fort Hays experiment station (Hurt 1981, p. 69), were involved in combating dust storms. Finally, the governments including the state level and the federal level participated in combating dust storms. Then the effort of combating dust storms and wind erosion began.

Roughly, I deemed the case of combating the dust storms during the 1930s as a semi-successful case of scholar-based governance by evaluating its performance from 1932 to 2007. Hansen and Libecap (2004) pointed out that there was not comparable level of wind erosion in the 1930s although there were similar droughts in the 1950s and the 1970s, because of soil conservation districts, established after 1937. Borchert (1971) also argued that since the 1930s contemporary basic changes in American had been accelerated: fewer, bigger, and more fragmented farmers, public controls and subsides, consolidation of urban business and services, and greater management. Dust storms, however, were returned in the Dust Bowl in the 1990s. Until 2007, although the wind erosion and dust storms in the Dust Bowl had been ameliorated and the frequency of dust storms has been reduced, dust storms still often happen in the Dust Bowl. Worster (2004) argued that the Conservation Reserve Program (CRP) established in 1985 had helped somewhat the agricultural interests in this region by emphasizing voluntarism and self-reliance in crop production. The Dust Bowl, however, was still losing population and had not gained a renewed sense of the prosperity existing before the drought, and mixed economy conservation might not be enough to restore the prairies and plains. Francis (1998) also pointed out that the fundamental challenge in the Dust Bowl remained the same. If a "Dust Bowl" type event reoccurred, it would be more devastating to today's agricultural economy. Let's analyze how the seven principles were satisfied from 1932 to 2007.

(1) *Many scholars involved in the activities of combating dust storms in the Dust Bowl were field-based scholars.*

As shown above, they had an independent social identity, a high social status, high enthusiasm and a capacity of participation, practical knowledge, high social capital with other social actors, and high social responsibility and spirit of practice. They also respected other social actors and their local knowledge and ideas.

(2) *A bottom-up federal structure of organizations was applied, and concrete purposes were set up.*

For example, in order to facilitate the development of a unified conservation program, the United States Department of Agriculture (USDA) consolidated all braches working on the wind erosion problems. The shelterbelt project showed not only the characteristics of their bottom-up structure of the activities of combating dust storms but also their concrete purposes. To better organize the upcoming planting season, the states were divided into different districts by the Forest Service to focus on more concrete affairs, respectively. Although the Congress did not support the shelterbelt project at the beginning, it changed ideas when the grass-roots support for the project began increasing. Especially, People in Kansas showed the greatest enthusiasm among the three Dust Bowl States (Hurt 1981, p. 132).

(3) *It applied the democratic and collaborative management and almost consensus method at the grassroots level and used different selective incentives.*

For example, because wind erosion was so serious, most farmers and ranchers began to be willing to adopt the appropriate measures to bring their soil under control by the spring of 1934 (Ibid, p. 68). The project organized by the Soil Erosion Service (SES) and the Civilian Conservation Corps (CCC) used contract method. Even to the shelterbelt project, the farmers were required to sign a contract with the Forest Service. In this contract, the farmers promised to donate the land, build and maintain fences around the plantings, cultivate the seedlings, and help control rodents. Then, the Forest Service in turn agreed to supply the stock, plant the seedlings, and help cultivate and build fences whenever funds were available (Ibid, p. 131). Their participation in the related programs was also encouraged. For instance, in a massive soil listing program in Kansas, the farmers received ten cents per acre for emergency listing of their blowing fields (in 1936, its was increased to twenty cents per acre) and forth cents per acre if they had to hire the work done. In Hamilton County, "farmers were allotted one gallon of fuel and one-sixteenth gallon of oil for each acre worked with tractors, and ten pounds each of grain and hay for every acre worked with horses" (Ibid, pp. 69–70). In thirty-nine western Kansas counties, the farmers were first given 60% of the money necessary to list their field in advance, and then they received the remaining 40% after the work was satisfactorily completed. In the shelterbelt project, the farmers, however, were prohibited from removing the trees through the arrangement that the Forest Service retained the right to inspect the trees at anytime (Ibid).

(4) *There were only relatively stable local scholar-entrepreneurs.*

Although in the Dust Bowl era farmers who had some scholars' characteristics began to learn how to resolve the problems of dust storms, they could not form an independent group to lead other farmers to resolve the problems. For some external scholars, once their external support and impetus from the government was stooped, their influence would decrease. Thus, although the activities of combating dust storms of the 1930s were relatively successful, this success did not continue. During the mid of the 1940s, a few dust storms occurred in the southern Plains. Most of them were local in nature, and only scatted fields were damaged. On the newly broken cotton lands in the sandy land south and west of Lubbock in Texas, however, dust storms were common. Some of them were severe, and the wind erosion scoured many fields. By late December 1948, the wheat land in western Kansas and eastern Colorado as far south as the Oklahoma Panhandle were blowing. Until mid-January of the 1950, soil blowing became worse again in the southern Plains (Hurt 1981). One reason of this phenomenon was that soil conservation was not continued very well by the Dust Bowl farmers, especially by some suitcase farmers, because of the lack of stable field scholars as entrepreneurs. Theoretically, the farmers, however, were the only people could soon recognize the seriousness of the new wind erosion menace and began to continue the soil conservation procedures learned in the 1930s (Ibid). Hansen and Libecap (2004) even argued that these small farmers liked to cultivate more of their land and were less likely to invest in erosion control than larger farmers.

(5) *In addition to environmental protection, farmers obtained more benefit, and this perception was reinforced as the program was developed.*

For example, in the case of using contour plowing, there was little contour farming outside the demonstration project by the spring of 1935. In late May rains, however, the tests indicated that one inch of additional moisture was conserved, and subsoil moisture was more than a fort deeper than on lands not contoured. One Texas farmer in north of Amarillo benefited from contouring by raising 160 acres of feed on contour plowed land. No one else as his neighbors, however, got this benefit without doing that. The reason why the farmers embraced the Soil Conservation Service (SCS) program was that they were geared to practicality and low cost. Also, because the farmers had the necessary implements, they could do most of the work themselves (Hurt 1981, p. 85). In the shelterbelt project, for example, by 1938, trees planted in 1935 were already eighteen to twenty feet tall and were already protecting crop land from wind erosion by the 1938 planting season.

(6) *An experiment-extension method was adopted.*

For example, to spread the use of basin lister, the Fort Hays Experiment Station in western Kansas first organized a demonstration of using the basin lister during the spring and summer of 1938. It was watched by three thousand farmers in thirty counties. The experiment station taught farmers how to convert their standard lister into the "damming" model (Ibid). Also, in order to stimulate the farmers to agree to follow the conservation practices, the SES, county agents, and the CCC established some demonstration projects in which they provided the necessary technical expertise and equipment to encourage farmers to participate in a conservation program. The SES first established demonstration project on private lands where cooperating farmers signed five-year contracts to do that. The federal government originally allotted $ 35,000 for the project, and the first demonstration unit was a 15,195-acre project located at east of Dalhart, Texas. When the amount was doubled by October 1934, the project was expanded to 30,000 acres. Then, similar projects were soon established in other states (such as Colorado, Kansas, Oklahoma, and New Mexico) and several other Texas Panhandle locations. The CCC also established fourteen demonstration projects in the Dust Bowl, and the farmers within twenty-five miles of the CCC camps could sign cooperative agreements for soil conservation work (Ibid).

(7) *Although there was relatively stable external support (including financial, technical, and institutional), especially from government, the support was not stable in a long run.*

For example, although many government officials and departments at different levels (such as Presidents, especially Franklin Delano Roosevelt, Congress, the Federal Emergency Relief Administration, the SES, the CCC, the SCS, the USDA, the different state governments, the County governments, the County commissioners, and agricultural agents) were all involved into the activities of combating dust storms in the 1930s, their participation did not continue very well after the 1930s (Francis 1998; Hansen and Libecap 2004; Worster 2004).

Case 8: The Gascoyne Basin-Australia (Unsuccessful)

The annual rainfall in the Gascoyne Basin is about 200 mm (Williams 1978). Considering erosion and desertification of this area, I deem it as an unsuccessful case of combating desertification from 1876 to 1980. Through carefully reanalyzing the case studies of the Gascoyne by various researchers, particularly by Williams et al. (1978), I found that all the seven Principles of successful scholar-based governance were not satisfied. Furthermore, the arguments stated by these scholars were often similar to the ideas of the seven principles of successful scholar-based governance. Especially, many of their arguments shown in their recommendations and lessons learned were consistent with the ideas of Principles 1–7. For example, William, Suijdendrop and Wilcox said:

> The two assessors appointed to this first inspection were experienced officers of the Department of Agricultural and the Department of lands and Surveys, long resident in region of arid Western Australia, who were conversant with the Gascoune Basin, who knew many of the pastoralists, but who were not involved directly in the original survey. (Williams et al. 1980, p. 60)

Here, they highlighted the importance of stable local scholars' participation, social capital, local knowledge, and collaboration, etc., although these were not realized in the Gascoyne Basin. Watson (2003) also pointed out that research, development, and training agencies could help individual managers do better management decisions. The analysis of lessees' participation and attitude by Williams et al. (1980) also showed the importance of democratic and collaborative management. Their arguments proposed that these governance characteristics, however, were not satisfied in the Gascoyne Basin. That is, *Principles 1, 2, 3, and 4 were not satisfied.*

The activities of combating desertification also could not combine environmental protection and economic development. Mabbutt argued:

> In the Gascoyne valley of Western Australia one-third of the graziers may have to leave their properties because of rising costs and falling wool prices combined with the inherently low carrying capacity of their pastures and their remoteness from markets. (Mabbutt 1977, p. 430)

That is, *Principle 5 was not satisfied.*

Also, their analysis about experiments and extension could be deemed as the analysis about the experiment-extension method. For instance, Williams, Suijdendrop, and Wilcox suggested:

> Research, extension and administration should be centered on a few critical aspects of rangeland management. These are regeneration of shrubs, livestock-control and education. Enthusiastic co-operation between pastoralists, their organization, financial institutions and range management personnel is necessary for success in this resource-limited area. (Williams et al. 1980, p. 71)

Here, they stressed the importance of enthusiasm, spirits of practice, collaboration, the experiment-extension method, etc. These recommendations, however, had not been applied at that time including the experiment-extension method. Thus, *Principle 6 also was not satisfied.*

Their argument also highlighted the importance of external support for combating desertification in the Gascoyne Basin. For example, Williams et al. (1980, p. 71) argued that it might "prove in the interests of the State to provide financial incentives and compensation in order that an agreed cooperative rangeland program be supported at station level." The financial, technical, and institutional support, however, was not enough according to their findings. Thus, *Principle 7 also was not satisfied.*

Case 9: The Region of Coquimbo, Chile (Unsuccessful)

The study area of this case is situated in the region of Coquimbo, 31° to 32° S. Although the importance of technology was realized, its desertification before 1980 was very serious (Government of Chile 1980). Thus, the governance of combating desertification before 1980 (mainly from 1900 to 1980 according to the case study by Government of Chile 1980) in Coquimbo can be deemed as an unsuccessful case of scholar-based governance. Now let's analyze how the seven principles of successful scholar-based governance were violated.

(1) *There were no stable field scholars who significantly took part in the activities of combating desertification.*

Technology was introduced in this region both in Prehispanic and Spanish Periods, while the environmental protection was neglected. Furthermore, the lack of knowledge and stable field scholars was the reason of the failure of combating desertification. For example, the study by the Government of Chile (1980) argued, "Technical assistance to the *comuneros* has in general been rare, ill-planned and sporadic," although "the mind of *comunero* is open to technological innovations" (p. 110). Because of the lack of local scholars, even if the inhabitants of rural areas realized that local people should have more knowledge and better education, they could not be helped by local scholars.

(2) *Its activities of combating desertification were not organized through a federal structure and did not have a concrete organizational purpose.*

There were two agrarian units: the large *hacienda* and the agricultural community. The *haciendas* were in the hands of single owner through inheritance and maintained. These owners often only cared about their own interests. As to the communities, the case study by the Government of Chile argued:

> The comuneros are generally individualistic and competitive. The community deals with the cleaning of drains and irrigation canals, moving livestock, eliminating toxic plants from grazing lands, building social centres, etc. Each farmer manages his agricultural enterprise in an individualistic way. This aspect explains the failure of some co-operative systems introduced since 1960. These rural co-operatives did not improve the organizational structures to the level required to meet the hard living conditions of the region; contrary to expectations, they generated conflicts by superimposing new organizations on existing traditional practices. (Government of Chile 1980, p. 110)

Also, uncertainty concerning boundaries of communities caused continual conflicts between neighboring communities. In summary, these managements and conflicts showed that there was no a federal organization for combating desertification and a concrete purpose of combating desertification.

(3) *Its governance did not apply democratic and collaborative management and the federal mechanism of awards and sanctions.*

This can be seen from the analysis about Principle 2. Especially, the hierarchical land-ownership and social systems hindered the democratic and collaborative management. For instance, the hacienda received the lands from the Crown, transmitted through inheritance, and maintained undivided as a singer owner; while the peon (the seasonal agricultural laborer), the tenant, and inquilino (the permanent laborer) were the three classes connected with the haciendas (see Borros Arana 1886; Government of Chile 1980; Pascal 1968).

(4) *There was no stable local scholar-entrepreneurship.*

Although the General Assembly *comuneros* could be deemed as the local leaders of communities, they were not scholar-style leaders (Government of Chile 1980). Thus, in these communities, there was no stable local scholar-entrepreneurship.

(5) *There were no predictable and reinforced benefits.*

For example, the Government of Chile argued:

> …developmental programmes of agriculture and livestock-breeding have been conceived to yield profit on a short-term basis, and no attention has been given to measures tending to preserve natural resources. The programmes have also been of short duration and have not lived up to the expectations of the rural population. (Government of Chile 1980, p. 52)

> *Comuneros* are the poorest people of the region. Their farms are small and barely profitable because they are located in the most degraded areas. (Ibid, p. 104)

(6) *Its governance did not apply the experimental-extension method.*

For example, although there were "several examples of programs conceived in the great hope that they wound permanently solve existing problems," because of the lack of using experimental-extension method, they soon failed (for detailed analyses of these examples, please see Ibid, p. 110).

(7) *There was no stable external support.*

As stated above, the lack of external knowledge and technological support was one of major reasons why its activities of combating desertification failed. Also, there was no external financial support. For example, although it was not difficult to introduce technological innovations in this region, say the use of small motor pumps to tap underground water, most farmers had "inadequate means to meet the purchase debt and then the fuel costs" (Ibid, p. 110). The institutional support was also not enough.

In summary, the Coquimbo case was an unsuccessful case of scholar-based governance before 1980. As stated above, all the seven principles of successful scholar-based governance were not satisfied, and this was reflected by the recommendations for further activities of combating desertification in this region by the Government of Chile (Government of Chile 1980, p. 112). Also, this was echoed by a case study of the Norte Chico, a larger area comprising the regions Atacama and Coquimbo (Schneider 1982).

Many achievements, however, had been obtained after 1980 through trying hard to satisfying these principles. For example, relying on the collaboration among the Corporacion de Fomento de la Producion (CORFO), the University of Chile, the Servicio Agricola y Ganadero (SAG), and the Instituto Nacional de Investigaciones Agropecuarias (INIA), more than 48,000 ha of Coquimbo had been planted with forage shrubs until 2004, particularly *Atriplex nummularia* (more than 90% of the area) which has good forage yields, strong resistance to extreme drought, resistance to grazing with fast resprouting, resistance to disease, and easy propagation (Mulas and Mulas 2004). In 2008, an award, granted the Universities of La Serena (Chile) and La Regina (Canada), was even given to a joint UNDP-EU program in the Alhuemilla Agricultural Community, for it good practices in the sustainable management of water in rural communities competition (Shneider et al. 2008).

Case 10: The Luni Development Block, Rajasthan, India (Unsuccessful)

Desertification is often deemed as an important contributing factor to the decline of the earliest civilization in India, and it still presents a challenge to India society (Mann 1982). This case study focuses on the activities of combating desertification before 1982, mainly from 1900 to 1982. The following account draws mainly on the work by the Government of India (1980), Dhir (1982), Mabbutt and Wilson (1980), Malhotra and Mann (1982), and Mann (1982).

The Luni Development Block was located in the Rajasthan Desert with the Jodhpur district. It encompassed 106 villages in a total area of 1989 km^2. Desertification in this block was very serious. Its reasons included (1)"degradation of natural vegetation," (2)"landform changes resulting from increased surface instability caused by cultivation severe overgrazing," (3)"soil fertility changes ensuing from wind and water erosion processes," (4)"surface-salinity increase owing to interference with natural drainage," (5)"change of groundwater conditions," and (6) "finally the reversal of desertification following the development of groundwater use" (Government of India 1980). Mabbutt and Wilson (1980) argued that secondary salinization resulting from the clearing of deep-rooted perennial vegetation for cropping was a widespread, although localized, problem in many dry-farming areas. Although there were no adequate analyses how desertification in this region was combated to help us study whether the seven principles of scholar-based governance were satisfied in this program, the recommendations by these studies can help us do this. The recommendations by the Government of India (1980, pp. 167–168) were consistent with the seven principles of successful scholar-based governance. Let's analyze them successively.

First, its recommendations about social education and transfer of technology to the field and intensification of research highlighted the importance of the participation of local field scholars (*Principle 1*) and stable local scholar-entrepreneurship (*Principle 4*). Malhotra and Mann (1982) also suggested that it was necessary to have "intensive microcosmic anthropological studies pertaining to traditional relations of production" in order to "provide the basis for the evolution of the traditional institutions and for envisaging plans for their change and their response to change" (p. 310).

Second, its recommendations about relief planning, especially on the central assistance and the states' complementary relief funds, showed the importance of federal organizations and clear purposes of the program (*Principle 2*). Dhir (1982) argued that careful planning was urgently required to minimize "further loss in the productivity of the renewable natural resources of India's dry lands" (p. 331).

Third, its recommendations about international collaboration and cooperation among different institutions highlighted collaborative management stressed by *Principle 3*. Also, its recommendation of population control emphasized the importance of democratic management (*Principle 3*). Jodha also suggested:

> The individual land user's obligations in terms of adherence to conservation practices should be specifically incorporated in the land laws. These laws can be enforced with liberal recourse to penalties in the case of default, and the granting of rent remission as reward. (Jodha 1982, p. 348)

This showed the importance of the mechanism of awards and sanctions (*Principle 3*).

Fourth, its recommendations on animal husbandry and nomadism, rational land use, harnessing solar and wind energy, and development of water resources suggested the importance of the combination between economic development and environmental protection (*Principle 5*).

Fifth, its recommendation about harnessing solar and wind energy emphasized the application of the experiment-extension method (*Principle 6*). For example, it said:

> The use of solar and wind energy has been amply demonstrated at CAZRI (the Central Arid Zone Research Institute—the author) in Jodhpur, and the successful manufacture of implements such as solar water heaters, solar cookers, solar driers, and sail-wing type windmills provides a lead in this direction. (Government of India 1980, p. 168).

For the detailed information about CAZRI, please see Mann (1982).

Sixth, its recommendations on the central assistance, the states' complementary relief funds, international collaboration, transfers of technology to the field, etc., showed the importance of external support of the program (*Principle 7*).

In summary, these recommendations told us that the activities of combating desertification in this region did not satisfy the seven principles of successful scholar-based governance.

Case 11: The Turan Program, Iran (Unsuccessful)

In 1975, the Department of the Environment (Tehran, Iran) embarked on a comprehensive ecological research program in Turan in order to provide better information for dealing with pastoral and rural development problems in the arid parts of Iran (Spooner 1984; Spooner and Mann 1982). The Turan Protected Area in Iran included 1.8 million hectares on northeastern margin of the central deserts. The reason why it was selected for protection as a Wildlife Refuge in 1976 and a Biosphere Reserve in 1977 was its potential richness of vegetation which was judged an excellent basis for the development of alternative methods of conservation. It, however, did not produce predictable results at predictable time (the Department of the Environment 1980). Let's analyze whether the management of this protected area satisfied the seven principles of successful scholar-based governance from 1975 to 1984.

(1) *There were no stable filed scholars who participated in this program.*

Because of the introduction of a new irritation technology—the qanat, many cities flourished in the period of economic and political expansion of the Achaemenian empire (sixth to fourth centuries BC). Until the next comparable technological innovation-mechanization, there was a long period of technological stability and consequent stability of land use patterns between these two processes arbitrarily from 750 BC to AD 1960 (Department of the Environment 1980). Mechanization along with the population growth, however, caused increased desertification. One of the reasons of this was that this process was an authoritarian and technocratic process. Government almost controlled everything. For example, the case prepared by the Department of the Environment, Tehran, Iran, for the United Nations Conference on Desertification argued:

> In 1962 a program of land reform was begun; in 1963 all forests and rangelands were nationalized, and in 1967 the nationalization programme was extended to the country water resources. These administrative and legislative measures-once again exogenous factors—had far-reaching effects on man-land relationships throughout those parts of Iran venerable to desertification. (Ibid, p. 185)

Obviously, there were no many local stable field scholars who participated in this program. Furthermore, the use of technology caused mechanization which allowed the use of groundwater for irrigation by means of cheap motorized pumps. The knowledge of environmental protection, however, was neglected. Thus, mechanization led to increased desertification. This area was not carefully studied before the policy was made and implemented (Ibid). Furthermore, information on the history of Turan and its traditional knowledge was limited (Dennell 1982; Department of the Environment 1980; Nyerges 1982). Thus, it can be concluded that there was no adequate knowledge. Furthermore, if there were no enough stable field scholars who took part in the program, and then it is not necessary to talk about Principles 1.1–1.6.

(2) *Its organizational structure was not a federal type, and there were no concrete organizational purposes.*

First, as stated above, because its management was authoritarian and controlled by government, its organizational structure was centralized (Department of the Environment 1980). The traditional and rural world was also distrusted and ignored. Martin (1982) argued that the lack of communal organization contributed a lot to environmental destruction in Tauran. He also carefully studied its reasons (please see pp. 165–168).

Second, their organizational purposes were not concrete and often led to contradictory results. For example, after analyzing administrative and legislative measures, the case study provided by the Department of the Environment argued:

> These measures were radial and far-sighted but their effects have not always been restricted to the benefits that were planned. For example, certain details of the Land Reform Law combined with the availability of tractors allowed the extension of dry farming onto unsuitable soils (including vertical ploughing of hillsides), which led to increased soil erosion. Nationalization of rangelands, although it had definite social benefits, has reduced the pastoralists' flexibility, and perhaps also their sense of ecological responsibility-essential features of efficient land use in arid zones-and removed an element of personal investment. It is noteworthy that one result of both of these measures was to reduce the need for investment as a basis for exploitation, and that degradation ensued. (Department of the Environment 1980, p. 185)

(3) *Its management was not democratic and collaborative, and federal mechanisms of awards and sanctions were not applied.*

First, its management was authoritarian rather than democratic as stated above. For instance, the Department of the Environment (1980, p. 208) argued, "Villages used to be fortified, and the area was controlled by leaders who sought to concentrate the ownership of resources in their own hands."

Second, its management was not collaborative. For example, the case study provided by the Department of the Environment said:

> Interest in the desert generally increased through the sixties among intellectuals and planners, and in the seventies a number of government institutions have established special departments or programmes responsible for the desert areas. The responsibilities of these agencies vary in their emphasis on natural and human factors, and suffer generally from the lack of an overall coordinating and planning authority, specifically and actively concerned with vulnerable areas. (Ibid, p.185)

Although the Turan Programme sought to bring together representatives of all the academic disciplines (especially the natural, human biological, and the social sciences) and sought out scholars both in Iran and abroad, there were no enough local scholars. Furthermore, these scholars' expertise and local knowledge were also limited (Ibid).

Third, the analysis of grazing pressure by the case study provided by Department of the Environment (1980, p. 200) showed the management of this problem lacked the federal monitoring mechanism of awards and sections.

(4) *There was no stable scholar-entrepreneurship.*

This can be concluded from the analysis of Principle 1.

(5) *Benefits of the program were not expectable and reinforced.*

The purpose of this program was to increase economic productivity and ecological viability (Spooner 1984). This purpose, however, was not realized. For example, the Department of the Environment argued:

> This economic review suggested that the outlook for transhumant pastoralism in Turan is uncertain unless productivity can be increased and shepherding be made more attractive. Economically, the self-employed resident mixed farmer does relatively well. However, the viability of this latter adaptation during the coming decades will depend on the interest of the younger generation and the rate of migration to the cities. Apart, therefore, from arguments concerning the ecological efficiency of these two adaptations, there is room for serious doubt about the survival of either unless they are included and encouraged in long-term management and development programmes. (Department of the Environment 1980, p. 216)

Breckle (1982) pointed out, "Grazing is also greatly disadvantaged, both because of low productivity and because of salt" (p. 291).

(6) *It did not apply the experiment-extension method.*

Evidence of this method was not found.

(7) *There was no adequate external support, especially financial and technological.*

Please also see the analysis by Department of the Environment (1980).

Case 12: The Grater Mussayeb Project, Iraq (Unsuccessful)

Desertification and dust storms are frequent in Iraq, which is a major region of dust storms in the Middle East (Middleton 1986). The Greater Mussayeb Project (GMP) was initiated by the government of Iraq in 1953 and located in the center of the Mesopotamian Valley, approximately 90 km south of Baghdad (Spooner 1984). Before 1980, this project had three stages (Government of Iraq 1980): the first stage of operation (1956–65), the period of rehabilitation (1965–68), and the period of new land reclamation and development (1968–1980). From the perspective of combating desertification, the implementation of this project disturbed the original balanced ecosystem and caused more desertification. Thus, this can be deemed as an unsuccessful case in combating desertification. Let's analyze how the seven principles were violated from 1953 to 1980.

(1) *There were no enough stable field scholars who significantly took part in this project.*

Although data on initial basic resources were collected and the irrigation and drainage systems were technically designed and installed, they were only done by some external researchers and government officials. Also, there were no enough local field scholars who took part in them. For example, when canal deterioration and increased salinity reduced production, many settlers chose to leave, and the project never had enough technical staff to help farmers locally. Especially, most allottees were originally pastoral nomads who had no intention of becoming farmers. Thus, they could not utilize the land satisfactorily with their own technical means. After the rehabilitation in 1965, more people who had university degrees and higher education were involved in the projects. However, there were no enough local scholars. In 1969, Dr. D. Al-Hardan, former Director General of the Greater Mussayeb Project Authority, pointed out that, at the village level, offices were almost nonexistent. Furthermore, even these external technical scholars were limited. For the first stage of operation (1956–65), the analysis by the Government of Iraq argued:

> As technical staff and manpower were limited, this under-utilization of available land caused maintenance of certain portions of the irrigation and drainage network to be neglected. As a result, canals and drains were rapidly clogged by weeds and by waterborne and wind-blown silt. Other factors such as the lack of extension services, and inadequate technical infrastructure, and insufficient agricultural machinery also contributed to the initial failure of the project. (Government of Iraq 1980, p. 187)

(1.3) *There was no adequate knowledge supporting this project.*

For example, for the design of the GMP irrigation and drainage, Nugteren's report pointed out, "no experimental data were available on water requirements of the proposed new crops"; and "no semi-detailed land classification was made available" (Ibid). Spooner (1984) argued that the reason of the failure of the program only ten years later (by 1964) was that the technology was not fitted to the existing social forms, or a society was not generated to be structurally adaptive to the requirements of the new technology. The inadequacy of animal health control also influenced the field of livestock production. The analysis of the Government of Iraq even argued:

> More difficult and urgent than the technical matters, however, are the sociological problems. The adoption of a system which relies on the combined presence of special managerial skills, a well-organized administration, teams of engineers, researchers, economists, technicians, and social workers, and last but not least—a farmer population receptive new methods and advice, often means a revolutionary change in life-style for technical personnel as well as for the farmer population. Both groups are required to live for protracted period on new agricultural settlement areas, often cult off from their social and cultural roots and without apparent professional, material, or spiritual compensation. The authors of this study believe that the factors which thus motivate staff members and settlers alike are too little understood. (Government of Iraq 1980, p. 212)

Furthermore, as already stated above, many allottees were illiterate (e.g., in 1965, illiteracy among farmers remained as high as 95%) and had no knowledge and experience of farming. Even if more and more schools were established in different stages of this project, these scholars were only primary and intermediate schools. There were no local institutes which were developed to study how to settle human beings and how to combat local desertification. Also, "there existed neither adult education nor any form of vocational training" (Ibid, p. 193).

Because both Principles 1 and 1.3 were unsatisfied, it is not necessary to analyze Principles 1.1, 1.2, and 1.4–1.6.

(2) *It did not have a federal organizational structure and concrete organizational purposes.*

The Government of Iraq (1980, p. 205) argued:

> In 1956–57 the project started with a small administrative unit placed under the department of Irrigation of the Development Board and divided into two sections: irrigation affairs with one civil engineer and three surveyors, and agricultural and co-operative affairs with a section head, two extension officers and four cooperative supervisors.
>
> The tasks of the administration were limited to the provision of water requirements and planting of forests, windbreaks, and new orchards (principally date palm and grapes). (Government of Iraq 1980, p. 205)

From above citation, it can be concluded that this project did not have a federal organization structure. This can be further explained by the lack of the application of the experiment-extension method. During the first stage of operation, although the organizational purposes were clear, they were too abroad (please see Ibid, pp. 177–178). In the period of rehabilitation (1965–68), the purposes were set more concretely, but they were still too abroad (please see Ibid, p. 191).

(3) *This project neither applied democratic and collaborative management nor built the federal mechanism of awards and sanction.*

First, “lacking the necessary authority” (Government of Iraq, 1980, p. 177) and the undemocratic management were important reasons of the project’s failure. For example, one of the most important reasons why most housing was rejected by most settlers was its undemocratic management. Because the government and designers did not listen to the farmers, these houses had a lot of drawbacks such as lack of intimacy, the small size, too modern for the very traditional population, too far from the farms, and the involvement of costly materials in completion of the houses.

Second, its management was not collaborative. For example, although the government hoped that when allottees joined a multi-purpose agricultural cooperative society, providing credit, marketing and supply facilities, and the isolation of the farmer’s families could help to develop the communal and cooperative spirit, this hope did not realize. Furthermore, although additional assistance was obtained from UNDP through FAO in the period of rehabilitation, this collaboration was not enough. Especially, the good collaboration among different scholars was not found. The Government of Iraq (1980, p. 212) argued that “more coordinated and detained advance research” was “of great importance” in the future.

Third, the federal mechanism of awards and sanctions was not found.

(4) *There was no stable scholar-entrepreneurship.*

This can be seen from the analysis of Principle 1

(5) *This project did not guarantee the expectable benefits, and the benefits were not reinforced after the project was implemented.*

One of the reason of setters’ rejection of government housing was this (for detailed analysis, please see Government of Iraq 1980, p. 189). For the agricultural economy, it was the same. The Government of Iraq (1980) pointed out, “After a few years of operation, the farmers found it increasingly difficult to obtain a satisfactory return from their efforts” (p. 191), and “yields in Greater Mussayeb Project remain low” (p. 209).

(6) *It did not apply the experiment-extension method very well.*

Although the importance of agricultural research and extension service was realized, the experiment-extension method did not apply very well. For example, Dr. D. Al-Hardan in 1969 pointed out:

> Extension service is included as a section under the branch of the agriculture co-operation Directorate. In effect, agricultural extension has a secondary character and it has not been the object of comprehensive development. On the village level, offices are almost non-existent. The existing agricultural division was not designed for carrying out an extension programme. It deals chiefly with regulations and procedures, instead of

> agricultural extension. The agricultural co-operation division does not perform extension activities, except for by occasional and rare farm visits by extension field technicians. (Al-Hardan, as Cited in Government of Iraq 1980, p. 206)

The Government of Iraq (1980) further argued that the lack of an effective extension service undoubtedly resulted in many of the Greater Mussayeb Project's failures and impeded the project's overall development. It particularly suggested that "the practice of eroding the ranks of extension-service personnel to meet ad hoc needs of other project sections should be discontinued" (p. 206).

(7) *It lacked necessary external support, especially financial and technical support.*

The lack of technical support can be found from the analyses of Principles and Principles 1.1–1.6. The lack of financial support was also obvious. For example, the Government of Iraq (1980, p. 189) argued, in the first stage of operation (1956–65), "it was obvious that the farmer, with his poor financial and technical means, could not utilize the land satisfactorily." The budget allocation to the Greater Mussayeb Project over the years 1965 to 1974 was also analyzed by the Government of Iraq (1980, p. 207). Financial support for local farmers and scholars, however, was not found.

Case 13: The Eghazer and Azawak Region, Niger (Unsuccessful)

Niger is roughly included within the Saharan, Sahelian, and north Sudan zones. It is a main source area for the dust in the Sahara (Brooks and Legrand 2000) as the world's largest source of dust in the atmosphere (Swap et al. 1996). This case study draws mainly on the work by the Government of Niger (1980). The case area, the Eghazer and Azawak Region in Niger with an area of a little over 100,000 km^2, is located within the north Sahel nomadic zone. Because its desertification from 1900 to the 1970s was not controlled, this region can be deemed as an unsuccessful case of combating desertification. Now let's analyze how the seven principles were violated.

(1) *There were no stable local field scholars who significantly took part in combating desertification.*

Some technologies were introduced in this region since the colonial era, but significant scholar-participation and technological support were not found. First, although the use of technology increased production of cattle and other animals, adequate knowledge and technology of environmental production were not introduced and developed. Second, there were some external scholars attending combating desertification, even American scholars, but there were no enough local scholars. Third, although there were some scholars who took in combating desertification, their participation was not stable.

(2) *The activities of combating desertification were not federally organized and had no concrete organizational purposes.*

First, there were no systematic activities for combating desertification. Second, although there were different organizations involved in combating desertification (some originations were centralized, some were colonial authorities, some were military originations, and some might also be spontaneous organizations), they were not federally organized. Third, some concrete objectives were made for some small and short-term projects (e.g., a three-year program from 1976 to 1978), but concrete purposes were not made to systematically combat desertification.

(3) *Its management was not democratic and collaborative and there was no the federal mechanism of awards and sanctions for the activities of combating desertification.*

First, colonial management, conflicts among different groups, military intervention, centralized chieftaincy, etc., showed that its management of combating desertification was not democratic and collaborative. Second, systematic and federal mechanism of awards and sanctions for the activities of combating desertification was not built (see Government of Niger 1980).

(4) *There was no stable local scholar-entrepreneurship.*

These can be seen from the analysis of Principle 1.

(5) *The principle of expectable and reinforced benefit was not applied.*

Important and systematic activities of combating desertification, which benefit could be expected and reinforced by the later implementation, were not organized from 1900 to the 1970s. Furthermore, as the Government of Niger (1980, p. 144) argued, sometimes although "the strategy and techniques for the struggle against desertification" were known, their application often conflicted with "the short-term requirements of the authorities" that were "responsible for protecting the lives of the populations and feeding them."

(6) *The experiment-extension method was not applied.*

The experiment-extension method was unintentionally used (e.g., building market, pumping stations, and administrative centers after drought) (please see Government of Niger 1980, pp. 141–141), but this method was not systematically used to combat desertification.

(7) *There was no stable external financial, technological, and institutional support.*

Although there was some external support, it was not enough and stable. For example, as stated above, there was a three-year program, but its finical, technological, and institutional support to this region was limited and unstable (for detailed analysis, please see Government of Niger 1980, pp. 142–143). Furthermore, to combat desertification, a three-year period was really too short.

Case 14: The Oglat Merteba Region, Tunisia (Unsuccessful)

The total area of Tunisia is 155,000 km^2. Its four-fifth of the entire area (about 120,000 km^2) is in arid zones (Government of Tunisia 1980). The Oglat Merteba region in southern Tunisia is on the pre-Sahara with only 100–200 mm of rain per year (Grainger 1984). This region, covering a 20,000 area, is a representative example of the ecosystems of the pre-Saharan region. This case study draws mainly on the work of the Government of Tunisia (1980) and focuses on the management before the 1980s. It may cover many years before the 1980s, but mainly focus on from 1900 to the 1980s. Now let's analyze how the seven principles were satisfied.

After studying this case, the Government of Tunisia (1980) gave some recommendations for future activities to combat desertification. These recommendations were consistent with the fundamental principles of successful scholar-based governance. For example, its recommendation about intensified research, local "specialist training" and "mass-education" (p. 50) showed the importance of *Principle 1*. After studying traditional grazing practices and their adaptation to modern conditions in Tunisia and the Sahelian countries, Nonikoff (1976) also highlighted the importance of problems such as "a long-term study comparing the advantages and drawbacks of subsistence economy versus market economy," "working out new grazing or cultivation methods adapted to highly erodible soils," to study "grazing practices," and "methods close to stubble-much faring" (p. 68). Also, the recommendations about "integrated managements" and "rational planning" by the Government of Tunisia (1980, p. 50) partly strengthened the importance of *Principle 2*. Furthermore, its recommendation about raising the level of income for the rural population in order to "reduce the intensity of exploitation of the environment" (p. 50) highlighted the importance of *Principle 5*.

In summary, the lack of activities done according to these recommendations was the reason why its desertification governance was failed. Based on comparing these recommendations and other analyses of this case, it was found that there was no any fundamental principle of scholar-based governance which had been satisfied by the activities of combating desertification in the Oglat Merteba region in Tunisia.

Case 15: The Golodnaya Steppe, Uzbekistan (Unsuccessful)

Arid and sandy deserts occupy around 80% of Uzbekistan's 447,000 km^2 (Toderich et al. 2001). This case about the Golodnaya Steppe mainly draws on the work by U. S.S.R. Ministry of Reclamation and Water Management (1980) and Matley (1970). The Golodnaya Steppe was conquered by Russia in the 1960s and then belonged to the areas of Turkestan. Since 1924, when Turkestan A.S.S.R. was divided into a number of republics and autonomous oblasts, "the Golodnaya Steppe had been divided between the Uzbek S.S.R. and the Kirgiz A.S.S.R. (renamed the Kazakh S. S.R. in 1936), with a small area in the Tadzhik A.S.S.R. The main part of the irrigation system was always in the Uzbek S.S.R., but the Pakhta-Aral sovkhoz was in the Kazakh S.S.R."(Matley 1970, pp. 343–344). In 1956, it was entirely within the Uzbek S.S.R., except for the small area in the Tadzhik S.S.R. In 1991, the Uzbek SSR became independent and was renamed the "Republic of Uzbekistan."

My case study mainly focuses on the time from 1869 to 1980 under the Russian government. However, because this region now is within Uzbekistan, it is deemed as a case study in Uzbekistan. To Matley (1970), this Russian irrigation venture in central Asia neither reached its planned targets, nor did it solve the problem of soil Stalinization. Thus, from the perspective of combating desertification, this venture can be deemed an unsuccessful case. Now let's see how the seven principles of successful scholar-based governance were violated.

Both in the Soviet period and the Czarist period, many scholars were involved in this project. For example, even in 1869, the first Russian governor of Turkestan, General Konstantin P. von Kaufman, "ordered a topographical survey of the Golodnaya Steppe" (Ibid, p.332). In Soviet period, the application of improved techniques and fertilizer, the large-scale use of machinery, etc., showed the importance of scholars' participation. Both the participation of local scholars and their knowledge, however, were limited. For example, for the first five settlements by 1901, Matley (1970, p. 334) argued, "The inhabitants were mainly former canal construction workers and other migrants who had little interests in farming, and the settlements had not stable agricultural population." Matley (Ibid, p. 337) also pointed out, "in 1974 the Russians were facing serious technical and economic problems in their attempts to irrigate the Golodnaya Steppe." In summary, both *Principles 1 and 1.3 were not satisfied.* Also, from these analyses and the case studies by U.S.S.R. Ministry of Reclamation and Water Management (1980) and Matley (1970), it was found that *Principles 1.1, 1.2, and 1.4–1.6 were not satisfied.*

Because of the strong authority of the Tsarist and Soviet governments, the federal organizational structure and democratic and collaborative management were impossible (U.S.S.R. Ministry of Reclamation and Water Management 1980; Matley 1970). The direct control by the governments, the use of unpaid labor, unconcern about the willingness of the local population, the rejection of private individuals' and boogiemen's participation, etc., showed this (Matley 1970). Thus, *Principle 2 and Principle 3 were not satisfied.* From the analyses of Principles 1–2, it also could be concluded that there was no stable local scholar-entrepreneurship. Thus, *Principle 4 was not satisfied.*

Although the application of new technology, machinery, fertilizers, etc., sometimes improved the agricultural yields, the cost of building and maintaining irrigation systems and the management cost were high. Furthermore, "irrigation and drainage of new lands" were "not enough to ensure an increase in cotton production" (Ibid, p. 346). Thus, *Principle 5 was not satisfied.*

The experiment-extension method was used sometimes (e.g., the Prince Masal'skiy experiment and its extension), but its application was not very well (for detailed analysis, please see Ibid, p. 336). Thus, *Principle 6 was not satisfied.*

Both the Czarist and Soviet governments supplied some financial and institutional support to this project, but the support was not adequate and stable. Furthermore, civil war and the attendant disorders often prevented the work from being carried out. For example, Matley (1970, p. 341) argued, "During the war the cotton-growing regions of Central Asia suffered from lack of investment and attention as food production became of prime importance." That is, *Principle 7 was not satisfied.*

Appendix I
Case Studies by Changing Fields with Changing Temporal Extent

In this appendix, seventeen cases in China are first analyzed, and then three case studies in other countries—People's Planning in Kerala in India, the Gal Oya irrigation project in Sri Lanka, and the Saemaul movement in Korea—are successively proposed. Of the seventeen Chinese cases, they are mainly arranged according to their occurrence time. When appropriate, some similar cases are arranged together regardless of their time of occurrence.

I.1. Historical Echoes in Ancient China (221 BC–1905)—The Village Governance System

Case 1: The Village Governance System in ancient China (221 BC–1905)
Ancient China here means China before 1905, the year the system of empirical examinations was abolished. Numerous studies have touched on the governance problem in the countryside in ancient China,[1] but the problem which I want to explore here is the scholars' role in its village governance system from 221 BC (the foundation year of Qin Dynasty) to 1905. In particular, I focused on the village self-governance system from 1067 to 1905. Yu (2002) pointed out that there were three phases of village governance from the Qin Dynasty to the Qing Dynasty. From 221 BC to 595 (The fifth Year of Kaihuang of Emperor Wen in the Sui Dynasty), although some leaders of townships could be chosen by villagers themselves, they were often appointed by government. From 596 to 1070 (The third year of Xining of Emperor Shenzong in the Song Dynasty), the system of villages and townships was gradually changed to the Zhiyi System. Zhiyi was a kind of

[1]One literature review was done by Yu (2001).

L. Yang, *Knowledge-Driven Governance*,
https://doi.org/10.1007/978-981-13-2910-4

corvée in the Song Dynasty. From 1071 (the fourth year of Wang Anshi's Reform) to 1905, the self-governance system under the county level was built.

Basically, the major actors of governance in ancient times could be grouped into five classes: (1) farmers (*nong*, means the farmers; or *min*, means the common people) who were individuals and often organized in families, (2) clan (*zongzu*) organizations, (3) the village administration system (*xiangxun xiangzheng guanli zhidu*), especially the neighborhood administrative system (*baojia zhidu*) after the Song Dynasty, (4) scholars (*Shi*, or *shen*, or combined as *Shishen*), and (5) government (neighborhood administrative systems were its agents in villages). The village governance generally liked a game among these five actors. Sometimes it might also include other kinds of actors such as businessmen and religious organizations, but these five were the major actors.

From a power, authority or right perspective, the village governance in ancient China was also like a game of authority, power, and rights possessed by different organizations or actors. Farmers represented the farmer or citizen authority, power, or rights; the clan organization represented the clan or consanguineous authority, power, or rights; the *baojia* organization represented the administration authority, power, or even rights; scholars represented the scholar or knowledge authority, power, and rights; but the government represented the imperial authority, power, or even rights.

In a village, collective goods were often provided by families, clans, and the whole village. Families, which were based on marriage and kinship, supplied the collective goods for the whole families through the collective action organized by family members. Sometimes a large family might include different smaller families. Clans, which were based on the same kinship and made up of different individuals and families, supplied the collective goods for the whole clans through collective actions organized by clan organizations.

How were collective goods produced and collective actions organized at the village level? At this level, although all the five aforementioned actors took part in different programs or actions, and sometimes they acted could independently, generally individuals, families, and clans could be deemed as one party, representing the common people. The neighborhood administration system was the agent of governments in the village, and it could be deemed as a party which represented governmental behaviors and actions, although it was also strongly influenced by clan organizations and scholars. Sometimes when there was another governmental agent (e.g., the upper government), it also could be deemed as a part of the common people, because the workers in the neighborhood administration system must be locals and generally were recommended by the general public (e.g., in Qing Dynasty). In summary, at this level, village games happened mainly among the farmers as the common people, the government, and the scholars. In these games, scholars often played a very important role. From my point of view, this is also a type of scholar-participation governance.

Who were these scholars? Although many researchers agree that scholars were important actors in the ancient countryside, they often have different ideas about who could be scholars. Some argued that scholars in the countryside were some intellectual landlords and retired officials (e.g., Hu 1948), but some claimed that these scholars were related to *gongming* (scholarly honor or official rank in feudal China), *xuepin* (characters as a scholar), *xuexian* (the official rank as a scholar), and *guanzhi* (official position), etc. (e.g., Zhang 1991, p. 1). Yu (2001), however, distinguished scholars into three types.

The first type is *gongminlei* (the type of *gongmin*) who became *Shishen* because they got Gongming, a scholarly honor or official rank in feudal China. Thus, they might be retired officials and also might be some who only got a scholarly honor but never became government officials. For example, in the Shaozhuang Village (he gives it an academic name Yuecun, which means the Yue Village) in Baoguo Town, Hengshan County, Hunan Province, according to the record of pedigree of the *Li* clan (the family name is *Li*), while Qianlong was emperor in Qing Dynasty, there were six people who got different scholarly honors. But they only passed the imperial examination at the provincial level and could not be government officials, so they worked in the village as teachers. Because they had knowledge and always did good things for people, they had a very high social status and were respected by the locals. Families were very proud if they could invite these men to attend their affairs like weddings and funerals. When there were entanglements and conflicts in the village, they were often invited to resolve. Even the government strongly depended on them to do things in the village such as registering and measuring land.

The second type is *Furenlei* (the type of *furen*) who were respected by the villagers because they were rich. One example was *Zhaojia Laoye* (Milord Zhao, a milord whose family name was Zhao) in the Shaozhuang Village. When he was a child, his parents died, so he survived by grazing cows for his clan. While other children in his clan studied, he could only overhear. However, because he was very wise, although he did not know the Chinese characters, he remembered many sentences and essays in the classic literature. He could not get a scholarly honor or official rank in feudal China. When in his forties he became a very famous rich man and always treated the villagers kindly. Once, when a *juren*, a successful candidate in the imperial examinations at the provincial level in the Ming and Qing dynasties, haughtily came to his village, he gave him a lesson using his wisdom. All the villagers were proud of him. After that, when people had problems, no matter what kind, (the entanglements and conflicts among clans, or government's difficulties), he was able to help resolve all of those problems.

The third type is *nengrenlei* (the type of *nengren*) who were respected and given local authority because they had some special knowledge and capacity needed by the villagers. An herbalist doctor, Li Jifang in the Baiguo town, was a good example. Because he was not only a very famous and highly skilled doctor but also a welldoer, people relied on his judgment. Even the government officials listened to his ideas when they encountered law cases about the Baiguo town and related places.

I agree with most of the ideas stated above by doctor Yu, but I also want to highlight that although wealth or official position could help people be the *Shishen* in villages, the basic requirement to be a real scholar was that these people must have advanced knowledge as stated above by these three types of scholar cases. This means whether a rich or a retired official could be a scholar in the ancient countryside first depended on their background knowledge. If they were retired officials, but they had no real knowledge respected by villagers, they could not be the real scholars in the village. Sometimes even if they could use their former capacity (good or bad) to influence affairs in the village, they were more like government officials than scholars. However, if they had respected background knowledge, they could be deemed as scholars, even if they were not retired officials. Also, if some people had only wealth but had no particular knowledge, they could not be a respected scholar, although their wealth could help them get more respect, authority, and high social status. Furthermore, in addition to knowledge advantages, respected scholars were the ones who were living in the village or had special relationships with the village such as kinship. They also had high social status and were respected by the locals, had a relatively independent social identity and social responsibility, and listened and helped people resolve their problems. All these clearly tell us that ancient scholars had the same characteristics described by the seven principles of successful scholar-participation governance. Like Confucians, these ancient scholars had a strong spirit of social responsibility and mission. For example, Zengzi (or Tseng Tzu) said:

> The scholar may not be without breadth of mind and vigorous endurance. His burden is heavy and his course is long. Perfect virtue is the burden which he considers it is his to sustain, — is it not heavy? Only with death does his course stop, —is it not long? (*Confucian analects* 1900, pp. 74–75).[2]

Now let's analyze how this ancient scholar-participation governance worked. According to the case of the Shaozhuang Village, the basic governance mechanism of ancient scholar-participation governance was also consistent with the seven principles. First, scholars, as information providers or entrepreneurs, could help villages resolve many problems, especially the conflicts among different clans, and assist in public affairs issues such as constructing bridges and maintaining roads. For example, the scholars in the Shaozhuang Village participated in many common welfare efforts such as mediating conflicts, organizing public projects, developing village education, and maintaining schools, the examination hall (*gongyuan*), and

[2]In Chinese: 曾子曰: “士不可以不弘毅, 任重而道远。仁以为己任, 不亦重乎?死而后已, 不亦远乎?” The English is revised from *Confucian analects* (1900) translated by James Legge.

the temple. When there was government intervention, on the one hand, scholars could help farmers deal with problems with government and give them some advice depending on their knowledge about governmental laws, policies, teachings by some sages, etc. They also pursued local interests with farmers together. Sometimes, they did not hesitate to antagonize government. Generally, these conflicts were no threat to the central government, and sometimes they could even be used by central or other high-level governments to balance local governmental officials' behaviors. On the other hand, these scholars also helped the government get more information about the locals, particularly farmers. They often helped government explain its policy to farmers and helped local governmental officials implement governmental policies and programs as agents. Also, they might directly give suggestions to government or work for it. A handbook for magistrates, *Mu-Ling-Shu*, even clearly taught the magistrates how to appropriately treat gentry's members and said:

> In administering the affairs of a district, the high families [*shen*] are not to be disturbed. They should be met in a virtuous manner and received courteously and are not to be repressed with power and prestige. Also the scholars [*shih*] are at the head of the people, and since the laws and discipline of the court cannot be exhaustively explained to the people, and since the scholars are close to the people and can easily gain their confidence, and learned and virtuous scholars are exactly the ones to rely upon in persuading the people to follow the instructions of the officials. Therefore, they should be loved and treated with importance. When they happen to come because of public affairs, if they are sincere and self-respecting, they should be consulted on the problems of whether bandits exist in their villages, what the jobs of the villagers are, and whether the customs of their places are praiseworthy….[3] (Chang 1955, p. 32)

For instance, the scholars in the Shaozhuang Village helped the government organize military training, collect taxes, etc. The functions of the neighborhood administration system as the local agent of government strongly depended on the scholars' support. For example, milord Zhao (*Zhaojia laoye*) in the Shaozhuang Village could even change "*baozhang*," the leader of the neighborhood administration system. Thus, *Principle 1 was satisfied.*

Although this ancient scholar-based governance was generated in a so-called autarchic system ruled by emperors, it helped ancient China keep its tradition of self-governance by the locals below the county level over thousands of years. The organization of this governance was a bottom-up system or at least had similar characteristics. It was also a federal system. Various social units (such as individuals, families, and clans) had complex relationships with scholars. This governance system was also problem-driven and interest-driven. Only if scholars could help local citizens or government officials resolve the problem, would they be

[3]Wang Feng-sheng, "On Gentry," in *Mu-Ling-Shu* 牧令书, compiled by Xu Zhichu (or Hsü Chih-ch'u) 徐致初, 1848; cited from *The Chinese Gentry* (Chang 1955, p. 32).

believed and trusted. From this perspective, this governance also had some democratic attributes, although these attributes were limited and put under an unequal and non-individual social environment. Moreover, in order to resolve problems, scholars themselves often tried very hard to listen to different people's ideas and balance them. The primary method they used was getting consensus through mediation, persuasion, education, etc., rather than using punishment and threats. Thus, *Principles 2, 3, and 4 were satisfied.* The problem-driven and interest-driven methods also showed that *Principle 5 was satisfied.* Furthermore, this governance used the experiment-extension method. Although knowledge, wealth, and governmental position could help scholars be a *shishen*, their authority and capacity in solving problems were built on their reputations. These reputations could only be gradually established based on their different behaviors and actions in previous situations and based on their understanding and perception of farmers and officials. Thus, *Principle 6 was satisfied.* Finally, from a historical perspective, this governance system in the countryside was also endogenous. Sometimes even if the dynasty was changed, the governance still remained. That is, *Principle 7 was also satisfied.*

Some irrigation examples demonstrated by Chang (1955) were very vivid descriptions of scholar-based governance in ancient China, so I excerpt several paragraphs below:

> Many examples can also be given to show the gentry's part in the promotion of irrigation projects. For instance, in Cho-hsien, Chihli, a *sheng-yüan* of the countryside, who took an active part in such matters as bridge and road construction, establishment of shrines and temples, relief work, and charity schools, also promoted an irrigation project for his locality. In the Chia-ch'ing period, he started a canalization project which irrigated more than 3000 *mou* (about 500 acres). The gazetteer, which was compiled in the Republic period, states that the villagers were still benefiting from this project.
>
> In Ch'ing-yüan, Chihli, a chü-jen who was an influential teacher and was also active in welfare activities became interested in irrigation. During a drought in the Kuang-hsü period he wrote an essay on relief measures in which he pointed out the need for an irrigation project. He secured the cooperation of the local people in promoting and carrying through an irrigation project covering several hundred thousand *mou*. He himself advanced the cost of the labor.
>
> In both these examples, the gentry promoting the project were concerned with the whole locality rather than with their privately owned land. These two examples also illustrate the usual pattern of gentry action in which the upper gentry functioned in a wider area than the lower gentry. (Chang 1955, p. 58)

Some examples of the gentry's participation in promotion of irrigation projects had been summarized in Table I.1.

Table I.1 Examples of the Gentry's participation in promotion of irrigation projects

Period	General statement in Gazetteer	Explanation in attached memorials, reports, or articles
Chia-ching 23 (1818)	The governor sent two special officials with instruction to dredge the Wu-sung River	The governor's memorial reported the gentry's petitions and indicated that the work was executed by the gentry and financed by the local people
Tao-kuang 3 (1823)	A *hsien* magistrate dredged the Cha-chiang	The attached article pointed out that the original request came from gentry members and that the work was financed by local inhabitants
Tao-kuang 4 (1824)	A *hsien* magistrate dredged the Mao-hu	An attached letter from a gentry member to the magistrate suggested the need for the work and the method of doing it
Tao-kuang 7 (1827)	The governor instructed an intendant to dredge the Wu-sung River	The governor's memorial mentioned the censor's suggestion of branch irrigation projects to be directed by gentry and elders and financed according to the practice that landowners supplied provision and tenants supplied labor
Tao-kuang 15 (1853)	Magistrates of two neighboring districts dredged the Pai-lien-chin	An attached article and the compiler's annotation show the action originated with gentry landowners of the localities concerned, and that the work was carried through by the point effort of the magistrates and the gentry. The article states that the cost was met partly by the people themselves and partly by an official "contribution"
Tao-kuang 16 (1836)	The governor instructed three *hsien* magistrates to dredge the P'u-wei-t'ang	The attached memorial divides the credit among the officials, gentry, and commoners. A petition from the gentry of these localities discusses methods of carrying out the work and suggests the maintenance of a large dam. The petition was approved by the governor
Tao-kuang 26 (1846)	The prefect constructed a dam	The prefect's article attached stated that the gentry members of the prefecture had discussed this construction work and had requested the former prefect to take action. The

(continued)

Table I.1 (continued)

Period	General statement in Gazetteer	Explanation in attached memorials, reports, or articles
		advantages of the project had been presented to him by members of the upper gentry. The project was also supported by a financial commissioner who was a native of the locality and was at home because of a mourning period
T'ung-chih 3 (1846)	An intendant dredged creeks around and with Shanghai	The intendant's attached article referred to his consultation with gentry and elders on the planning of the work
T'ung-chih 5 (1866)	Hsien magistrates of Lou-hsien, Shanghai, and Nan-hui dredged rivers	The magistrates' attached reports give the decision and suggestions arrived at in the meeting of "gentry-managers." The reports indicate that the gentry managed the work and that officials acted as coordinators between the gentry of the different districts
T'ung-chih 8 (1869)	*Hsien* magistrates of Lou-hsien, Shanghai, and Ch'ing-p'u dredged rivers and built dikes	The explanation reveals that in this case the officials instructed "managers of the various units" of the locality to carry out the work. The work was to be financed according to the practice that the landowners supplied provisions and tenants supplied labor. The technical plans which were followed were drawn up by a local man
T'ung-chih 9 (1870)	*Hsien* magistrates of Hua-t'ing, Ch'ing-p'u, Ching-shan, and Shanghai dredged ponds and rivers	A notation states that the magistrates gathered the city and countryside "managers" and decided how the work would be financed
T'ung-chih 9 (1870)	*Hsien* magistrates of the prefecture dredged rivers and ponds	In one case the gentry petitioned for the work to be done and the officials acted as supervisors. Attached is an article by Ling Chieh-his, a gentry member of a neighboring prefecture, who was an authority on irrigation problems of this and other areas of southeast China. In the article, Ling discusses technical problems of the work on a large lake and certain

(continued)

Table I.1 (continued)

Period	General statement in Gazetteer	Explanation in attached memorials, reports, or articles
		rivers. In another case, a joint report of the *hsien* magistrates states that the "managers" are taking care of the work. These "managers" were gentry again. In a third case, the *hsien* magistrates reported on the gentry's completion of the dredging of the Wu-sung River together with a petition from the gentry. The gentry asked that it should not become an established practice to expect them to carry out similar work on this river in the future, as their localities received little benefit from it

Adapted from Chang (1955), pp. 59–61

I.2 Variant Echoes from Transforming China (1905–2007)—Various Social Experiments

Generally, many historians deem the Opium War (1839–1842) as the starting point of transforming China. From the perspective of scholar-based governance, however, I like to choose 1905, the year when the system of empirical examinations was abolished. There are two reasons. First, beginning in that year, traditional scholars in China, especially in the countryside, suddenly began to gradually lose almost all of their influence. Second, although after the Opium War China began to become a semicolonial and semifeudal country, this influence mainly occurred in cities and littoral areas at the beginning. Dr. Yu (2001) also had the same idea.

From 1905 to present, there are over 100 years which can be divided into three periods: Late Qing Dynasty (1905–1911), the Republic of China (1911–1949), and the People's Republic of China (1949–present). During these periods, many things happened in the countryside. Since I could not study everything in detail, the method used here is to draw a broad-brush outline from the perspective of scholar-based governance to help us understand this tortuous transformation.

I.3 The Last Struggle of the Old Dynasty—Rural Autonomy Movement in the Late Qing Dynasty (1905–1911)

Case 2: Rural Autonomy Movement in the late Qing Dynasty (1905–1911) Before 1905, in Chinese rural society, scholars were the bond between government and common farmers. After the system of empirical examination was abolished, government cut its connection with traditional scholars. The relationship between government and traditional scholars suddenly was totally changed. Scholars no longer were the cooperators of government, nor could they help government to govern the countryside. Government began to change its attitude to traditional scholars, especially under the pressure of modernized reform. Traditional scholars' social status was also suddenly changed. They were no longer highly respected by farmers, and farmers could no longer depend on scholars to resolve local problems and the relations with government. Scholars themselves also encountered tremendous changes in their lives and no longer had resources and time to do the work they did before. This tremendous change was so huge that no body could suddenly be seasoned with this. In the countryside, government almost lost its control, and traditional scholars lost their authority and social responsibility in public affairs. Farmers themselves lacked adequate capacity to realize self-governance. With the movement of many traditional scholars to cities to find new lives and the descending station of the countryside along with the growth of new cities and their new elites, the differences between the countryside and the cities became larger and larger, and finally the countryside almost went to the dangerous fringe of non-intelligence and governance vacuum. Under such conditions, the countryside was often controlled by local tyrants or despots, gangsters, ruffians, etc.

To resolve these serious problems, the government of the Qing Dynasty organized a rural autonomy movement along with a local autonomy movement in the whole country. For example, a new statute, titled "*Chengzhenxiang Defang Zizhi Zhangcheng*" (the Statute of Local Autonomy in Cities, Towns, and Villages), was enacted in 1908. This law gave the right to farmers in the countryside to elect their own governors. This movement had several serious problems, however.

First, after the abolishment of empirical examinations, the loss of the scholars resulted in the government losing its basic channel to communicate with and rule within the villages. Even if the central government had a good policy, nobody at the grassroots level could help implement this policy. All these meant that at this time, neither scholar-based governance nor vertical government governance was possible. Thus, *Principles 1 and 4 were not satisfied.*

Second, this movement was organized using top-down method, and the mobilization at grassroots level was not enough. The basic way adopted to resolve rural problems in this movement was still the strict state-control method although it was named local autonomy. For instance, during this movement, the neighborhood administration system was compressed under the town level. That is, the central government wanted to realize its directly administrative control at the town level, but this broke the Chinese long tradition where the administrative power never goes down below the county level. Thus, *Principles 2 and 3 were not satisfied.*

Fourth, it did not apply the experiment-extension method and also did not provide any benefits to the villagers. Thus, *Principles 5 and 6 were not satisfied.*

Fifth, because the national capacity of Qiang Dynasty had largely declined and its legitimacy became suspect, it lacked enough authority and resources to really implement its policy at that time. That is, *Principle 7 was not satisfied.*

All these tell us this can be deemed as an unsuccessful case of scholar-based governance. During the field study in Hunan Province, Dr. Yu (2001) even argued that he could not find any concrete records about this movement in the counties.

I.4 Rural Vicissitude and Experiments in the Republic of China (1911–1949)—Various Attempts in a Turbulent Era

Case 3: The Movement of Local Autonomy in the Republic of China (1911–1949)

After the Qing Dynasty was overthrown by the 1911 Revolution, China entered its new period of the Republic of China (1911–1949). Due to the weakness of Kuomintang and many other reasons, its governance was stratocracy in the early period, and Kuomintang only began to relatively control the whole country after 1927. The governance capability during this period was very weak, especially when the whole nation was organized to resist Japanese invasion after 1937. Based on inertia and experience of the late Qing Dynasty, government continued the Movement of Local Autonomy. Under the rule of the Shihkai Yuan, his government enacted "*Zizhi Tiaoli*" (the Autonomy Ordinance). Even the Northern Warlords did some experiments on local autonomy. In the period of the Republic of China, government encountered almost the same problem in the Late Qiang Dynasty and tried to adopt the same method—the government control method—to resolve the problem. Almost because of the same reasons, these experiments were not successful, however. Dr. Yu argued that according his field study, although several counties all had records about this movement, these records could only be found on the level of "qu" (an area classification between a county and a township) and Xiang (township). Under the township level, the traditional neighborhood administration system was left intact.

There were several differences between this new movement and the movement in the Late Qing Dynasty. First, this movement was promoted by the new governments, although some of them were the Warlord governments. Second, this movement stressed civil rights, although they were still very limited. Third, during this period, although the clan authority had rapidly developed when the government control and scholars' governance shrank in order to protect farmers, violence and uncivilized religious authority were still very popular in villages. All the other aforementioned problems in the Late Qing Dynasty still existed, however.

In 1935, a national conference was organized to talk about the local autonomy problem and the neighborhood administrative system. Some scholars said, because farmers were not all trained in the world of conformism, they had no capability to realize autonomy. Even the elections were neglected by them. Some scholars, on the contrary, insisted on the idea of local autonomy. Therefore, the government decided to do both—train farmers and implement autonomy policy (Yu 2001). But these problems were not resolved finally because their endeavors still completely depended on farmers' autonomy and strict state control rather than scholar-based governance.

In summary, *all the principles were not satisfied, and this was an unsuccessful case of scholar-based governance.*

During this period, several important experiments were worthy of consideration, however. They were the three Rural Education Movements led by scholars and the Farmers' Association Movement led by the Chinese Communist Party. Three representative Rural Education Movements were Xingzhi Tao's (or Hsing-chih T'ao) experiments in Xiaozhuang (or Hsiao-chuang) normal school (1927–1930) in Jiangsu Province, Yangchu Yan's (or James Yen) rural reconstruction experiment in Ding County (or the Ting Hsien Experiment, 1929–1937) in Henan Province, and Shuming Liang's rural reconstruction experiment in Zouping County (1931–1937) in Shangdong Province. I explore these cases as follows.

Case 4: Tao's Xiaozhuang Normal School Experiment in Nanjing, Jiangsu (1927–1930)

Tao's fundamental education thoughts were "the union of doing, learning, and teaching" and "life education" (shenghuo jiaoyü). Xiaozhuang Experimental Village School, located in Xiaozhuang Village outside of Nanjing, was Tao's most venturesome experiment. His basic purpose was to develop non-formal education at the village level and then to improve the quality of farmers. Many other scholars in that era believed many problems in villages were caused by the low quality of farmers. If their quality could be improved, many problems could be readily solved. Although his school only lasted about three years and then was closed by Kuomintang's troops, his experiments succeeded in many aspects. After carefully studying his experimental method, I found that the organizational styles of his experiment were consistent with some of the principles of successful scholar-based governance, and it can be deemed as a semi-successful case.

His school was located in a village. Although the teachers were trained with the more distant aim of rural education and village renewal, the potential teachers were required to joint village farmers in their daily work (Keenan 1977, p. 101). As the leader, Tao also had a very strong social responsibility and spirit. He even could smoke the calumet together with farmers and his followers. He deserted his own position in China's small middle and upper class and often dressed like the farmers. As a doctor, he also got much respect from other people. All these meant that *Principles 1 (including Principles 1.1–1.6) and 4 were weakly satisfied* (because their role in the village was still limited and many of them still could not become the local scholars within 3 years).

The governance method of his school even had some democratic characteristics. For example, Keenan argued:

> In a dispute over well water, for example, the school arranged a public discussion to settle the problem; villagers were proved to share experience and knowledge, regardless of age or personnel predetermined social status. A teenager ran the meeting, and school personnel participated from the floor to set and atmosphere in which all opinions could be considered equally. (Keenan 1977, p. 101)

Its governance also had some characteristics of a bottom-up federal organization, and experiment-extension governance. For instance, Keenan points out:

> As the teachers' school expanded, a central elementary school for in-service education was established in the village. Branch schools, extending only through the compulsory four grades, grew out from the central school in network that connected several small villages, each of about 100 families. Adult classes and special sessions for working young people were available at the elementary schools. Gradually other social functions were provided. The normal school established a village hospital, and a 'self-defense league' for protection against bandits. (Ibid, p. 101)

Thus, *Principles 2 and 3 were satisfied, and Principle 6 was also weakly satisfied.*

From a pure education perspective, Tao's experiment was very successful. Although his school was closed in 1930 by Kuomintang's troops, his models and ideas were deemed as the central elements in the education system of the People's Republic of China. And this is one of the important reasons why the People's Republic of China can gain great achievement in rural education. But his experiment only focused on the farmers' education and did not include political and economic elements, especially, the economic development and the governance structure in the village. Thus, *Principle 5 was not satisfied.* Furthermore, the fact that his school was closed by Kuomintang's troops also showed that his experiment did not get enough external support. Thus, *Principle 7 was not satisfied.*

Case 5: Yan's Rural Reconstruction Experiment in Ding County, Henan (1929–1937)

Like the Xiaozhuang Experiment, Yan's Ding County Experiment still focused on rural education, but its scope was broader. Yan tried to use four kinds of education and three types of methods to realize his program. The four kinds of education were (1) *wenyi jiaoyu* (literature and art education) which tried to teach the common people literature, arts, rural drama, and science, publish newspapers, make farmers literate, construct radio station, and so forth; (2) *shengji jiaoyu* (education of livelihood means) which trained farmers with means of livelihood, constructed related schools, trained *biao zhuanjia* (exampled experts), especially organized *hezuoshe* (rural cooperation), extended a new breed of cotton, and meliorated the breed of pigs and chickens, etc.; (3) *weisheng jiaoyu* (education of people's health) which mainly focused on the establishment of institutions of health protection, including *baojianyuan* (the person who was with responsibility for people's health care) in village, *baojiansuo* (an institute) which took charge of the people's health care within a *qu* (an administration unit between the town and the county) including about 30 thousand people, and *biaojianyuan* (the highest institute in the county) which took charge of people's health protection in the whole county; (4) *gongming jiaoyu* (civil education) which focused on the training of national spirit. Three types of methods used were (1) *xuexiaoshi jiaoyu* (education in classrooms), (2) *jiatingshi jiaoyu* (education in homes and families), and (3) *shehuishi jiaoyu* (education through the whole society) which mainly focused on the continuing education of schools' graduates and the four kinds of education to other farmers through the schoolfellows' association and tried to make these graduates become the cadremen in the rural reconstruction movement.

His experiment even involved political reforms. The new administration system had three levels. The first was *gongming fuwutuan* (civil service groups) which were grassroots organizations and functioned as gathering, training, organizing, and applying different resources of the people. The second was *xiangzheng jianshe weiyuanhui* (the committee of rural reconstruction in towns) which had 6–12 committee members and the teachers in elementary schools were automatically the members and the secretaries. Meanwhile, *xiangzhen gongming dahui* (the civil congress in towns) was also organized to supervise the committee members. The third was *xiangzheng weiyuanhui* (the county administrative committee). He reformed the old administration system and established a new system which only had five branches including civil administration, finance, education, economy, and police.

In summary, Yan and his fellows played the role of scholars in the countryside not only as information providers but also as new scholar-entrepreneurs. These scholars were mainly external elites, however. If they could educate new scholars who always lived in county or they decided to live there forever or at least always keep their strong supports for the country, this experiment could survive. If these conditions could not be satisfied, once these external elites left the county, his experiment could not be continued. This was also proved by history. Thus, I conclude that *Principles 1.1–1.6 were satisfied*, but *Principles 1 and 4 were weakly satisfied* (because although Yan's experiment lasted 8 years, many scholars still could not become stable local scholars).

The federal structure, concrete and stratified organizational purposes, democratic management, and the system of awards and sanctions in this experiment showed that *Principles 2 and 3 were satisfied.*

Although Yan's experiment tried very hard to combine economic development and other social, political, and cultural developments (including farmers' education) together for the benefits of farmers, it could not touch land reform, which in fact was the fundamental benefit to the farmers. So although the conditions of Ding County were much changed through scholar-based governance of his experiments, the basic rural problems still could not be resolved. Thus, *Principle 5 was only weakly satisfied.*

Yan got some external support including governmental support (even the President's support) and some international support (especially from the USA) (Hayford 1990). For example, depending on the governmental support, in Hunan

Province, Yan even dismissed about 2/3 corrupted officials instead of 5000 scholars and scientists. But these supports were not enough, and so his experiments could not be extended to many other counties. Thus, *Principle 6 was not satisfied, but Principle 7 was weakly satisfied.*

Case 6: Liang's Rural Reconstruction Experiment in Zouping County, Shandong (1931–1937)

Liang's experiment in Zouping County was similar to Yan's experiment in Ding County. His experiments included education, economy, and political affairs. For example, he organized a villagers' self-defense organization in Zouping. The most famous event of his experiment was successfully helping farmers to plant a new breed of cotton and organizing cooperation organizations. Impelled by Liang, over one hundred scholars who had over a bachelor's education devoted themselves to the rural reconstruction movement in Zouping County. These scholars were entrepreneurs in this county and they really did very successfully in the early period. This experiment was even supported by the provincial government (especially chairman Fuju Han). But also like Yan's experiments, because of the Japanese invasion and without provincial government support again, his experiment was stopped (Alitto 1986). *Its satisfaction of the seven principles was the same as Yan's experiment*, so I also deem it as a semi-successful case of scholar-based governance.

In sum, as great social experiments, these three experiments along with others done by other scholars such as Jinghan Li and Yanpei Huang showed us some aspects of the new scholars' social functions in the countryside. In addition to improving the farmers' educational level and helping the natives do some practical things, their attendance in countryside activities reconstructed the scholar's function and social status to some extent, and partly reconstructed scholar-based governance in these villages. Compared with traditional scholars, they were new scholars using new ways to help new farmers resolve new problems. These experiments also had their inherent shortcomings, however. First, from the villagers' perspective, most of the leaders of these new experiments were from the outside rather than local scholars. Second, as already stated above, although their attendance could improve the farmers' capacity for self-governance, their experiments could not get enough external support. These experiments, however, became the important heritage of scholarship about Chinese rural problems and strongly influenced the Chinese rural policies such as the New Life Movement (*Xinshenghuo Yudong, 1934–1949*) by the Kuomintang Government. Even now, these experiments are still influential to researchers' ideas and government's policy selection and policymaking. Although many researchers traditionally deem these experiments as the rural education experiments, I prefer to deem them as new experiments of scholar-based governance in modern China.

Case 7: The Farmers' Movement and the Farmers' Association (1921–1949)

If the rural reconstruction movements can be deemed as social experiments in the legitimate system of Kuomintang's Government (although those scholars had a relatively strong and independent social identity), the Farmers' Movement and the Farmers' Association led by the Chinese Communist Party can be deemed as a social experiment out of the legitimate system of Kuomintang's Government. Many rural policies in the People's Republic of China after 1949 were influenced by this experiment.

Different from pure scholars, the members of the Communist Party in the early period had two identities. First, they were scholars and always were highly educated. Sometimes even if they were not scholars, they at least were literate people. For example, although all the leaders of the farmers' associations in *Yuebei* (*the north of Yue village*) were originally farmers, they already learned much new thought and were literate people (Yu 2001). Second, they were revolutionists. The aforementioned pure scholars who organized the above experiments generally wanted to reform the villages in the legitimate system and under the control of Kuomintang's Government through rural education; the scholar-revolutionists, however, not only wanted to change the rural society through education but also pursued some more exhaustive reforms including economic, cultural, political, social, etc. Furthermore, they clearly claimed to share the public authority with the legitimate government. For example, in "*xiangxiangxian nongxianhui zhixing-weiyuanhui he qunongming xiehui weiyuanzhang lianxihuiyi xuanyan*[4]" (one enouncement of the farmers' association in *Xiangxiang* County) directly argued that their only purpose was to eliminate local tyrants, evilgentry, bullus, and evilgentry. The enouncement of *Xiangtai County* (the hometown of Zedong Mao) also argued that the purpose of farmers' association was to realize the farmers' self-liberation and oppose imperialistic pressure and oppressing organizations.

The principal members of farmers' associations were rural farmers, and they claimed that all policies should be decided and implemented by the farmers themselves. According to an investigation in 1926, there were 54 counties which established the farmers' associations, and the total members in these associations were 1,070,000, in which the owner-farmers (*zigengnong*) accounted for 8.15%, the semi-ownerfarmer (*ban zigengnong*) 13.82%, the tenant farmer (*diannong*) 19.31%, the farmhand (*gunong*) 26.42%, craftsmen (*shougongyezhe*) 7.44%, small businessmen (*xiaoshang*) 2.21%, and teachers in elementary schools (*xiaoxue jiaoshi*) and others (*qita qunzhong*) 2.15% (HSCXBW 1995, p. 192). Farmers' associations even had a slogan "all power should be held by farmers' associations (*yiqie quanli gui nonghui*)." Using different methods, they partly controlled or at least shared the administrational and jurisdictional power with the legitimate system, and then limited the power of the head of the counties, reduced the taxes, measured land, equally reallocated land, confiscated public property and temple property, organized different economic organizations, punished local tyrants, evil gentry, bullies, and organized many public affairs such as constructing road, bridges, reclaimed

[4]In Chinese: 《湘乡县农民协会执行委员会和区农民协会委员长联席会议宣言》.

wasteland, and afforestation. They even opposed superstition and old etiquette, liberated women, and set up new elementary schools and evening schools for farmers. They also opposed old clan authority and many methods of excruciation. They even established a new farmer militia and took over the armed forces controlled by landowners and government.

As new scholar-revolutionists, their leaders' behaviors were often consistent with basic behavioral principles of successful scholar-based governance. Most of them were local scholars, and almost all of them were originally farmers. They had high education levels, had more knowledge about the outside world, and directly went to villages working together with farmers. At the *qu* and *xiang* levels, they mainly depended on their knowledge and capacity of resolving problems obtained leadership, social status, and respects (Yu 2001). They worked, lived, ate, and slept together with farmers as well as wearing the same clothes, shoes, caps, etc.

Their methods of governance were also consistent with governance characteristics of successful scholar-based governance. These can be listed as follows:

(1) They directly went to the grassroots level in villages to organize their movements;
(2) They developed federal organizations through snowballing methods and set up concrete purposes for the organizations;
(3) They used democratic methods to resolve problems, they often liked to reach consensus among farmers through educational and persuading methods, and they built the mechanism of awards and sanctions;
(4) They further considered farmers' benefit and interests;
(5) They helped farmers gradually understand their movements, devoted all their efforts to do almost every possible thing with zealous passion and used the experiment-extension method to develop and ameliorate their movements.

Compared with the experiments organized by some pure scholars, their experiment was more exhaustive, radical, and broad although they still could not resolve the land problems. To Kuomintang's Government, their experiment was dangerous even astounding, and it often strongly opposed and even violently suppressed it. So if scholar-revolutionists lacked enough capacity to oppose governmental intervention, their experiment would fail. For example, in about June 1927, in Hengshan County, Hunan Province, about 100 communists and related people were killed by governmental troops, and about 2000 people were put into jails. Thus, in this experiment, *Principles 1–6 were satisfied, but Principle 7 was not.* Totally, it was a semi-successful case of scholar-based governance.

I.5 Rural Vicissitude and Experiments in the People's Republic of China (1949–2007)—From Extremeness to the Middle Way

Roughly, the experiments after 1949 can be grouped into two phases: from 1949 to 1977, and from 1978 to present. The reason why I choose 1978 as a dividing line is that this year is the starting point of China's new opening up and reform policy. If the experiments before 1978 can be deemed some extreme or even crazy experiments, the experiments and policies after 1978 began to return to the middle way (i.e., become more rational and cautious).

From 1949 to 1977, under the Soviet system of planning economy and highly even extremely centralized political arrangement, at least three groups of important events strongly influenced the governance system in the countryside: (1) the Farmers' Association and Land Reform (*Nongmin Xiehui he Tudi Gaige,* 1950–1953); (2) the Cooperative Movement (*Hezuohua Yundong*, 1949–1955), the Great Leap Forward movement (*Dayuejin Yundong,* 1958–1960), and the People's Commune Movement (*Renmin Gongshehua Yundong*, 1958–1978); and (3) the Intellectuals' Ideological Remolding Movement (*Zhishi Fenzi Sixiang Gaizao Yundong*, 1951–1952), the "Anti-rightist" Movement ("*Fanyou*" *Yundong*, 1957–1958), the Four Cleanups Movement (*Siqing yundong*, or The Socialist Education Movement, 1963–1965), the Great Cultural Revolution (*wenhua dageming*, 1966–1976), and the Rustication Movement of Educated Youth (*shangshan xiaxiang yundong*, or the Movement of Going Up to the Mountains and Down to the Villages, 1955–1981, mainly 1968–1978).

After 1978, I chose to study the case of the Household Contract Responsibility System and Development from 1978 to 2007.

Case 8: The Farmers' Association and Land Reform (1950–1953)

Before 1949, especially during the period of the war of Resistance against Japanese Invasion (1937–1945) and the Civil War (1945–1949), the land reform in fact had been done in the liberated areas (*jiefang qu*). After the funding of new China, this movement was continued in many new liberated areas, where there lived about 264 million farmers. According to the Land Reform Law of the People's Republic of China (*Zhonghua Renmin Gongheguo Tudi Gaige Fa*), this movement was gradually implemented.

Basically, this movement had two steps: the first step was conducting different *jianxiban* (workshops) about land reform by the Land Reform Training Institute. This training program was organized from the top-down. The central government trained the representatives at the provincial level, and then these trained representatives from different provinces went back their provinces and trained other representatives on down to the village level. These trained representatives used different ways to find the villages' real situation to prepare the land reform through the method of working in sites (*dundian*). Generally, they also had many characteristics of scholars in successful scholar-based governance. For example, they had high social status, high social responsibility, and high social capital with the locals, etc. The second step was organizing the farmers' association (*nongmin xiehui*). These farmers' associations were very similar to the ones talked about before, but now the government control was stricter, and the people often adopted the class-struggle (*jieji douzheng*) method to resolve problems and conflicts, especially between poor farmers and land owners or rich farmers. For example, in the Shaozhuang Village, some trained representatives were called by farmers as "*Liu Zhengfu*," "*Wang Zhengfu*," and "*Ren Zhengfu*" (Liu, Wang, and Ren were their family names, while "*Zhengfu*" meant the government) (Yu 2001). That is, *Principles 1.2 to 1.6 were satisfied, but Principles 1, 1.1, 2, and 4 were only weakly satisfied.*

Why could this movement relatively succeed although sometimes it was brutal and bloody due to the class-struggle? In addition to the aforementioned analysis about Principles 1 (including 1.1–1.6), 2, and 4, the first reason was that the governance system of these representatives partly applied democratic and collaborative management. That is, *Principle 3 was weakly satisfied.* The second reason was that the land reform satisfied poor farmers' fundamental interests and hopes, and they often completely supported this reform. Thus, *Principle 5 was satisfied.* Third, this movement was also gradually implemented and extended relying on previous experiments. That is, *Principle 6 was also weakly satisfied.* Fourth, at the beginning of the new China, the new government had unparalleled authority and rigorous organizations to implement its policy, and both government officials and farmers had unimaginable passion to support this policy. That is, *Principle 7 was satisfied.*

After this movement, the government began to exert strong control over the villagers, and grassroots governmental organizations were established. Particularly, when the land reform was finished, the farmers' association was gradually substituted with the new administrative regime in villages, and most of the cadres in the farmers' association became the new cadres in the new administrative system. For example, in Hengshan County, the farmer association's work was stopped in November 1953, and a new governmental system which employed many old cadres in the farmer association was set up in spring of 1954 (Yu 2001). By introducing the concept of "class (*jieji*)" to the villagers, however, the clan and family authority was dramatically decreased in villages, and the difference among various classes was strongly perceived. This was also found in the Shaozhuang Village (Yu 2001).

Case 9: The Cooperative Movement (1949–1956)

The Cooperative Movement led by the Communist Party was a rural centralization movement. Through this movement, the Communist Party tried to transfer the rural individual economy (*nongye geti jingji*) based on private ownership to centralize an economy (*jiti jingji*) which relied on collective ownership.

This movement had three phases. The first was from 1949 to 1953, mainly focusing on the mutual aid group (*huzhu zu*) along with the experiments on elementary forms of agricultural cooperatives. During the winter of 1952 to the spring of 1953 there was a kind of hot-headed and rash-advance tendency (*jizao maojin qushi*), however. On April 3, 1953, the government posed a new guideline called "steadily advance (*wenbu qianjin*)." But this guideline and new amendment were opposed by Zedong Mao in October and November 1954.

In December 1953, the central government made a new decision which promoted the agricultural cooperatives to a new developmental phase from the experimental phase, and this movement went into its second phase from 1954 to the first half year of 1955. During this period, the elemental cooperatives (*chuji she*) were widely established and developed all over the country. In 1956, after the investigation from Hunan Province, Zedong Mao suggested quickening the step, but Zihui Deng (the minister of Ministry of Agriculture) did not agree. Mao then said that Deng's idea was the one of right tendencies (*youqing*).

The third phase was from the second half of 1955 to the end of 1956. During this period, the cooperative movement rapidly developed, and about 96.3% families attended the elementary agricultural cooperatives and about 87.8% attended the advanced cooperatives until the end of 1956. Then, the whole society in the countryside changed to a socialist collective system from the old individual system.

During this movement, scholars were often excluded, strict governmental control and top-down methods were applied, farmers' real interests were only weakly considered, the experiment-extension method was abandoned, and external support except for strict governmental control was not guaranteed. In summary, all *Principles 1–7 were not satisfied*, and it was an unsuccessful case of scholar-based governance.

Case 10: The Great Leap Forward Movement (1958–1969)

The centralized movement did not stop at the stage of the Cooperative Movement. In January 1958, Mao criticized the "against rash-advance (*fan maojin*)" in Nanning Conference (*nanning huiyi*) and argued that this tendency was the Right Tendency (*youqing*) and was hindering progress. Opposing "against rash-advance (*fan maojin*)" (*fan fan-maojin*), he started "the Great Leap Forward (*da yuejin*)." After this, the government made some plans which set very high, unpractical objectives of agricultural production, and the newspaper even crazily propagated the "high yield satellite (*gaochan weixing*)" and the slogans such as "the crop from the field is as large as guts can yield (*ren you duoda dan, di you duoda chan*)." Because of this and the overestimation of agricultural production, at the Beidaihe Conference (Beidaihe Huiyi) in 1958, the central government even proposed a plan to ask every province to change their major task from the agriculture to industry, and even the farmers in the countryside were organized to attend to steel-making work. Although from the Zhengzhou Conference in November 1958 to the early period of the Lushan Conference (Lushan huiyi) in July 1959, Mao, and the Communist Party began to correct the Communism Vogue (*goongchan feng*) and Flatulent Vogue (*fukua feng*), Mao started up the animadversion to Dehuang Peng during the later period, and then the movement of resistance against the right tendencies was extended. Also, the left-tendencies in economic development were extended. Until the winter of 1960, because of the correction of the left-tendencies on the rural work by the central government and Zedong Mao, the Great Leap Forward Movement was finally stopped. From the standing point of scholar-based governance, *all Principles 1–7 were violated* in this movement, and it was an unsuccessful case.

Case 11: The People's Commune Movement (1958–1983)

The People's Commune Movement even made the conditions worse. In 1958, the Chinese Communist Party published a decision about establishing the people's commune in the countryside (*zhonggong zhongyang guanyu zai nongcun jianli renmin gongshe de jueyi*). During the first half of 1958, in order to construct irrigation works and improve agricultural production, many small cooperatives (*xiaoshe*) were successively united into large cooperatives (*dashe*), and people spontaneously called them "communist communes (*gongchan zhuyi gongshe*)," "large cooperatives (*dashe*)," "people's communes (*renmin gongshe*)," "collective farms (*jiti nongzhuang*)," etc. During the first ten days of August 1958, when Mao inspected Hebei, Henan and Shangdong provinces, he said that the people's commune was good, and its advantage was that it could unite the workers, farmers, businessmen, students and scholars, and soldiers together, making it was very convenient for leadership. Also, he pointed out two attributes of people's communes: The first was they were large, and the second was they were public (*yida ergong*). After that, people's communes were rapidly extended all over the country.

During August 17–30 in 1958, in the enlarged conference of the Political Bureau of the CPC Central Committee, the decision of the people's communes was published. This decision argued that the people's commune was the inevitable trend of development. After this, within only several months, the farmers who attended the people's communes reached over 120 million families, accounting for over 99% of the total population.

Yu (2001) classified the whole period of this movement into four phases. The first period was the funding period from 1958 to 1960. During this period, people's communes were widely established all over the country. Their governance system also had several general characteristics. First, they had a huge scale. For example, in Hengshan County in 1958, although there were 147.6 thousand families and the total population was 569.9 thousand, the total number of communes was only 25. This meant every commune had 5900 families and about 22,700 people. In 1959, the number of communes was reduced to 17, and then every commune had over 80 families and about over 33,500 people. Second, their political structure was militarized. People were organized through the military system to attend agricultural activities, and most of people were organized into militia. For example, in Hengshan County, almost all of the people from the age 16–45 took part in the militia organization, which was not only a military organization but also a method of social organization administration. Third, people's lives became the collective life. People ate together in the collective eatery, and sometimes they were even required to separate males and females. Fourth, the cadre was bureaucratized. For example, in Xiangtai County, even the grassroots managers were included into the governmental administrative system.

The second period was from 1960 to 1967 which was the period of putting into final form the design of people's communes. But because of the tragedy in the first period experiments, some new methods were adopted to correct the extreme policies. The main characteristics of these experiments were the separation of power between communes (*she*) and teams (*dui*, a smaller organization in countryside), reestablishing *qu* (between the town and the county) system, reducing the scale of communes, widely establishing the organization of the Communist Party in the villages, reinforcing the control of grassroots cadres, stopping the public eatery system, and resuming the family as the basic accounting unit.

The third period was the revolutionized period. Because of the Great Cultural Revolution, the organization of communes was substituted with the revolutionary committee (*geming weiyuanhui* or *gewei hui*). Even farmers' daily life was organized under the organization of the Communist Party, and the economic function of communes and teams was reduced.

The fourth period was the disaggregating period from 1979 to 1983. During this period, the people's communes were gradually disaggregated.

In summary, scholars' participation, federal organizations, concrete purposes, democratic and collaborative managements, effective mechanisms of awards and sections, stable scholar-entrepreneurs, and farmers' interests were all neglected by this movement. Even the experiment-extension method was not applied very well, and enough external support was not supplied except for strict governmental control. That is, all *Principles 1–7 were violated*, and it was an unsuccessful case of scholar-based governance.

Cases 12–16: The Intellectuals' Ideological Remolding Movement (1951–1952), the "Anti-Rightist" Movement (1957–1958), the Four Cleanups Movement (1963–1966), the Great Cultural Revolution (1966–1976), and the Rustication Movement of Educated Youth (1955–1978)

Roughly, the land reform and movements such as the cooperative movement and the people's commune movement reflected the direct interactions between governments and farmers, but the movement mentioned here mainly reflects the interaction between scholars and governments.

The relationship between scholars and governments can have three classical forms: Situation 1—a strong government and a weak scholar group; Situation 2—a government and a scholar group that counterbalance each other; and Situation 3—a weak government and a strong scholar group. In Situation 1, scholars' behaviors will be strongly influenced by the government. The most serious condition is that scholars are completely controlled by the government and even killed by government officials when the government and scholars are in extreme conflict each other such as in the burning of books and burying of Confucian scholars alive in Qin Dynasty. In Situation 2, the government and scholar groups often influence and depend on each other. In Situation 3, scholar groups may have more influence on the government. Particularly, when scholars become new scholar-entrepreneurs of farmers, they can organize farmers to oppose the government. Numerous farmers' revolts in ancient China were the good examples of this situation (Yang 1999).

In ancient times, especially after the Han Dynasty (202, BC 220), or more recently after the Sui Dynasty (581–618), because most of the government officials were originally scholars, Chinese society as a whole was governed by scholar-based governance. The only difference between scholars and government officials was when scholars took part in government they were formal government officials. After the Late Qing Dynasty, due to the reduced status of ancient scholars, this traditional system was partly stopped. In the Republic of China, because of weak government, scholars still had a strongly independent social identity, although they often were stopped or killed or assassinated by the government. In that period, because of the status of the out-party and them needing the help of scholars, the Communist Party could also treat scholars kindly. After the founding of new China, the People's Republic of China, the government's policy on scholars dramatically changed after a honeymoon period, because of ideological reasons and the imitation of the Soviet Union system. The following movements were the classical movements which strongly influenced the fate of Chinese scholars under this system.

The Intellectuals' Ideological Remolding Movement was the movement in which scholars were required to carry out self-education and self remolding particularly of their thought led by the new government. The purpose of this movement was to make the thought of scholars become Marxist thought. This movement was first started among the teachers and students in over 3000 colleges and universities in Beijing and Tianjing in September 1951. In 1952, it was extended to all the literature domain and intelligentsia, including scholars in the countryside. Although in one aspect, this movement actually helped scholars conquer some bad thoughts of imperialism, feudalism, and bureaucratism, it lifted the curtain on the attempt of government wanting to control scholar's thought and to establish a new government-controlled society.

After about five years, another tragedy which strongly influenced the Chinese scholars' fate began, and it was the Anti-Rightist movement. From April 1956 to April 1957, the government made a new policy of "all flowers are in bloom and contention of a hundred schools of thought (*baihua qifang, baijia zhengming*)" encouraging scholars to give advice to government to improve its work. Generally, the studies show that the original intention of this movement might be good. Scholars corresponded to the government to express their discontent and criticize government, and media also attended to this movement. Different ideas then grew up suddenly, and scholars thought the Communist Party was so great that it could self-criticize. After May 1957, the situation was dramatically changed, however. Because Mao and other leaders of the Communist Party thought there were some people who wanted to overthrow the new government under the cover of the new Movement of Rectification Campaign (*zhengfeng yundong*), they began to criticize that some scholars were rightists. Some people even argued the new Movement of Rectification Campaign in fact was a cabal (*yinmou*) which tried to draw a snake out of its hole (*yinshe chudong*). From then, the Anti-Rightist movement which tried to find rightists among scholars began.

Because of the simplified criteria of rightists and the inherent disadvantages of the Soviet Union system, this movement was extremely enlarged. According to the statistics in 1978, during the Anti-Rightist Movement in 1957 and the Movement of Making Up a Missed Lesson of Anti-Rightists (*fanyou buke*) in1958, totally about 550 thousand people were deemed as rightists. Some scholars even estimated the number might be larger. Most of these people in fact were not. Numerous scholars were falsely and unjustly charged. Until 1958, because the major task of the government transferred to economic development, this movement was gradually stopped. During this movement, many scholars, especially some scholars in the grassroots level, were persecuted to death. Until 1978, only about 100 thousand old-rightists were still alive. In 1977, Yaobang Hu was appointed as the president of Party School of the Central Committee of the Communist Party of China (zhongyang dangxiao) and began to redress the mishandled cases and depurge the rightist (pingfan). His work was highly praised by scholars. In 1989, his death became the fuse of the Tiananmen Event. After this movement, independent scholars were almost excluded from the political stage and almost lost all their influence on the government. Then nobody had the courage to criticize the government, and the government became more and more centralized.

The Four Cleanups Movement (or the Socialist Education Movement) was organized during 1963 to 1965. In the countryside, the first purpose of this movement was to cleanup the accounts (*zhuangmu*), storehouses (*cangku*), property (*caiwu*), and workpoints (*gongfen*). In a later period, however, the purpose was developed to cleanup politics, economy, thoughts, and organizations, which were called by a joint name "four cleanups (*siqing*)." During this movement, many cadres of the grass roots were strongly punished. Although some rectifications were adopted in the first half year of 1965, some more leftist ideas were developed. Finally, it became a reason for the Great Cultural Revolution which was really an unprecedented world-shaking catastrophe for Chinese scholars.

During the Great Cultural Revolution, in addition to its catastrophic influence on politics, economy and culture, numerous scholars in the countryside were persecuted to death or killed themselves. This meant that government tried to not only control scholars' behaviors, thoughts, and ideas but also corporeally eliminate them. Then the society in the countryside almost totally became a government-controlled society. Even common farmers were included in governmental actions through collective actions, or they were taken advantage of by government to oppose scholars.

The Rustication Movement of Educated Youth was also an important movement worthy of consideration. This movement started in 1955 and then gradually swept the whole country. From 1955 to 1961, it was the phase of experiments, and the purpose was to resolve the unemployment of the youth in cities. From 1962 to 1965, the government began to organizationally develop this movement. But the large-scale organization began in 1968 when Mao Zedong wanted to put a brake on the Movement of the Red Guards, and this was the third phase. Mao said the educated youth should go to the countryside to accept reeducation by poor and middle-class farmers. Almost all students in junior middle schools (*chuzhong*) and high schools in 1968 (they were enrolled in 1966, 1967, and 1968 and called "*lao sanjie*"), about 2.6 million, went down to the countryside. During the Cultural Revolution, about 14 million youth go down to the villages. These students either, like farmers, directly went to villages and lived with farmers or took part in the Production and Construction Corps (*shengchan jianshe bingtuan*) living like soldiers. Generally, researchers argued that the reasons why Mao wanted so many youths to go to the countryside was that he believed that the whole country in that era, especially during the Cultural Revolution, had been controlled by the people in authority taking the capitalist road (*zou zibenzhuyi daolu de dangquanpai*), and he thought the youth should go down to the countryside to know more about the farmers. There were also another two reasons. First, Mao wanted to dismiss the Red Guards during the Cultural Revolution through this new movement; second, the government wanted to use this movement to resolve the unemployment problem in cities, especially during the first and the second phases of this movement. All these reasons were reasonable, but I also want to add another one. Because of different political movements on scholars, the countryside almost had no the educated people to help government and farmers do some necessary work, and there were even no people who could do the work such as accounting and receiving and paying out money. Considering these, this movement in fact was also a blood transfusion of scholars to villages. Because most of these youths had no experience in the countryside and could not be educated as the criteria of local scholars had talked before, this movement improved some situations and sometimes developed education in the countryside, but its negative effects were larger than its positive effects. Furthermore, this movement made all the youths go to the countryside, stopping their higher education and research and then eliminated educated scholars in society from the perspective of the whole country. All these influenced the science and academic development and led to a tremendous loss to the whole society.

In summary, these aforementioned movements dramatically changed the fate of numerous Chinese scholars and were unprecedented catastrophes not only for Chinese scholars but also for the whole Chinese nation and country. From the perspective of the governance system, this movement reinforced government control on the whole society and the farmers through eliminating scholars from various methods (teachers and scholars were even called "Stinking Number Nine, *chou lao jiu*"). They totally distorted the whole society and whole country and finally made the Chinese society and culture go to the fringe of destruction. Obviously, all *Principles 1–7 were violated in these movements*, and all of them were unsuccessful cases of scholar-based governance.

Fortunately, after the death of Zedong Mao in 1976, especially after the 3rd Plenary Session of the 11th Central Committee of the CPC (*Zhongguo Gongchandang shiyijie san zhong quanhui*) in December 1978, most of the above policies have gradually stopped or changed, all the things were returned to the middle way because of the new opening up and reform policy. From then on, China has been experiencing a rapid development in every domain, especially in the economic domain, and this is often called as "China's Miracle." From the perspective of the countryside, the most important event worthy of carefully analyses is the experiment and extension of the Household Contract Responsibility System and Development. This policy still works now, and my case studies the period from 1978 to 2007. If the history from 1949 to 1978 was a period when scholars were gradually, sometimes even bloodily, excluded from the governance system, it has been a process of gradual renaissance of scholar-based governance from 1978 to 2007.

Case 17: The Household Contract Responsibility System and Development (1978–2007)

Although it had some other historical contributions such as land reform, the 30-year centralized movement put the countryside on the fringe of danger. Particularly due to reinforcing government authority, organizing farmers using administrative power, and excluding scholars, the interactions between farmers and the government became more and more direct and sometime even ridiculous. Partly because of lacking information, knowledge, and support from scholars, farmers became easier to be controlled by government, and sometimes they even blindly followed the rules set by the government. Their capabilities of resistance again the government's intervention and self-governance were both reduced. Also, partly because of lacking the information, knowledge, and balance from the scholar, government's policies became more and more radical and extreme, and the damages of bad policies were dramatically increased due to their direct control on farmers without scholar's checks and balances. Under such a situation, a new reform was gradually gestated. The classical event starting this reform occurred in a very small and poor village (at that time, called *shengchandui*, meant a production team), named Xiaogang,[5] Fengyang County, Anhui Province.

Fengyang was the hometown of Yuanzhang Zhu, who was the founding emperor of the Ming Dynasty (1368–1644). Like all the rural counties in China, this county also experienced all of the aforementioned movements. After several years of development during the early period of new China, from 1960 to 1978, the situation of this county gradually deteriorated. Particularly in 1978, the number of the people fleeing from the county due to famine suddenly increased to 13 thousand from several thousand during the previous years. The Xiaogang Village, as a small village in Fengyang County, experienced a similar situation, even worse. Before the Cooperative Movement, there were 34 households, 175 people, 90 laborers, 30 cultivating beasts of burden, and 1100-mu land (mu is Chinese unit of area, about 1/15 of a hectare). The average yield of grains was 90 thousand *jin* (a Chinese unit, about 600 g) per *mu*. During the fortunate years, it could reach 100 thousand *jin*, and every one could get over 500 kg. So people had enough food every year. Because of establishing the advanced communes, however, this situation was dramatically changed. Before 1955, the average income per person was 80 *yuan* (as a rich middle farmer at that period). But in 1956, it became 55.50 *yuan*. The yield of grain even dramatically decreased to 30-thousand kg in 1957 from 82.5-thousand kg in 1956. Some people began to flee from the village because of famine. Until the summer in 1962, there were only 10 households, 39 people, 1 cattle, and 100 *mu* cultivated land. In this year, farmers first tried the method of the Household Production Contract (*baochan daohu*), but their experiment was criticized as the restoration of the old system (*fubi*). During the Culture Revolution, the situation became more serious because of cruel and ridiculous intervention of the

[5]The materials of this case mainly come from the County Annals of Fengyang (AFDBW 1999) and Xia Yurun's *Xiaogang Cun yu Da Baogan* (2005).

government. For example, during the 10 years of the Culture Revolution, the yield of grain was only 1–1.5 thousand kg per year. In 1976, the average income per person became only 22 *yuan*, and the average yield of gain per person became about 50 kg. This village became famous for its poverty, bachelors, and fleeing from famine.

In October 1978, Junchuang Yan was elected as the leader (*duizhuang*), HongChang Yan was elected as the vice header (*fu duizhuang*), and Lixue Yan as the accountant (*kuaiji*) of the village. Because of a large drought, many villagers began to prepare to flee. These cadres began to think about how they could increase the yield of grain production. There were 20 households, 115 people in that year. Because of the conflicts among people, they thought the organization of communes in fact was not a good way to improve the yield. First, they divided the production team into two groups (*zu*) using the method of "work contracted to groups (*baogan daozu*)." But it still was not good, because the people in the two groups often conflicted with each other. They asked the commune secretary of the CPC, Minglou Zhang, to divide the production team into four task groups (*zuoye zu*). Because Zhang knew farmers' hardship and also knew this village was very intractable, he broke the rule to agree their ideas. But the conflict and problems still could not be resolved, because of the problems such as working times (some people went to work early but others late), the accounting of working points, and the unreasonable division of labor. In November 8 1978, they secretly divided the production team into eight groups. Thus, most of the groups became "father-son groups (*fuzi zu*)" and "brother groups (*xiongdi zu*)." When the groups became smaller, the enthusiasm for production was increased. Because of a lack of labor, even the older family members and children went to work together. If there were no beasts of burden, farmers did the work by hand. Thus, the task of growing in fall was rapidly finished.

But this good time did not last long. Because of the conflict among brothers, wives of brothers and neighbors, new problems appeared. Hongchuang Yan, the vice leader of the village, was an emulative young man. Relying on his personal experience of contracting for a job, he knew if this situation was kept, there was no hope for farmers. One evening, he went to talk with Tingzhu Guang, the oldest farmer in the village. Guang said, "The farmland covered by contract (*zeren tian*) in 1962 was very efficacious. If we do want not to quarrel with each other, the only way is to divide the individual household and let each do it alone. But I am afraid that government does not allow that, and your cadres have no courage to do it." When three leaders met together and discussed the problem, Hongchan Yan and Lixue Yan said they also asked some older people, who all suggested dividing the land into the individual household and letting each do it alone. Thus, in November 24, three leaders decided to hold a meeting of householders. Among 20 households, 18 householders attended. They were Hongchang Yan, Junchang Yan, Lixue Yan, Guopin Yan, Youjiang Guan, Youshen Guan, Jiaqi Yan, Fuchang Yan, Xuechang Yan, Guoyun Han, Youzhang Guang, Jiazhi Yan, Tinzhu Guang, Likun Yan, Lihua Yan, Lifu Yan, Meichang Yan, and Jinchang Yan. Only Youde Guang and Guochang Yan could not attend, because they were in Jiangxi Province because of fleeing from famine.

The topic of this meeting was how to do the agricultural work. People argued for different ideas. One farmer, Jiazhu Yan first said, "If we do not want to quarrel, and want to have a bowl of food to eat, the only way is to divide the land to the individual household and let each do it alone." Tingzhu Guan also said, "during the early days of 1950s, the land was divided into the household, all the people worked in full swing, people strived to be the first to send agricultural tax paid in grain (*gongliang*) and sold the rest of the grain, and every household had enough food and clothing and had no conflict and contradiction, so it would be better to do as in that period, divide the land, and let each household do it alone (*fentian dangan*)." "Great, right, divide the land and let each household do it alone," suddenly almost all the people agreed in jubilation. Some people, however, said "Although this idea is good, the government does not allow us to do this, and there are no cadres who have the courage. If this thing is known by the upper officials, disaster will happen. The leaders will be paraded under escort through the streets as a light punishment or will be possibly put into jail until death. 'Every people eat the meat, only one will afford the punishment (*rou chi qian kou, zui luo yi ren*)'." At that time, the leaders were also feared. But they really had no other way because of the poverty. Finally, they decided to try it for one year to see if they thought it was still worthy of being put into the jail for doing something to let villagers have food to eat. Then, the leader, Junchang Yan, stood up and said, "Since all the people want to divide the land to let the household do it alone, we, as cadres, do not want to be cowards (*nao*)." The prolocutor, Hongchang Yan, also the vice leader, stood up and said, "If every villager wants to divide the land to let households do it alone, we must make a contract." After a longtime talking, they made three agreements:

(1) After dividing the land for every household, every household should first pay the agricultural national tax and withdrawing and retaining of collectivity using the first-yield grain of the second season in middle fall (*quqiu erji*), and nobody would be a coward and betrayer (*zhuang naozhong*).
(2) The policy is actually to divide the land to the household but ostensibly organized as larger groups (*mingzu anhu*); any person is disallowed to tell this to the upper officials and other people. Anyone who does that will be the public enemy of all villagers.
(3) If the leaders of production team were put into jail, all the team members should afford the farm work of their households and guarantee to bring up their children until they are 18.

When the contract of fixing a plot of land for each household was written, all the people swore and vowed and said they'd never break their promise. Then all the householders put their fingerprints or personal seal on the contract, and even the nephew of Youde Guang and Guochang Yan also put the fingerprints for their uncles (see Fig. I.1). After the meeting, the villagers immediately distributed beasts of burden, farm implements, and measured land in the same night. The entire land (totally about 517 *mu*) was divided for every household considering the good and the bad quality. Every two households got one farm cattle after converting into money. All the things were done before the morning of the next day.

Fig. I.1 The scene of 18 farmers putting fingerprints or private seals on the secrete contract (the sculpture in the memorial in the Xiaogang Village) and its original script. Adapted from Xia (2005)

However, their secret actions were immediately noticed by the commune, and the leaders of the commune persuaded the farmers to return to the group organization, but nobody agreed. Then the leaders criticized the Xiaogang Village in the meeting of the cadres and announced that if the farmers could not return to the group organization, they would not give them chemical fertilizer, agricultural loan, and seeds. When the farmers knew this, they all disagreed. They resolved the loan by themselves and used the farmhouse fertilizer to substitute for the chemical fertilizer. Only seeds, which could not be bought from the market and also borrowed from other villages, were still the problem, so they had to ask the commune. In April 15, 1979, when the head of the county went to the Liyuan commune to inspect its work, the cadres of the commune reported the problems of Xiaogang Village to him. He and other cadres of communes went to Xiaogang at once. After listening to the report of the leaders of the village and seeing the good growing of the crops and the farmers' enthusiasm, he said "now they become so poor (*qiong 'hui' diao le*), what kind of capitalism can they do? At most they only want to produce some grains to resolve the food problem. They had already divided the land, and the crops were planted by households. If they are reunited, it is even not good to cast accounts. Let them do one year, this may work. Our county has over 3000 production teams; this one village is irrelevant to the overall situation. Furthermore, anyhow the crops cannot run to other places, and the seeds could be given to them." Thus, this experiment in the Xiaogang Village was safeguarded. From 1979 to 1980, Chen and other leaders of County Committee of the CPC went to Liyuan and Xiaogang 13 times and helped farmers resolve many problems.

At the beginning phase, the leaders of the County Committee also were ambivalent. On one side, they thought this new method could improve the farmers' enthusiasm and they should protect the farmers' benefit and safeguard this new thing; on the other hand, they thought this thing was not a tribal matter and they had to adopt some steps to limit this method. They sent the working team to the villages to suggest the farmers reunite themselves through the method "two household forming one group." But their constraints were only said strictly. Until December in 1979, the number of households which attended the new economic experiments was increased to 7160 (about 7.5% of total number in the county) from 202 in August. In order to compare different production responsibility systems, according to the directions of regional Committee of CPC, the County Committee sent Tingmei Wu, who was a member of the policy research office of the Country Committee, to the Xiaogang Village to do an investigation. He wrote a very important report *One Dose of Absolutely Necessary Tonic—the Survey of "Work Contracted to Households" in the Xiaogang Production Team, Liyuan Commune, Fengyang County* (*Yiji Bibukeshao de Buyao—Fenagyang Xian Liyuan She Xiaogang Shengchandui "Baogan Daohu" de Diaocha*), which concretely and vividly described the achievement of this new experiment and highly appraised it. On January 1, 1980, when Chen Tingyuan went to Hefei, the capital of Anhui Province, to attend the agriculture meeting, he gave this report (about 7000 Chinese characters) to Li Wan, who was one of the leaders of Provincial Committee of the CPC. Wan said he read that report twice in one breath as reading a novel and was deeply moved by the experiment in the Xiaogang Village and praised this method and farmers' initiative spirit. This information was suddenly extended in Fengyang County; many villages began to imitate Xiaogang's experiment. At the end of

February 1980, the case of Shaoqi Liu, who was the former National Chairman of China and was persecuted to death during the Culture Revolution, was redressed during the 5th plenary session of the 11th Central Committee of the CPC. The farmers in Fengyang said, "The case Shaoqi Liu was redressed, and farmland covered by contract can be done!" Until April, about 23% household in Fengyang County attended the experiment of work contracted to households. The County Committee also supported the extension of this new experiment by adopting the close-door method. In March 18, Tingyuan Chen and Wanli Xu, who was the vice secretary of the County Committee, went to Banqiao Qu and required the Qu Committee to make an administration instruction. In April 25, with the support of the policy research office of the County Committee, the Banqiao Qu Committee of the CPC wrote an administration direction (*Guangyu Nongyeshengchan Shixing "Baochandaohu" Zerenzhi de Jidian Yijian*) which was published in an inner communicated magazine "*Qingkuang Fanying*" and was distributed to other Qu Committees. This was the first time the County Committee publicly supported the extension of "work contracted to households" using the written method. Shortly, after the extension of "work contracted to households" in Fengyang County, the Central Committee of the CPC published a notice (Guanyu Jinyibu Jiaqiang he Wanshan Nongye Shengchan Zerenzhi de Jige Wenti de Tongzhi), which showed that the Central Committee of the CPC agreed to extend "work contracted to households" in remote and poor areas. At the end of 1980, 80% households attended the new economic reform in Fengyang County. Until the second half of 1982, in Fengyang County, all households attended in this reform.

As a famous collective action, the experiment of the Xiaogang Village was undoubtedly successful. According to Tingmei Wu's record, the yield of grains in first year of "*da baogan*" reached 132,370 *jin*, which was similar to the total amount of previous five years from 1966 to 1970. The total yield of oil plants reached 35,200 *jin*. Especially, the farmers said that the total amount of peanuts over the last 20 years was not over the yield of this year. The number of breed pigs was 135 which were over the historical record of any previous year. During the last 23 years, the farmers never could completely accomplish the task of the grain requisition by purchase about 2800 *jin* per year, but this year they sold 24,995 *jin* to government, which was in excess of the required amount over seven times. The farmers also sold dried sweet potato 5000 *jin*. The totally sold peanut and sesame were 24,933 *jin* which was over the required task about 80 times, and this task was never accomplished before. In 1980, the yield even increased 67.7% compared with the yield in 1979. The farmer income also increased, and in that year, the average income per person reached over 400 *yuan*. In sum, the change was dramatic and was continued after 1980 (see Fig. I.2).

From then on, the experiment was gradually extended all over the country. According to statistical analysis, until the end of 1983, 99.5% of households in

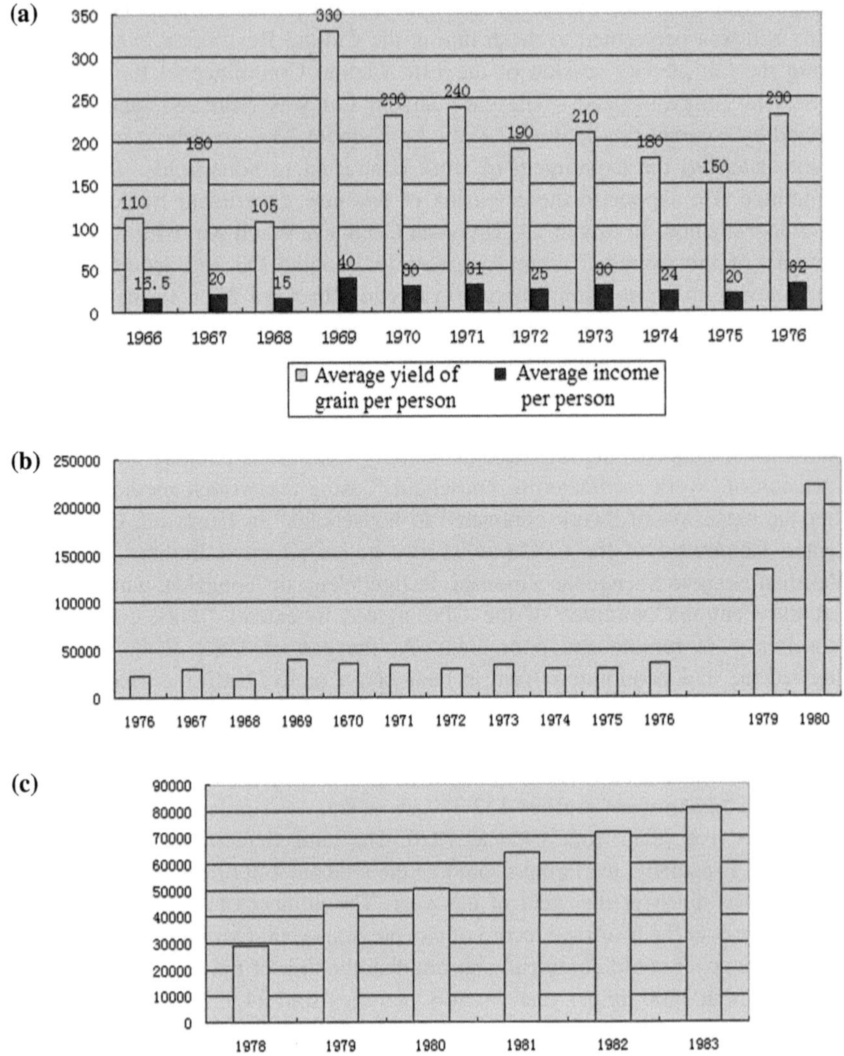

Fig. I.2 Statistics of the development in the Xiaogang Village from 1966 to 1983. **a** The statistics of average yield of grain per person and average income per person from 1966 to 1976 (*jin* and *yuan*). **b** The statistics of total yield of grain from 1966 to 1980 (*jin*). **c** The statistics of total yield of grain from 1978 to 1983. Adapted from Xia (2005)

China participated in the Household Contract Responsibility System, and among them about 97.8% adopted the method of the Xiaogang Village. The Xiaogang model then becomes the major method of Chinese agricultural management. Generally, scholars think the success of this experiment and its extension is a classical case in that the method was initially invented by the farmers and then accepted and extended by the government. But I think it is also a good case of scholar-based governance.

Before 1978, because scholars had been almost completely excluded from the governance system, government realized its complete control on farmers, and the whole countryside slide to the fringe of danger. Three leaders and the oldest farmer in the Xiaogang Village now stood up on the historical stage, and they actually played the role as new entrepreneurs. As stated before, they showed all the characteristics of scholars in successful scholar-based governance. As the farmers, they knew what the problems were and knew the farmers' hope and demands. As the elected leaders, they had more knowledge and information about the external world and the problems and also were respected by the leaders. In particular, they knew how to find the problem solutions through asking the oldest farmers and to get their support. They even had great courage to do a revolution, because they were already prepared to be put in the jail if their plan failed, and this also let them have some relatively independent social identity to oppose intervention from the upper government officials. If these leaders were put into jail, the farmer would lose new entrepreneurs. Fortunately, their new experiment was secretly supported by the leaders of the county and was even supported by the provincial leaders later and finally was supported by the national leaders such as Xiaoping Deng and Yaobang Hu. For example, in January 1980, when Li Wan authorized that the experiment in Xiaogang could be done five years during his inspection in the village, in Beijing, from January 11 to February 2, the experiment in Xiaogang was strongly criticized by most representatives who attended the meeting on the administration of the people's communes. The opponents argued that the Household Contract Responsibility System meant doing it alone and was the very dangerous capitalist method. Most of the leaders of national agricultural committee said all the things should be done according to the document of central committee of the CPC. This meant they in fact stand on the opposite side of the Xiaogang's experiment. In January 31, the leaders of the Political Bureau of the CPC Central Committee such as Guofeng Hua, Xiaoping Deng, Xiannian Li, Yaobang Hu, Qiuli Yu, Renzhong Wang, and Yilin Yao attended the meeting on the reporting of the above conference. Hua said that the farmers who participated in the experiment should be gradually organized together although he thought the household contract responsibility system was different from letting each household do it alone. Deng said that it was not very easy to evaluate this experiment because this was a big problem and he also had no information about that before. After this conference, however, Li Wan was promoted to Beijing and was elected as the secretary of the Secretariat of the CPC Central Committee in February in the 5th plenary session of the 11th Central Committee of the CPC and appointed as the vice premier in the 14th session of the Standing Committee of the 5th National People's Congress. After talking with Wan Li, reading many documents, and careful pondering, in April 2, 1980,

Xiaoping Deng called Yaobang Hu, Li Wan, Yilin Yao, Liqun Deng, etc., together and publicly supported Xiaogang's experiment. After that, he gave several other speeches to publicly support the experiment within several months. Yun Chen, Yaobang Hu, and other leaders also publicly showed their support. Thus, the external intervention of this experiment was excluded, and this was also an important reason for its success.

The aforementioned support was based on scholars' efforts. During this experiment, in addition to the local scholars growing from the farmers, some outside scholars also played a very important role, although they were already absorbed in the government system. For example, Tingmei Wu, as a scholar (he was also a member of the policy research office of the Country Committee), kept his relatively independent social identity and social conscience and highly praised the experiment in Xiaogang. In particular, he proved that the experiment in Xiaogang was not a capitalist movement and was consistent with theory of Marxism. At that era, this theoretical analysis was extremely important. This meant that he not only personally praised the experiment but also found the theoretical support for the experiment. As a researcher in this event, first he was an information collector who theorized the experiment. Second, he was a helper to the farmers through his report, praise, and theoretical analysis. Third, he was an information provider. For example, his report helped Tingyuan Chen continue his policy, and even Li Wan was moved by it. These helped the experiment get the support from the provincial leaders. The reason why this experiment was supported by the different leaders was also related to another important discussion, the Grand Discussion on the Criterion of Truth (*guanyu zhenli biaozhun wenti de taolun*). This movement drew a clear conclusion that practice is the sole criterion for testing truth (*shijian shi jianyan zhenli de weiyi biaozhun*). All these clearly show that the success of the Xiaogang experiment was also a case of successful scholar-based governance, although the scholars in this experiment did not have a completely independent social identity. That is, *Principles 1, 1.2–1.6, and 4 were satisfied*, but *Principle 1.1 was weakly* satisfied.

From the governance perspective, as described above, no matter whether we discuss the problem from the county level, provincial level, or the country level, this important reform in the countryside was consistent with characteristics of successful scholar-based governance. The organizational structure of this reform was a bottom-up federal form from the families to the village and finally to the whole country. Its purpose to let farmers have enough food to eat was very clear and concrete. At the grassroots level, this experiment was democratic and used the different multiple incentives as shown in the farmers' contracts. Collaborative management was also applied. The most important thing was that this experiment highly satisfied the farmers' hope and interests. Also, the experiment-extension method was applied. Furthermore, external support, especially governmental or institutional support, was guaranteed. In summary, all *Principles 2–7 were satisfied.*

I.6 Echoes from Other Countries (1970s–2007)

People's Planning in Kerala in India is a case in a large developing country, the Gal Oya Irrigation Project in Sri Lanka is a case from a small developing country, but the Saemaul Movement in Korea is a case in a semi-developed country. From the cultural perspective, Korea is a Confucian country influenced also by Buddhism and Christianity, India is Hindu and Moslem country, and Sri Lanka is a Buddhist country.

Case 18: People's Planning in Kerala, India (1970s–2007)[6]

As one of the poorest states in India, Kerala is located on the southwestern coast of this country and has 29 million inhabitants. After the implementation of social and redistributive policies by the successive governments of the state since the middle to late 1970s, particularly the implementation of the Kerala People's Campaign for Decentralized Planning since 1996, Kerala does a remarkable radical experiment in democracy (Isaac and Franke 2002). Their efforts of flattening out the distribution of income in Kerala were deemed remarkably successful (Fischer 2000; Isasc and Franke 2002). Although Kerala was not yet a model of sustainable development (Véron 2000), Franke and Chasin (2000, p. 38) argued, "Having learned their lessons from the past, Kerala's people are creating prospects for a sustainable future." Because a lot of scholars participated in this problem, let's analyze how the seven principles of successful scholar-based governance were satisfied from 1970s to 2007.

Scientific capacity building (Shrum and Ramanathaiyer 2000) and stable local field-based scholars played a very important role in this program. First, scholars from the people's science movement (KSSP) as local stable field-based scholars contributed a lot to this program (Isaac and Franke 2002). KSSP was established in 1962 by a number of scientists and social activists in Kerala "who were concerned that scientific information was basically inaccessible to the majority of the people of the state" (Fischer 2000, p. 162). Second, many other experts from the locality and outside were invited to participate in this program. For example, they were invited to attend the seminars to "come up with 'integrated solutions' for the various problems identified at the lower-level conventions" (Ibid, p. 160). Some volunteer experts were even invited by the State Planning Board to help workgroups prepare the final document. Third, many other kinds of scholars such as trained resource persons who served as discussion facilitators to guide what was best described as a "semi-structured discussion" (Ibid, p. 160) and the "wise farmers" who were also organized into a cadre of experts "to assist the local panchayats in their discussion" (Ibid, p. 167). These scholars not only had enough expertise but also had enough

[6]Because the projects People's Planning in Kerala, India and the Gal Oya Irrigation Project, Sri Lanka still continue, I chose the last years of the cited references as the end years of these two cases.

local, easily understood, and hands-on knowledge. For example, Fischer pointed out:

> Information about the local area is gathered by citizen groups, and specific development problems are identified. The citizens are then assisted in analyzing these problems on the basis of their own experience, and to the extent possible, they suggest solutions. (Ibid, p. 160)

Furthermore, the questions were often presented in local rather than scientific terminology, and local leaders are asked to imaginatively adapt successful models to specific local circumstances rather than only replicate it (Ibid). Thus, *Principles 1 and 1.3 were satisfied.* This does not mean they did that perfectly, however. Sharma (2003) claimed that the lack of technical, administrative, and financial know-how of the local bodies was a major bottleneck of decentralization, and it was even true in Kerala where literacy and government facilities were superior to those of other states.

The case described by Fischer (2000) also showed that these scholars had high enthusiasm and capability of participation, relatively independent social identity and high social status, high social capital, high social responsibility and spirit of practice, and respect for other social actors and local knowledge. For example, the fact that all of KSSP's members served as volunteers (Ibid) and worked very hard to help local people or governments to resolve the social problems showed their high enthusiasm and social responsibility. As local scholars, they had high social capital with local residents, governmental officials, and other social actors. For example, they obtained data by "door-to-door surveys of the villagers" (Ibid, p. 165). Because of two major efforts (a struggle over the building of a proposed hydroelectric dam in 1984 and involvement in a literacy campaign), they even got national attention. Fischer pointed out:

> So impressive were the results that KSSP was awarded the 'Alternative Nobel Prize' in 1996 by the Swedish foundation that has been giving the prize regularly now for some years. Because of the success of the project, coupled with the award, politicians and other organizations have found it difficult to ignore KSSP. (Ibid, p. 163)

These two efforts showed their capacity for participation and their spirit of practice. As a mass movement, the scholars involved in KSSP respected local residents and other social actors and their local knowledge. Also, the self-generated income permitted KSSP "to remain free of influences from outside government agencies and NGOs" (Ibid, p. 162). Thus, *Principles 1.1, 1.2, and 1.4–1.6 were satisfied.*

This program built a bottom-up federal organization and had concrete organizational purposes. For example, Fischer (2000, p. 160) pointed out, "This decentralized planning process constitutes a hierarchy of deliberations moving from the village to the block and district levels upward to the State Planning Board." He (Ibid) also argued that "The actual process commences with the formation of various groups to deal with specific issues, such as agriculture, schools, and environment." Thus, *Principle 2 was satisfied.*

Decentralized governance, bottom-up and deliberative planning, participatory democracy, and civil society building in People's Planning (Chettiparamb 2006;

Heller et al. 2007; Isaac and Franke 2002) showed that democratic management was applied in this program. The investment that Kerala had made in decentralization really had no parallels anywhere in the developing counties (John and Chathukulam 2003), although technical competence at the local level and the attitudes of actors were obstacles of decentralization in the first place (Mukundan and Bray 2006), and the state of Kerala only got a score of 2.00 according to a scale ranging between 0 and 5 of an ideal form of decentralization (John and Chathukulam 2003). Fischer pointed out, "35–40% of the planning activities would be formulated and implemented from below and allocated to the local level an equivalent share of the planning resources" (Ibid, p. 159). But the participation of the Left Democratic Front (LDF) brought people closer to their local government (Fischer 2000). The cooperative banks (see Isaac and Franke 2002, pp. 166–167), class coordination (Heller 2000), the extensive network of voluntary organizations and mass movement which motivated and brought together local representatives, officials in the various line departments, governmental and nongovernmental experts relevant to the local planning process (such as KSSP) and the mass citizenry (Fischer 2000, pp. 162–164), the synergies civil society, and local government bodies and state government (Törnquist 2000; Véron 2001) showed the internal collaborative management at different levels within India. Also, the involvement of the Swedish foundations and the like showed its international collaboration. Thus, *Principle 3 was satisfied.*

The analysis of different kinds of these scholars' involvements also showed that stable local scholar-entrepreneurship was also built in Kerala. KSSP was a good example. Thus, *Principle 4 was satisfied.*

The new policy in Kerala tried very hard to seek reconciliation of social, productive, and environmental objectives at the local level (Törnquist 2000; Véron 2001). Fischer argued:

> The groups engage in a detailed review of the proposals made at the seminar and draw up working plans, basing them on the necessary technical considerations, time frameworks of the overall process, cost-benefit estimates, and other details mandated by the State Planning Board. For each scheme, the task forces also assess the resources required, as well as the level of the available resources. In particular, they examine the extent to which the costs can be met through contributions from local financial and nonfinancial institutions, as well as through the labor of efforts of local volunteers. (Fischer 2000, p. 161)

Their participation in resource mapping and their success of improving social development in Kerala also showed this. Although the mismatch between social development and economic growth had attracted scholars' attention for some time and it was still a formidable change for Kerala to resolve the paradox—"social development coexisting with economic backwardness" (Kannan 2000, pp. 64–65), totally compared with the other cases, *Principle 5 in Kerala was satisfied.*

The experiment-extension method was applied. For example, KSSP's further participation in land literacy (see Fischer 2000, pp. 163–166) was based on their two successful efforts as stated above. A number of experiments tried by the Left Democratic Front (LDF) also showed this (for detailed analyses, please see Fischer

2000, pp. 158–159). Also, the replicability of the achievements of Kerala in other states was studied (Ramachandran 2000). Thus, *Principle 6 was satisfied.*

External scholars' participation as stated above and governmental support from the central government (e.g., it sponsored the literacy campaign) guaranteed its stable external technical, financial, and institutional support (see Isaac and Franke 2002). Also, because this program mainly was a self-governed one within the state of Kerala, its requirement for the external support was less. That is, *Principle 7 was also satisfied.*

Case 19: The Gal Oya Irrigation Project in Sri Lanka (1950–1992)

Relying on the case studies mainly by Perera (1986), Uphoff (1985a, b, c, 1986a, b, c) and Kasyanathan (1986), Ostrom (1990) deemed the Gal Oya Irrigation Project in Sri Lanka as a fragile case of self-governance. She pointed out:

> … I am unwilling to assume that these are robust institutions. Given the long history of central control, it would be difficult for farmers in Gal Oya to continue their organized efforts of a major change in the Irrigation Department were to place in office engineers who presumed that local farmers had little to offer. (Ibid, p. 181).

I found it is also a semi-successful case of scholar-based governance, however. Based on the found literature, this case study mainly focuses on the period from 1950 to 1992.

According to the game theoretical framework of product-institutional analysis (PIA), the major game parties of this case included farmers, institutional organizers (IOs) and ID officials. Before the IOs' participation the irrigation games were played mainly among different kinds of farmers and between farmers and ID officials as the representatives of government. Mutual distrust and animosity, however, had been reproduced over several generations (Ibid). Cooperation among farmers was actually minimal, and social relations among settlers coming from different areas of the country were often strained (Perera 1986). The conflict between Tamil speakers who were most of the cultivators in the tail areas and settled from nearby coastal areas and resettled Sinhalese who were most of the upstream cultivators (Upoff 1986a, b, c) made the entire situation even much worse and tense. Thus, the games among different farmers, especially the game between the farmers in the tail areas and the upstream areas, could not lead to effective self-governance. Also, mistrust and recriminations between farmers and ID officials were popular. Many field-level officials were often notorious for their corruption and thuggery, and farmers lost confidence in the competence or the trustworthiness of the ID's staff and even believed that local-level officials who had political and bureaucratic power behind them in fact were the main obstacle to efficient water management (Perera 1986). The ID officials, on the other hand, claimed that farmers were a part of the problem, and it was necessary to organize, educate, and discipline the farmers to do what they asked them to do because farmers could not use water responsibly and carefully (Ibid).

After the IOs' participation, however, people overcame perverse incentives that lead to suboptimal outcomes. Let's see how different principles of successful scholar-based governance had been satisfied.

Let's analyze these IOs' personal characteristics. First, these IOs were college graduates, and this social position and character would let them "be able to grasp organizing principles rapidly" and let them "have the status needed to deal effectively with ID officials" (Ostrom 1990, p. 168). Second, their farm background and the requirement—"if possible, from large settlements like Gal Oya" ensured that they could work with the farmers. Third, they had to be "willing to live in the remote project area" (Ibid, p.168). Ostrom also pointed out:

> The ARTI/Cornell team specifically rejected that model of external regimentation. Instead, they chose to facilitate the problem-solving capabilities of local farmers and officials by introducing "human catalysts" who were to work directly with farmers and officials at the field-channel level trying to solve problems. (Ibid. p. 172)

Fourth, they were given about six weeks of training in "how to approach and motivate farmers and in technical subjects related to agriculture and irrigation" (Ibid, p. 168) although some of them "were field with inadequate training" (Ibid, p. 171). Fifth, they were divided into small groups of four and five; and each group was responsible for the area served by one distributary canal. Sixth, "Each team divided its areas into smaller units using field canals as the primary basis for division" (Ibid, p. 168). Seventh, "Each group of IOs met weekly in order to learn from each other's experiences and bolster morale when necessary" (Ibid, p. 168). All these tell us that the IOs in fact had almost all the requirements of the scholars of successful scholar-based governance: the capacity and tendency for learning, the field knowledge except for the expert knowledge, the comparative advantages in information and knowledge, social responsibilities, and as field worker, researchers, etc. Although a four-year project was not long enough, when farmer organizations were deemed as legitimate and permanent tools for coping with the long-term problems by both farmers and irrigation officials, the local stable scholar-entrepreneurship was also relatively built. Thus, *Principles 1.1–1.6 were satisfied,* and *Principles 1 and 4 were weakly satisfied.*

Now let's analyze the characteristics of their organizations. Ostrom argued:

> An IO was expected, first, to meet each farmer sharing water from a field canal to discuss the types of agriculture and irrigation problems they faced and to complete a survey of relevant information about the area. After becoming familiar with the farmers and their problems, the IO was expected to meet informally with small groups of farmers sharing the same field channel to plan self-help strategies. Instead of establishing a predefined organization, the IO tried to form a working committee to solve particular problems, such as repairing a broken control gate or desilting a field channel. Furtherer, IOs identified problems beyond those that could be solved by the farmers working together, problems that needed to be articulated to ID officials and others. Once farmers were used to working together and had achieved benefits from group action, the IO would then help form a local organization and select, through consensus, a farmer-representative. This representative could articulate the interests of the other farmers on his field channel at larger meeting and report back to the others what had happened in larger arenas. (Ibid, pp. 168–169)

All these descriptions along with four tiers of organization—the "field-channel organization" (FCO), the "distributory channel organization" (DCO), the branch-canal or area level which followed after FCOs and DCOs, and the project-level committee—and the discussion "between the farmers and engineers about the plans for local rehabilitation" (Ostrom 1990, p. 169) organized by the IOs and their nonpartisanship of FCOS and DCOs showed that the federal organizations were relatively built in Gal Oya, the purposes of the IOs were also relatively concrete, and democratic and collaborative management were applied. Also, consensus was a dominant rule in making decisions, and farmers were organized on four mutually reinforcing levels and were given reorganization and encouragements. As a result, more communications, better understanding and mutual trust were facilitated (Kasyanathan 1986; Perera 1986). On the one hand, "By the time the design phase was initiated, the farmers had already begun to work together and had good ideas about how to rehabilitate their field channels" (Ostrom 1990, p. 169). Even attitude and behaviors of many powerful farmers in the area who had originally opposed "the nonpartisan nature of the organization and bypassing those who had been the elite" were changed, and the degree of conflict among different farmers had also declined (Ibid). Uphoff (1992) pointed out that headenders even voluntarily shut their own gates so that tailenders' crops would get enough water when the value of water peaks in dry season. On the other hand, through these discussions, the ID officials began to change their fundamental orientation toward the farmers and finally changed their attitude toward the farmers, and farmers also changed their "attitude and behavior of irrigation officials toward them and in the policies adopted by the ID" (Ostrom 1990, p. 169). Uphoff (1992) even argued that after bureaucrats found that farmer participation began to yield efficiencies in infrastructural maintenance, water usage, and crop production, they surrendered their jealously guarded prerogatives. The information of graduated sanctions, however, was not found, and the major weakness in Gal Oya was that "farmers were expected to undertake construction at the field-channel level without pay" (Ostrom 1990, p. 171). Thus, *Principles 2 and 3 were weakly satisfied.*

In this case, the gains of farmers were obvious, although they "were expected to undertake construction at the field-channel level without pay" which was deemed as "an unrealistic hope" by Ostrom (1990, p. 171). Particularly since 1950s, the quantity of paddy rice produced in Sri Lanka had grown steadily. The gains of the ID officials were also obvious. In addition to the payment by the ARTI/Cornell team, at least they resolved the problem that they wanted to resolve and reduced their efforts. Thus, Principle 5 was satisfied.

Because the ARTI/Cornell team first started development of IOs in a 5000-acre plot area near the head of the system and then gradually expanded it to 25,000 acres (Ostrom 1990), *Principle 6 was also satisfied.*

Ostrom argued that "the Gal Oya project demonstrates how external agents may help appropriators overcome perverse incentives that lead to suboptimal outcomes, even when traditions of mutual distrust and animosity have been reproduced over several generations" (Ibid, p. 172), and it is not necessary to maintain large numbers of catalysts in the field for a long time when "both farmers and irrigation officials come to view the resulting farmer organizations as legitimate and permanent tools for coping with the long-term problems involved in the governance and management of any complex irrigation system" (Ibid, p. 173). But her principle of "minimal recognition of rights to organize" was only weakly applied in the Gal Oya project. Thus, *Principle 7 was weakly satisfied.*

Case 20: New Village Movement in South Korea (1972–1980)

This case mainly draws on the work by Douglass (2000), Ha (2001), Lee (1990), Park (1998) and Turner et al. (1993). The New Village Movement, also known as the New Community Movement, is called the Saemaul Ŭndong or the Saemaul Movement in South Korea. Korea is a country strongly influenced by Confucianism, especially in the Yi Dynasty (1392–1910). The ancient governance of the countryside in Korea was very similar to China and also could be deemed as a kind of scholar-based governance. However, these scholars in Korea were yangban, a social class of civilian and military officials (Turner et al. 1993), who formed a council for each magistrate, relayed the feeling of villagers to the district office, and served as advisers to the chief officials (Ibid). At that time, they played the roles not only as information providers and agents of government but also scholar-entrepreneurs and pure game players. So Turner et al. argued, "Self-help movements were organized by yangban during the Yi Dynasty, as they sought to encourage the expansion of irrigation and to further cooperation in agricultural production among villagers (Ibid).

After new Korea was built in 1948, the government leaders gave little attention to the countryside under the target of industrial development and the expansion of export trade during 1950s and 1960s. Because of difficulties living in the villages, lots of farmers took part in a rapid movement to the larger cities to search for better jobs, and this made the proportion of farmers in the total population fall from 70% in 1950 to 46% by 1971. More seriously, because many better-educated young men and women moved out and many villages were left with a component of the older members, the supply of talented people who might rise to positions of leadership in their communities was reduced. Under such a situation, the new village movement was conducted. Although there are different explanations of this movement, most scholars agree that the success of this movement was due to the Saemaul leaders and the government's efforts through the village classification system and Saemaul leaders training institutes (for detailed analysis, please Lee 1990, pp. 22–23). Thus, I am willing to deem it as a successful case of scholar-based governance.

Before the new village movement, because of lack of leadership, the capability of self-governance of villages was weak, and external help became necessary. From the perspective of the villages, the first force of this movement came from an external agency, the Korean government. If the central Korean governments had tried to directly control the behaviors and actions of farmers, a Leviathan model would have been applied. They did not do this, however. An alternative way they chose was scholar-based governance, in which the new educated and organized scholars, the Saemaul leaders, played a very important role in this movement. Let's analyze how the seven principles were satisfied.

The characteristics of the Saemaul leaders were consistent with the requirements of scholars in successful scholar-based governance.

First, the Saemaul leaders were relatively stable field-based workers. Most of them had a residential attachment to their villagers dating back to their grandfathers' or fathers' generation and were engaged in agriculture. Even if the outsiders paid by the government might perform an important role, they could not be "true leaders." Furthermore, these leaders had higher levels of education than other villagers, even the former leaders of the villagers. Especially, most of the leaders who were over the age of fifty were familiar with modern forms of organization and technology because of the experience of military service as private or staff sergeants. The methods used in the training program included cases studies of Saemaul success stories, practical knowledge, and sharing of experiences through group discussions for the purpose of enlightening them on the essential features of the Saemaul spirit and motivating them for continued action. They were even required to wear the same uniforms, live together, follow the same schedules, and even change their values to develop a common perspective through intermingling rural leaders and people in white-collar occupations to minimize rural/urban and class distinctions, sing songs, and attend early morning jogs according to the daily routine (Turner et al. 1993).

Second, in addition to the above education and knowledge characteristics and advantages, generally these leaders were in their 40s—younger than the average age of the population by about four years, and they owned more land than the ordinary villagers and had substantially higher yearly incomes. Together, they were respected and trusted by the local people. Also, they exhibited a willingness to accept common people's suggestions and to lead them in a democratic fashion (Ibid).

Third, they must be infused with "Saemaul spirit' and eager to sacrifice their personal interests for the welfare of community. Also, they should be creative, innovative, persuasive, and leading by example to encourage their neighbors to tear themselves loose from harmful traditions and help them realize diligence, self-help, and cooperation, which were deemed as the components of the "Saemaul spirit" (Ibid).

All these showed that as a relatively independent group even though they were partly created and organized by the government program, the characteristics of the Saemaul leaders were consistent with our requirements of scholars in successful scholar-based governance such as stable field scholars, enough knowledge (local knowledge and non-local knowledge), capable of learning and teaching, with high social status, independent social identity and high social capital, respect for the locals and the other actors, and social responsibility and action-tendency. Thus, *Principle 1 and 1.1–1.6 were satisfied.*

The characteristics of governance of the Saemaul movement were consistent with successful scholar-based governance.

First, in addition to the organizational system created according to the model of the Central Consultative Committee at each tier of administration from central government and the province to the township, the Saemaul organizations at the grassroots level were democratic and independent from the direct control from the upper levels of governmental systems. The basic objectives of the Saemaul movement were to increase farmers' income and make the communities better to live through modernizing villages, and more concretely, were "to provide tile roofs for houses, and to improve village roads and other infrastructure, such as water wells and electricity" (Douglass 2000, p. 11). From this perspective, they were bottom-up and federal organizations with concrete and clear purposes. Thus, *Principle 2 was satisfied.*

Second, the management and activities of the Saemaul organization at the grassroots level were democratic with gaining consensus and a strong incentive system both at the village and the individual levels. The residents elected two people as Saemaul leaders: a man and a woman. Generally, "the man assumed the primary leadership role, leaving the female leader in charge of women's affairs" (Turner et al. 1993, p. 79). They also could be chosen at the residents' request, "while some were selected by the influential members of the settlements or by the village development committee; only a few were appointed directly by the township chiefs" (Ibid, p. 82). Five to fifteen popularly elected members formed a village development committee, and one of the leaders was chair to assist the work of the Saemaul leaders. Occasionally, the village chief also would be chair of the committee, but usually the village chief and the Saemaul chief were kept separate. As to their major task, Turner et al. told us:

> The most important tasks of the Saemaul leader and his or her development committee was to agree among themselves on a project, often suggested to them by the county or township officials, and then to persuade the villagers to accept and implement the plan. It stood a better chance of being adopted when it was compatible with village needs and resources and when it was oriented to toward the benefit of the entire community rather than a few individuals. In the process of gaining consensus, the leader and members of the development committee sought the support of influential elders in the community (often these respected people were the effective decision makers), and then they took the matter to a general assembly meeting in the village hall. Approval at this level implied that the villagers had consented to participate by contributing their labor, money, and sometimes materials. The Saemaul leaders were responsible for carrying out the plan, recruiting labor, and making job assignments. They attended meetings at the township office to give reports and to receive directions from the administrators there. (Ibid, 1993, p. 80)

At the individual level, the community residents were distinguished by the amount of their voluntary contribution to the project, primarily through donations of man-hours of labor, sometimes also including a share of the investment costs (Ibid, p. 83); even the Saemaul leaders were also distinguished by their contributions. At the village level, they were classified into different phases such as underdeveloped (basic), developing (self-helping), and developed (self-managing) before 1974, and the developing/self-sustaining group or the developed/self-sustaining group in 1974. In 1988, the welfare phase was added. The methods of incentives were also numerous, including the grants (as in the case of the cement and steel rods), the cash aid either as grants or loans, Presidential Awards, the order of Merit, medals, etc. Although this system stimulated competition among the villages, which might remain an obstacle to the desired cooperation among them, totally it made the system work very well. Together, it can be concluded that *Principle 3 was satisfied.*

Third, the average length of incumbency of the Saemaul leaders was four to five years, "with the longest tenures being characteristic of officials of the villages that had developed the most rapidly" (Turner et al. 1993, p. 82). Thus, the stable entrepreneurship was guaranteed, and *Principle 4 was satisfied.*

Fourth, the benefits of the movement to the villagers were obvious and were reinforced along with the program development. For example, the villagers' living conditions and incomes were improved, and then rural and urban income differentials were narrowing (Kim 2000). Park argued:

> Clearly, the decade of the 1970s was an exciting period for farmers in Korea, for the age-old thatched roofs, traditional lamps, carrying things on men's backs and women's heads had vanished. For the first time in the history of villages, vehicles were coming into villages, and plowing by draft cattle was being replaced by engine-powered tillers. Cash incomes were increasing annually, allowing them to afford the purchase of consumer goods. The villagers obtained a confidence that "we can do" and "we can live better." (Park 1998, p. 125)

Also, the governmental officials received award and salaries. Even the president got more support form the people and his followers. Although the Saemaul leaders did not receive salaries, they were the recipients of some material benefits. For example, they were given priority when they applied for loans for their farming activities, they were allowed discounted fares when they travelled on official business, and their children were given special consideration when they sought educational scholarships. These leaders were also favored when they tried to secure jobs in government organizations" (Turner et al. 1993, p. 84). Thus, *Principle 5 was satisfied.*

Fifth, this program began as an accidental experiment at the outset without the initial pattern of implementation. Turner et al. pointed out:

> In 1971, the government had a surplus of cement, and the central officials decided to distribute 335 bags to each village with the proviso that the inhabitants use it, along with private capital and contributed labor, for community projects. The experiment caught on in many places, and a year later, the 16,600 villages (out of a total of 35,000) that had been the most responsive through self-help were each given an additional 500 bags of cement, one tone of steel rods, and some cash for further development of their community ventures. Seeing what was happening in these successful villages, 6108 other settlements instituted projects of their own volition, without assistance from the government. This development pointed the way to a more organized program that would help to dispel the disenchantment of rural voters. (Ibid, p. 76)

The cooperation among individuals was started by some small-scale endeavors such as "helping a neighbor build a new house, assisting in the field at harvest time, or coming to the aid of a family during an emergency." This small-scale cooperation was then extended to larger scales such as "the building of roads and bridges, the construction of an irrigation system, erosion control, and the creation of community workshops—programs aimed at developing infrastructure that would improve the economy of the community and at the same time benefit individuals indirectly" (Ibid, p. 76). Also, the movement was conducted in a stage-by-stage approach: The initial phase was one of self-help "designed to improve the living conditions of individual households;" the second phase "concentrated on the development of economic infrastructure;" and in the third phase, "the villagers focused their attention upon generating increased income, an aim that involved an even higher degree of cooperation" (Ibid, p. 77). Ha (2001) pointed out that the community development of the Saemaul movement in rural regions in the early 1960s was also spread to towns and metropolitan regions. Thus, *Principle 6 was satisfied.*

Sixth, the external support was guaranteed in the Saemaul movement. Especially, the father of the Saemaul movement is the president Park Chung Hee, "who had grown up in a poor rural environment, was sympathetic to the problems of the farmers" (Turner et al. 1993, p. 75). Under his leadership, many government organizations and agencies were mobilized to support this movement. Also, his policy was followed and even amended by the following presidents. Thus, this program lasted from 1972 to the mid-1980s, and *Principle 7 was satisfied.*

References for Appendices

AFDBW, Anhuisheng Fengyangxian Difangzhi Bianzuan Weiyuanhui. (1999). *Fengyang xianzhi (County annals of Fengyang)*, Beijing: Beijing Fangzhi Chubanshe http://kjj.fengyang.gov.cn/zjfy/fyxz.php (In Chinese).

Ai, K. X. (1995). Lindi baohu lu zai hefang—Yijnhuoluo Qi shi ruhe baohu lindi ziyuan de. *Journal of Inner Mongolia Forestry*, 4, 18–19 (In Chinese).

Alexander, H. (1995). *A framework for change: The state of the Community Landcare Movement in Australia. The National Landcare Facilitator Annual Report*. Canberra: National Landcare Program.

Alitto, G. (1986). *The last Confucian: Liang Shu-ming and the Chinese dilemma of modernity* (2nd ed.). Berkeley: University of California Press.

Allen, B. H. (1988). Wildlife, recreation, and other resources. In H. F. Heady (Ed.), *The Vale Rangeland Rehabilitation Program: An evaluation* (Pacific Northwest Research Station Resource Bulletin PNW-RB-157) (pp. 65–85). Portland, OR, USA: US Department of the Interior, Bureau of Land Management and US Department of Agriculture, Forest Service.

Ashraf, M. M. (1981). Farm water management at Mona Reclamation Experimental Project: An overview. In *Farm water management: Report and proceedings of the international expert consultation on farm water management* (pp. 67–82). Held in Islamabad, Pakistan (26 September to 4 October).

Babaev, A. G., Batyrov, A., Freikin, Z. G., Khodzhakuliev, Y. S., Lavrinenko, V. T., Lavronenko, O. S., Nechaeva, N. T., Nikolaev, V. N., Orlovsky, N. S., Petrov, M. P., & Sarkisov, M. M. (1980). Integrated desert development and desertification control in the Turkmenian SSR. In M. R. Biswas & A. K Biwas (Eds.), *Desertification: Associated case studies prepared for the United Nations Conference on Desertification* (pp. 481–523). Oxford: Pergamon Press.

Babaev, A. G., Muradov, C. O., & Khachaturyan, A. A. (1999). International cooperation in desertification control. In A. G. Babaev (Ed.), *Desert problems and desertification in Central Asia the researches of the Desert Institute* (pp. 257–265). Berlin: Springer.

Bai, C. (1982). Zai Chengjisihan changmian de difang—fang Yijinhuoluo Qi. *Journal of Inner Mongolia Forestry*, 2, 19–20 (In Chinese).

Bai, X. F., Chen, H. W., & Peng, J. F. (2006). Study on spatial change of ecosystem services value: A case study of Yijinhuoluo County. *China Land Science*, 20(6), 16–20, & 30 (In Chinese).

Barros Arana, D. (1886). *Historia general de Chile*. Santiago de Chile (In Spanish).

L. Yang, *Knowledge-Driven Governance*,
https://doi.org/10.1007/978-981-13-2910-4

Bartlett, E. T., Bartolome, J. W., & Quigley, M. Q. (1988). Chapter 6: Cost and Benefit of the Vale Program. In H. F. Heady (Ed.), *The Vale Rangeland Rehabilitation Program: An evaluation* (Pacific Northwest Research Station Resource Bulletin PNW-RB-157) (pp. 86–102). Portland, OR, USA: US Department of the Interior, Bureau of Land Management and US Department of Agriculture, Forest Service.

Bartolome, J. W., & Heady, H. F. (1988). Grazing management 1962 to 1986. In H. F. Heady (Ed.), *The Vale Rangeland Rehabilitation Program: An evaluation* (Pacific Northwest Research Station Resource Bulletin PNW-RB-157) (pp. 54–64). Portland, OR, USA: US Department of the Interior, Bureau of Land Management and US Department of Agriculture, Forest Service.

Baumhardt, R. L. (2003). The dust bowl era. In B.A. Stewart and T.A. Howell (Eds.), *Encyclopedia of water science* (pp. 187–191). NY: Marcel-Dekker.

Bezlova, A. (2006, November 11). Keeping the desert at bay. *Asia Time Online*. Retrieved October 16, 2008 from http://www.atimes.com/atimes/China_Business/HK01Cb02.html.

Borchert, J. R. (1971). The dust bowl in the 1970 s. Annals of the Association of *American Geographers*, 61(1), 1–22.

Breckle, S-W. (1982). The significance of salinity. In B. Spooner & H.S. Mann (Eds), *Desertification and development: Dryland ecology in social perspective* (pp. 277–292). New York: Academic Press.

Brooks, N. & Legrand, M. (2000). Dust variability over Northern Africa and rainfall in the Sahel. In S. J. McLaren and D. R. Kniveton (Eds.), *Linking climate change to land surface change* (pp. 1–25). Dordrecht: Kluwer.

Bu, X. L. (2005, May 27). Huochu yibeizi, ganhao yijian shi. *People's Daily*. Retrieved May 5, 2008, from http://www.bjdj.gov.cn/Article/ShowArticle.asp?ArticleID=25706 (In Chinese).

Buhe C. L., Dong, F. L., & Sun, R. (2001). Ganli Chaotou xie rensheng—ji Naiman Qi Guoyou Sharihaolai linchang changzhang Zhao Jingshuang. *Journal of Inner Mongolia Forestry*, 7, 31–32 (In Chinese).

Campbell, A. (1992). *Landcare in Australia: Taking the long view in tough times, third annual report*. Canberra: National Soil Conservation Program.

Campbell, A. (1994). *Landcare: Communities shaping the land and future*. Sydney: Allen and Unwin.

Carle, J., & Ma, Q. (2005). Challenges of translating science into practice: poplars and other species in the Three North Region of China. *Unasylva*, 221(56), 31–37.

Carrera, J. A., Simon, E., & Fisac, M. (1981). *Regional management of Mediterranean ecosystems in Spain. Presented at the Symposium on Dynamics and Management of Mediterranean-type Ecosystems*. San Diego, California (June 22–26).

Chamala, S. (1995). Overview of participative action approaches in Australian land and water management. In S. Chamala & K. Keith (Eds.), *Participative approaches for Landcare: Perspectives, policies and programs* (pp. 5–42). Brisbane: Australian Academic Press.

Chambers, R., Pacey, A. & Thrupp, L. (Eds). (1989). *Farmer first-farmer innovation and agricultural research*. London: Intermediate Technology Publications.

Chang, C. (1955). *The Chinese gentry: Studies on their role in nineteenth-Century Chinese society*. Washington: University of Washington Press.

Chao, L. (2005, February 17). Joint forestry project to see 3000 gectares. *China Daily*. Retrieved May 16, 2008, from http://www.chinadaily.com.cn/english/doc/2005-02/17/content_416948.htm.

Chen, L. J. (2000, September 6). Hanzhi bei ju Maowusu. *People's Daily*. Retrieved May 16, 2008, from http://www.envir.online.sh.cn/info/np/file.asp?file=9-6-9.txt (In Chinese).

Chen, Z. T. (2008, May 2). Wang Xueding shuji tiaoyan weiwen Zhongkeyuan Naiman Shamohua Yanjiuzhan. *Zhongguo Kexueyuan Lanzhou Fenyuan wangzhan* . Retrieved May 16, 2008, from http://210.77.70.10/show.aspx?id=517&cid=26 (In Chinese).

Cheng, Z. L. (China Features). (2007, September 26). *Tree by tree, China rolls back deserts. Embassy of the People's Republic of China in the United Kingdom of Great Britain and Northern Ireland.* Retrieved May 19, 2008, from http://uk.china-embassy.org/eng/zt/Features/t367033.htm.

Chettiparamb, A. (2006). Bottom-up Planning and the future of planning education in India. *Journal of Planning Education and Research*, 26, 185–194.

China National Committee for the Implementation of the UN Convention to Combat Desertification (CCICCD). (1996, August). *China National Action Program to Combat Desertification* (Abstract). Retrieved by October 16, 2008, from http://www.desertification.gov.cn/xinxi/xdfa/0.htm.

Chinese National Ecosystem Observation & Research Network. (2006). *Naiman Desertification Research Station, Chinese Academy of Sciences.* Retrieved May 16, 2008, http://www.ceorn.org.cn/ceorn_en/web/stations/stationindex.aspx?NodeID=93.

Christie, E. K. (1986). *Pastoralism and ecology in arid Australia.* Melbourne: Longman Cheshire Pty. Ltd.

Ci, L. J. (2004). Large-scale sandstorm disasters–Their causes and prevention measures. *China Development Gateway.* Retrieved May 18, 2008, http://en.chinagate.com.cn/english/257.htm.

Confucian analects. (1900). In *Zhushi jiaozheng huaying sishu* (*Si shu*) (English & Chinese) (Trans. by J. Legge). Shanghai: Shangwu Yinshuguan.

Curtis, A. (1998). Agency–community partnership in Landcare: Lessons for State-Sponsored Citizen Resource Management. *Environmental Management*, 22(4), 563–574.

Curtis, A. & Lacy, T. D. (1996). Landcare in Australia: Does it make a difference? *Journal of Environmental Management*, 46, 119–137.

Curtis, A., Van Nouhuys, M., Robinson, W., & Mackay, J. (2000). Exploring landcare effectiveness using organizational theory. *Australian Geographer*, 31(3), 349–366.

Dennell, R. W. (1982). Dryland agriculture and soil conservation: An archaeological study of Check-dam Farming and Wadi Siltation. In B. Spooner & H.S. Mann (Eds.), *Desertification and development: Dryland ecology in social perspective* (pp. 171–200). New York: Academic Press.

Department of the Environment. (1980). The Turan Programme. In M. R. Biswas & A. K Biwas (Eds.), *Desertification: Associated case studies prepared for the United Nations Conference on Desertification* (pp. 181–251). Oxford: Pergamon Press.

Dhir, R. P. (1982). The human factor in ecological history. In B. Spooner & H.S. Mann (Eds.), *Desertification and development: Dryland ecology in social perspective* (pp. 311–331). New York: Academic Press.

Dorr, J. (2004). *Desertification of the Gobi. Geology comps papers, Carleton Geology Department, Carleton College.* Posted at Digital Commons, Carleton College. Retrieved May 17, 2008, from http://digitalcommons.carleton.edu/gcp/5.

Dou, Z. Y., & Song, B. M. (1995). Sanbei linglingjiu xiangmu zai Naiman jinzhan shunli. *Journal of Inner Mongolia Forestry*, 1, 11–12 (In Chinese).

Douglass, M. (2000). *Turning points in the Korean space-economy: From the developmental state to intercity competition, 1953–2000.* The Urban Dynamics of East Asia Project Discussion Papers.

Duffy, P. & Migongo-Bake, E. (2003). The application of EIA in the evaluation of sustainable desertification control activity: examples from Namibia and Inner Mongolia. *Impact Assessment and Project Appraisal*, 21(1), 65–73.

Durikov, M., & Annaklycheva, J. (2005). Realization of the National Action Program to Combat Desertification in Turkmenistan. In B. Faye & P. Esenov (Eds.), *Desertification combat and food safety: The added value of camel producers* (pp. 55–59). Amsterdam: IOS Press.

Environmental Protection Service, Government of Israel. (1980). The Negev—a desert reclaimed. In J. A. Mabbutt & C. Flotet (Eds.), *Case studies on desertification* (pp. 263–265). Paris: Unesco.

Evenari, M. (1989). *The awakening desert: The autobiography of an Israeli scientist*. Berlin: Springer-Verlag.

Evenari, M., Shanan, L., & Tadmor N. (1982). *The Negev: The challenge of a desert*. Cambridge: Harvard University Press.

Fan, S. Y. & Zhou, L. H. (2001). Desertification control in China: Possible solutions. *Ambio*, 30 (6), 384–385.

Feng, Q, Cheng, G. D., Endo, K. N. (2001). Towards sustainable development of the environmentally degraded River Heihe basin, China. *Hydrohgica lSciences-Journal-des Sciences Hydrologiques*, 46(5), 647–658.

Fischer, F. (2000). *Citizens, experts, and the environment: The politics of local knowledge*. Durham and London: Duke University Press.

Forestry of Gansu. (2005). Jiakuai fangsha zhisha shifanqu jianshe bufa—Zhangye fangsha zhisha shiji shifanqu jianshe zongshu. *Forestry of Gansu*, 2, 37–38 (In Chinese).

Forest Division of Yulin Prefecture, Shaanxi Province. (1998). Introduction to the control and development of sand area of Yulin. In traditional knowledge and practical techniques for combating desertification in China. Zhongguo Shamohua Fangzhi Wang (desertification.gov.cn). Retrieved May 19, 2008, from http://www.desertification.gov.cn/xinxi/book2/ctk-12.htm.

Franke, R. W, & Chasin, B. H. (2000). Is the Kerala model sustainable? Lessons from the past, prospects for the future. In G. Parayil (Ed.), Kerala: The Development Experience Reflections on Sustainability and Replicability (pp. 16–39). London: Zed Books.

Grainger, A. (1984). *Desertification* (2nd ed.). London: Earthscan - International Institute for Environment and Development – IIED.

Government of Chile. (1980). Desertification in the region of Coquimbo, Chile. In J. A. Mabbutt & C. Flotet (Eds.), *Case studies on desertification* (pp. 52–114). Paris: Unesco.

Government of India. (1980). Desertification in the Luni Development Block, Rajasthan, India. In J. A. Mabbutt & C. Flotet (Eds.), *Case studies on desertification* (pp. 147–175). Paris: Unesco.

Government of Iraq. (1980). Desertification in the Greater Mussayeb Project. In J. A.Mabbutt & C. Flotet (Eds.), *Case studies on desertification* (pp. 176–213). Paris: Unesco.

Government of Niger. (1980). Desertification in the Eghazer and Azawak Region. In J. A. Mabbutt & C. Flotet (Eds.), *Case studies on desertification* (pp. 115–146). Paris: Unesco.

Government of Pakistan. (1980). Desertification in the Mona Reclamation Experimental Project, Pakistan. In J. A. Mabbutt & C. Flotet (Eds.), *Case studies on desertification* (pp. 214–252). Paris: Unesco.

Government of Tunisia. (1980). Desertification in the Oglat Merteba Region, Tunisia. In J. A. Mabbutt & C. Flotet (Eds.), *Case studies on desertification* (pp. 11–51). Paris: Unesco.

Government of the United States of America. (1980). Vale Rangeland Rehabilitation Programme, Oregon. In J. A. Mabbutt & C. Flotet (Eds.), *Case studies on desertification* (pp. 269–272). Paris: Unesco.

Ha, S.-K. (2001). Developing a community-based approach to urban redevelopment. *GeoJournal*, 53: 39–45.

Hansen, Z. K., & Libecap, G. D. (2004). Small farms, externalities, and the Dust Bowl of the 1930s. *Journal of Political Economy*, 112 (3), 665–694.

Hayford, C. W. (1990). *To the People: James Yen and village China*. New York: Columbia University Press, 1990.

Heady, H. F. (1988a). The Vale BLM District. In H. F. Heady (Ed.), *The Vale Rangeland Rehabilitation Program: An evaluation* (Pacific Northwest Research Station Resource Bulletin PNW-RB-157, pp. 1–19). Portland, OR, USA: US Department of the Interior, Bureau of Land Management and US Department of Agriculture, Forest Service.

Heady, H. F. (1988b). The Vale Rehabilitation Program. In Heady. H. F. (Ed.), *The Vale Rangeland Rehabilitation Program: An evaluation* (Pacific Northwest Research Station Resource Bulletin PNW-RB-157, pp. 20–35). Portland, OR, USA: US Department of the Interior, Bureau of Land Management and US Department of Agriculture, Forest Service.

Heady. H. F. (1988c). *The Vale Rangeland Rehabilitation Program: An evaluation* (Pacific Northwest Research Station Resource Bulletin PNW-RB-157). Portland, OR, USA: US Department of the Interior, Bureau of Land Management and US Department of Agriculture, Forest Service.

Heller, P. (2000). Social capital and the developmental state: Industrial workers in Kerala. In G. Parayil (Ed.), *Kerala: The development experience reflections on sustainability and replicability* (pp. 66–87). London: Zed Books.

Heller, P., Harilal, K. N., & Chaudhuri, S. (2007). Building local democracy: Evaluating the impact of decentralization in Kerala, India. *World Development*, 35(4), 626–648.

Ho, P. (2000). The Myth of Desertification at China's Northwestern Frontier: The Case of Ningxia Province, 1929–1958. *Modern China*, 26(3), 348–395.

Hobbs, P. R., Sayre, K. D., & Ortiz-Monasterio, J. I. (1997). *Increasing wheat yields sustainability through agronomic means* (Paper 98-01). CIMMYT, Mexico.

HSCXBW, Hunan Sheng Changsha Xianzhi Bianzuan Weiyuanhui. (1995). Changsha xianzhi (County annals of Changsha). Shanghai: Sanlian Shudian (In Chinese).

Hu, Q. (1948). Liangzhong quanli jiafeng zhong de baozhang. In X. Fei, & H. Wu (Eds.). *Huangquan he shenquan*. Shanghai: Shanghai Guancha Chubanshe: 135 (In Chinese).

Huntsinger, L., & Heady, H. (1988). Perceptions of the Vale Program. In H. F. Heady (Ed.), *The Vale Rangeland Rehabilitation Program: An evaluation* (Pacific Northwest Research Station Resource Bulletin PNW-RB-157, pp. 103–133). Portland, OR, USA: US Department of the Interior, Bureau of Land Management and US Department of Agriculture, Forest Service.

Hurt, R. D. (1981). *The dust bowl: An agricultural and social history*. Chicago: Nelson-Hall.

Ibáñez, J. J., Sánchez Díaz, J., Alba, S. de, López Árias, M., Boixadera, J. (2003). *Collection of soil information in Spain: A review in 2003* (European Soil Bureau Research Report No. 9).

IIMI. (1999). Impact assessment of Resource Conservation Technology (RCT) to enhance wheat Ppoductivity. In *Farmers' perceptions and observational evidences from RCT trials in Pakistan's Punjab Province Rice-Wheat cropping system*. Lahore: International Irrigation Management Institute.

Inner Mongolia Daily. (1990). Damo zhonghun—ji Yijinhuoluo Qi xinjie zhishazhan yuan dangzhibu shuji Wang Yushan. *Journal of Inner Mongolia Forestry*, 3, 32–34 (In Chinese).

Institute of Botany, Chinese Academy of Science. (2003). Zhongguo Kexueyuan Zhiwu Yanjiusuo Eerduosi Shadi Caodi Shengtai Yanjiuzhan. Retrieved May 12, 2008, from http://eco.ibcas.ac.cn/pub/liangshichu/OSERS/member.htm (In Chinese).

Institute of Sand-Control of Shaanxi (ISCS). (2008). Institute of Sand-Control of Shaanxi (ISCS). Retrieved May 15, 2008, from http://www.iscsx.com/shanxi%20zhisha/index.html.

Isaac, T. M. T. & Franke, R. W. (2002). *Local democracy and development The Kerala people's campaign for decentralized planning*. Lanham: Rowman & Littlefield Publishers, INC.

Jodha, N. S. (1982). The role of administration in desertification land tenure as a factor in the historical ecology of Western Rajasthan. In B. Spooner & H.S. Mann (Eds.), *Desertification and development: Dryland ecology in social perspective* (pp. 333–350). New York: Academic Press.

John, M. S., & Chathukulam, J. (2003). Measuring decentralization: The case of Kerala (India). *Public Administration and Development*, 23, 347–360.

Kannan, K. P. (2000). Poverty alleviation as advancing basic human capabilities: Kerala's achievements compared. In G. Parayil (Ed.), Kerala: The Development Experience Reflections on Sustainability and Replicability (pp. 40–65). London: Zed Books.

Kasyanathan, N. (1986). The farmer organization component in final impact assessment study of the Gal Oya Water Management Project. *Manuscript, ARTI Research Study Series*, Colombo, Sri Lanka.

Keenan, B. (1977). *The Dewey experiment in China. Educational reform and political power in the early republic*. Cambridge, MA: Harvard University Press.

Kharin, N. (2002). Vegetation degradation in Central Asia under the impact of human activities. Dordrecht: Kluwer Academic Publishers.

Kim, E.-S., Park, D. K., Zhao, X. Y., Hong, S. K., Koh, K. S., Suh, M. H., & Kim, Y. S. (2006). Sustainable management of grassland ecosystems for controlling Asian dusts and desertification in Asian continent and a suggestion of Eco-Village study in China. *Ecological Research*, 21, 907–911.

Kim, Y.-W. (2000). Territorial development policy—Spatial development. Proceedings from t*he OECD Korea Workshop on Territorial Policies and Issues*, Seoul, March 8–9, 2000, 73–95.

Lee, M (1990). *Policy and rule configuration: Korean rural development movement. Saemaul Undong*. Indiana University: Workshop in Political Theory and Policy Analysis.

Li, D. X. (2008). Shahai zhongde nongchaoer—Ji Tongliao di 2 jie renda daibiao Bao Ligao. *Naiman Qi Renda Changweihui Wangzhan*. Retrieved May 20, 2008, from http://www.nmrd.gov.cn/dbgz2005051901.htm (In Chinese).

Li, Y. C. (June 17, 1999). *Zai "Kaifa Jihuashu Fangzhi Huangmohua Shiyong Jishu Zuijia Shijian Jiang" Banjiang Yishi Shangde Jianggua*. Retrieved May 21, 2008, from http://www.desertification.gov.cn/gongyue/speechbyliyucai.html (In Chinese).

Li, Y. H., & Qian, L. P. (2006). T*he Great Western development strategy: Controversy and a case study of Zhangye, Gansu Province. 3rd CPN Conference Proceeding* (pp. 192-201). Retrieved October 16, 2008, from http://chinaplanningnetwork.org/english/2006PDF/YUHUI%20LI.pdf.

Li, S. W., Zhang, H. (1991). Bianchui lühua chuan jiahua—Hetian junfenqu zhishu zaolin jishi. *Forestry of Xinjiang*, 3, 11-12 (In Chinese).

Liu W. F. (September 14, 2005). Battling the desert in Minqin. *China Daily*. Retrieved May 21, 2008, from http://china.org.cn/english/environment/142016.htm.

Liu, H., Cai, X. M., Geng, L. H., & Zhong, H. P. (2005). Restoration of pastureland ecosystems: Case study of western Inner Mongolia. *Journal of Water Resources Planning and Management*, November/December, 420–430.

Liu, Z. X., & Zhu, Z. M. (1998). Jianshe shansha liangqu, shisian zaizao Naiman. *Inner Mongolia Publicize*, 6, 37 (In Chinese).

Lockie, S., & Vanclay, F. (1997). Critical Landcare: Introduction. L. Stewart & F. Vanclay (Eds.), *Critical landcare* (pp. 1-7). Wagga Wagga: Centre for Rural Social Research, Charles Sturt University.

Lü, S. M. (November 23, 2007). Gansu Songhe Cun zonghe zhili shahuan chengxiao xianzhu. *Xinhua News*. Retrieved May 8, 2008, from http://news.xinhuanet.com/misc/2007-11/23/content_7131500.htm (In Chinese).

Ludwig, J. A. & Tongway, D. J. (1995). Desertification in Australia: An eye to grass roots and landscapes. In D. A. Mouat & C. F. Hutchinson (Eds.), *Desertification in developed countries* (pp. 231-237). Dordrecht: Kluwer Academic Publishers.

Luo, F., Qi, S. Z., & Xiao, H. L. (2005). Landscape change and sandy desertification in arid areas: a case study in the Zhangye Region of Gansu Province, China. *Environmental Geology*, 49, 90–97.

Ma, T. S. (1999, March 12). Ecological agriculture is a necessary choice of sustainable development of rural economy. *China Economic Times*, p. A4.

Mabbutt, J. A. (1977). Perspectives on desertification. *Economic Geography*, 53 (4), 429–432.

Mabbutt, J.A. & Wilson, A. W. (1980). S*ocial and environmental aspects of desertification*. Tokyo: United Nations University Press.

Malhotra, S. P. & H. S. Mann. (1982). In B. Spooner & H.S. Mann (Eds), *Desertification and development: Dryland ecology in social perspective* (pp. 305–310). New York: Academic Press.

Mann, H. S. (1982). The Central Arid Zone Research Institute (India). In B. Spooner & H.S. Mann (Eds.), *Desertification and development: Dryland ecology in social perspective* (pp. 293–303). New York: Academic Press.

Martin, M. A. (1982). Conservation at the local level: Individual perceptions and group mechanisms. In B. Spooner & H.S. Mann (Eds.), *Desertification and development: Dryland ecology in social perspective* (pp. 145–169). New York: Academic Press.

Matley, I. M. (1970). The Golodnaya Steppe: A Russian irrigation venture in Central Asia. *Geographical Review*, 60 (3), 328–346.
McGarry, D. (2005). The "current status" of conservation agriculture in China: A report for the People's Republic of China/global environmental facility (OP12)/Asian Development Bank "partnership on land degradation in dryland ecosystems" (first draft). In *Asian Development Bank*. Retrieved May 23, 2008, from http://www.adb.org/Projects/PRC_GEF_Partnership/current-status.pdf.
Michaelowa, A., Rawat, V. R. S. (2007). CDM afforestation and reforestation baseline methodologies: An analysis of the submission and approval process. *Hamburg Institute of International Economics (HWWI)*. Retrieved May 17, 2008, from http://www.hwwi.org/uploads/tx_wilpubdb/HWWI_Research_Paper_4-9.pdf.
Middleton, N. J. (1986). *The Geography of dust storms*. Unpublished doctoral dissertation, University of Oxford, Oxford.
Mitchell, D. J., Fullen, M. A., Trueman, I. C., & Fearnehough, W. (1998). Sustainability of reclaimed desertified land in Ningxia, China. *Journal of Arid Environment*, 39, 239–251.
Mukundan, M., & Bray, M. (2006). The decentralization of education in Kerala State, India: Rhetoric and Reality. In J. Zajda (Ed.), *Decentralisation and privatisation in education: The role the state* (pp. 111–131). Dordrecht: Springer.
Mulas, M., & Mulas, G. (2004). The strategic use of atriplex and opuntia to combat desertification. *Short and Medium- Term Priority Environmental Action Programme (SMAP), Desertification Research Group, University of Sassari*. Retrieved May 18, 2008, from http://desa.uniss.it/mulas/desertIN.pdf.
Network for desertification information and data exchange. (2008). Root causes, development and trend of desertification in Yinjinhuoluo County, Inner Mongolia, China. *Yijinhuoluo County Network for Desertification Information and Data Exchange*. Retrieved May 11, 2008, from http://nic6.forestry.ac.cn/sts/cy/5/5.html.
Nie, J. J. (2005, May 25). Songhe Cun zhisha yugong Shi Shuzhu: Yibezi zhiwei qunzhong zuo yijian shi. *Xinhua News*. Retrieved May 16, 2008, from http://xhby.net/xhby/content/2005-05/25/content_802066.htm (In Chinese).
Ningxia Shangwu Ting. (2005, July 1). Ningxia Wuzhong Shi Yanchi Xian shahua mianji jianshao 300wan mu. *Wuzhong Shi Difang Shangwu Zhichuang*. Retrieved May 21, 2008, from http://wuzhongshi.mofcom.gov.cn/aarticle/dongtai/200506/20050600123384.html (In Chiense).
Nonikoff, G. (1976). Traditional grazing practices and their adaptation to modern conditions in Tunisia and the Sahelian countries. In A. Rapp, H. N. Le Houérou, & B. Lundholm (Eds.), *Can desert encroachment be stopped*? A study with emphasis on Africa. Stockholm: Ecological Bulletins No. 24.
Nyerges, A. E. (1982). Pastoralists, Flocks and Vegetation: Processes of Co-adaptation. In B. Spooner & H.S. Mann (Eds). *Desertification and development: Dryland ecology in social perspective* (pp. 217–247). New York: Academic Press.
OFWM. (2000). *Zero Tillage for enhancing wheat productivity, Recommendations for 1997 2000*. Punjab, Lahore: Director General Agriculture, Water Management Punjab.
O'Hara, S. L. (1997). Irrigation and land degradation: implications for agriculture in Turkmenistan, central Asia. *Journal of Arid Environments*, 37, 165–179.
Oñate, J. J., Cummings, C., Gómez, A., Peco, B., & Sumpsi, J. M. (2001). Report on published on unpublished references on policy implementation and desertification in Spain at the local, regional and national level. In J. Jenkins & G. Wilson (Eds.), *MEDACTION MODULE 2: Effects of past policies in the target Areas. Deliverable 16: National team reports on published and unpublished references on policy implementation and desertification at the local, regional and national level in project research countries* (2001, August). London: King's College London, Department of Geography.
Ostrom, E. (1990). *Governing the commons: The evolution of institutions for collective action*. Cambridge University Press.

Park, J.-H. (1998). T*he Saemaul Movement: Korea's approach to rural modernization in the 1970 s*. Seoul: Korea Rural Economic Institute.

Pascal, A. (1968). Relaciones de poder en una localidad rural. *Santiago: Instituto de Capacitación e Investigación en Reforma Agraria (ICIRA)* (in Spanish).

Paz, J.-M, Sa′nchez, J., & Visconti F. (2006). Combined use of GIS and environmental indicators for assessment of chemical, physical and biological soil degradation in a Spanish Mediterranean region. *Journal of Environmental Management*, 79, 150–162.

People's Daily. (2000, April 19). Be mentally prepared for long-term battle. *People's Daily*. Retrieved May 2, 2008, from http://english.people.com.cn/english/200004/19/eng20000419_39277.html.

People's Daily. (2000, June 27). Keji renyuan zhujin cun, Songhe Cun renjin shatui. *People's Daily*. Retrieved May 5, 2008, from http://www.envir.gov.cn/info/2000/6/627657.htm (In Chinese).

People's Daily. (2000, July 18). Bai ChunLan he tade "lüse zhuangyuan". *People's Daily*. Retrieved May 4, 2008, from http://www.people.com.cn/GB/paper464/1029/148601.html (In Chinese).

People's Daily. (2007, September 1). Quanguo Daode Mofan huoxuan ren (jingye fengxian): Shi Shuzhu. Xinhua News. Retrieved May 25, 2008, from http://news.xinhuanet.com/newscenter/2007-09/01/content_6642748.htm (In Chinese).

Perera, J. (1986). The Gal Oya farmer organization programme: A learning process? In IIMI (Ed.), Participatory Management in Sri Lanka's Irrigation Schemes, International Irrigation Management Institue. Sri Lanka: Digana Village via Kandy.

Perez-Soba, A. & Barrientos, F. (1986). The "LUCDEME" Program in the southeast of Spain to combat desertification in the Mediterranean Region. In R. Fantechi & N. S. Margaris (Eds.), *Desertification in Europe: Proceedings of the information symposium in the EEC Programme on Climatology*, held in Mytilene, Greece, 15–18 April 1984. Dordrecht: D. Reidel Publishing Company.

Qi, L. (2006, March 16). Quanguo fangsha zhisha zonghe shifang qu jianshe qidong. *Huzhou Linye Xinxi Wang*. Retired May 13, 2008, from http://www.lzly.gov.cn/Article/ShowArticle.asp?ArticleID=570 (In Chinese).

Ramachandran, V. K. (2000). Kerala's development achievements and their replicability. In G. Parayil (Ed.), Kerala: The development experience reflections on sustainability and replicability (pp. 88-115). London: Zed Books.

Ren, X. Z. (1990). Wang Yushan tongzhi gei women liuxia de…. *Journal of Inner Mongolia Gorestry*, 3, 34 (In Chinese).

Saeed, M. M., Ashraf, M., Asghar, M.N. (2003). Hydraulic and hydro-salinity behavior of skimming wells under different pumping regimes. *Agricultural Water Management*, 61, 163–177.

Saigal, S. (2003, June). Turkmenistan: Issues and approaches to combat desertification (Discussion Draft). *TA 5941-REG: Combating desertification in Asia*. Retrieved May 11, 2008, from http://www.global-mechanism.org/dynamic/documents/document_file/turkiadc.pdf.

Scheckter, Y. & Galai, C. (1980). The Negev—a desert reclaimed. In M. R. Biswas, & A. K. Biwas (Eds.), *Desertification: Associated case studies prepared for the United Nations Conference on Desertification* (pp. 255–308). Oxford: Pergamon Press.

Schneider, H. J. (1982). Drought, demography, and destitution: Crisis in the Norte Chico. *GeoJournal*, 6(2), 111–119.

Schubert, S. D., Suarez, M. J., Pegion, P. J., Koster, R. D. & Bacmeister, J. T. (2004). On the cause of the 1930s dust bowl. *Science*, 303, 1855–1859.

Sharma, R. (2003, September 6). Kerala's decentralisation idea in practice. *Economic and Political Weekly*, 3831–3850.

Sheng, G. Q., & Yang, G. S. (1997). Baguo shamo shi lüzhou—Neimenggu Naiman Qi renmin zhengfu fu qizhang Bao Qinghe. *Zhongguo Tongyi Zhanxian*, 8, 39–41 (In Chinese).

Shi, M. J., Cheng, S. L, & Zhang, Q. Y. (2006). Scenario analysis of environmental policy for combating desertification in Northern China: A bio-economic model approach. *Journal of Natural Resources*, 3, 465–472 (In Chinese).

Shi, M. J., Zhang, Q. Y., & Wang, T. (2005). Better access to new technologies and credit service, farmers' land use decision, and policy for poverty alleviation and rangeland conservation. *JARQ*, 39 (3), 181 – 190.

Shneider, Catalina and Valeria Foncea. (2008, June 3). Award given to joint UNDP-EU program in Chile. *United Nations Development Programme*. Retired May 4, 2008, from http://content.undp.org/go/newsroom/2008/june/proyecto-para-combatir-la-desertificacin-gana-premio-en-chile.en.

Shilongwei of Wuwei City. (2008, March 31). Minqin Xian Xuebai Xiang Songhe Cun Cunmin Huodong Zhongxin jianjie. *Wuwei Dangjian*. Retrieved May 5, 2008, from http://www.wwdj.gov.cn/Article/ShowArticle.asp?ArticleID=168 (In Chinese).

Shrum, W., & Ramanathaiyer, S. (2000). Knowledge, democratization and sustainability: The 'Kerala model' of scientific capacity building. In G. Parayil (Ed.), Kerala: The development experience reflections on sustainability and replicability (pp. 157–177). London: Zed Books.

Spooner, B. 1984. *Ecology in development: A rationale for three-dimensional policy*. Tokyo, Japan: United Nations University.

Spooner, B., & Mann, H. S. (1982). *Desertification and development: Dryland ecology in social perspective.* New York: Academic Press.

Stephens, P. H. (1937). Why the dust bowl? *Journal of Farm Economics*, 19(3), 750–757.

Sun, R., & Buhe, C. L. (2000). Fengyu linye xing—Ji Naiman Baxiantong linchang changzhang Li Shan. *Journal of Inner Mongolia Forestry*, 10, 32–33 (In Chinese).

Sun, R., Du, F., & Buhe, C. L. (1998). Tade xin xizhe linchang—Ji Naiman Xinzhen linchang changzhang Liu Zhong. *Journal of Inner Mongolia Forestry*, 4, 31–32 (In Chinese).

Sun, R., & Liu, Y. Y. (1997). Shaihai mingzhu—Ji Naiman Qi Shelihu shuiku zhuren Liu Yuanbin. *Dangjian Yu Rencai*, 1, 20 (In Chinese).

Swap, R., Ulanski, S., Cobbett, M., & Garstang, M. (1996). Temporal and spatial characteristics of Saharan dust outbreaks. *Journal of Geophysical Research*, 101(d), 4205–20.

Tang, J. (2007, May 15). *Aohan Qi zoushang huangjin zhilu, Beifang Xinwen Wang*. Retrieved May 9, 2008, from http://www.northnews.cn/news/2007/200705/2007-05-15/85565_2.html (In Chinese).

Tebie Baodaozu. (2007, April 17). Zhizhuo jianren—Longren pinge de xinnian lidu. *Gansu Economic Daily*. Retrieved May 10, 2008, from http://www.gsjb.com/Get/sdbd/20070417082758.htm (In Chinese).

Thornes, J. (1995). Global environmental change and regional response: the European Mediterranean. *Transactions of the Institute of British Geographers, New Series*, 20 (3), 357–367.

Toderich, C. H., Goldshtein, R. I., Aparin, W. B., Idzikowska, K., & Rashidova, G. S. (2001). Environmental state and an analysis of phytogenetic resources of halophytic plants for rehabilitation and livestock feeding in arid and sandy deserts of Uzbekistan. In S. –W. Breckle, M. Veste, & W. WUcherer (Eds.), *Sustainable land use in deserts* (pp. 154-165). Berlin: Springer.

Törnquist, O. (2000). The new popular politics of development: Kerala's experience. In G. Parayil (Ed.), Kerala: The development experience reflections on sustainability and replicability (pp. 116-138). London: Zed Books.

Turner, J. E., Hesli, V. L., Bark, D. S., & Yu, H. (1993). *Villages astir: Community development, tradition, and change in Korea*. London: Praeger Publishers.

United Nations Environment Programme (UNEP). (2002, December). Increasing productivity on fragile soils: Comprehensive desertification control using sustainable moisture conservation, crops and animal husbandry technologies in Naimanqui Banner County, China. In *Success stories in the struggle against desertification: A holistic and integrated approach to environmental conservation and sustainable livelihoods*. Retrieved May 12, 2008, from http://www.unep.org/desertification/successstories/SSS%20Evaluation%20report%20.pdf.

Uphoff, N. T. (1985a). Fitting project to people. In M. M. Cernea (Ed.), *Putting people first* (pp. 359–395). Oxford: Oxford University Press.

Uphoff, N. T. (1985b). *Summary of January 1985 trip report on farmer organization program in Gal Oya, Sri Lanka*. Manuscript, Cornell University.

Uphoff, N. T. (1985c). People's participation in water management: Gal Oya, Sri Lanka. In J. C. Garcia-Zamor (Ed.), *Public participation in development planning and management: Cases from Africa and Asia* (pp. 131–78). Boulder: Westview Press.

Uphoff, N. T. (1986a). Activating community capacity for water management in Sri Lanka. In D. C. Korten (ed.), *Community management: Asian experience and perspectives* (pp. 201–219). West Hartford, Conn.: Kumarian Press.

Uphoff, N. T. (1986b). *Local institutional development: An analytical sourcebook with cases*. West Harford, Conn.: Kumarian Press.

Uphoff, N. T. (1986c). *Getting the process right: Improve irrigation water management with farmer participation*. Boulder: Westview Press.

Uphoff, N. T. (1992). *Learning from Gal Oya possibilities for participatory development and post-Newtonian social science*. Ithaca and London: Cornell University Press.

U.S.S.R. Ministry of Reclamation and Water Management. (1980). U.S.S.R. Golodnaya (Hungry) Steppe: A Case Study of Desertification. In Biswas, M. R. & Biwas, A. K (Eds.). *Desertification: Associated case studies prepared for the United Nations Conference on Desertification* (pp. 427-480). Oxford: Pergamon Press.

Véron, R. (2000). Sustainability and the 'new' Kerala model. In G. Parayil (ed.), *Kerala: The development experience reflections on sustainability and replicability* (pp. 212–229). London: Zed Books.

Véron, R. (2001). The "new" Kerala model: Lessons for sustainable development. *World Development*, 29(4): 601–617.

Wang, B. (2005). Hetian: Yongyuan de lüzhou. *Forest and Humankind*, Vol. 12, 56–58 (In Chinese).

Wang, D. L. (2008, February). Governmental guiding, projects motivating and public participation to mobilize resources for combating land desertification in Ningxia. *Ningxia Forestry Bureau*. Retrieved May 13, 2008, from http://www.un.org/esa/sustdev/sdissues/desertification/beijing2008/presentations/wang.pdf (In Chinese).

Wang, G. J. (2003a). Zhili huangmohua de chenggong moshi—Neimenggu Aohan Qi shengtai huangjing jianshe huigu. *Neimenggu Aohan Qiwei Xuanchuanbu*. Retired May 14, 2008, from http://www.aohan.gov.cn/html/cgms.htm (In Chinese).

Wang, P. Z. (2003b). Naiman Qi zhenbo yiqi daofa linmu an. *Journal of Inner Mongolia Forestry*, 5, 26 (In Chinese).

Wang, S. J. (2003c, February 28). Aohan Qi shuqi zhisha yimian qi. *People's Daily*. Retrid May 16, 2008, from http://iea.cass.cn/jianbao/GRDB003_1714.htm (In Chinese).

Wang, Z. F. (2003d). Linhai chonglang ren—Ji Baxiantong linchang changzhang Li Shan. *Journal of Inner Mongolia Forestry*, 3, 43-44 (In Chinese).

Wang, Z. F. (2003e). Huangmo shenchu de shengtai jingji yuanqu—Ji Neimenggu Naiman Qi Baiantong linchang. *Forestry of China*, 11, 34 (In Chinese).

Wang, X. Z., & Li, G. L. (1982). Shahai xinlü—Yijinhuoluo Qi jixing. *Journal of Inner Mongolia Forestry*, 51, 22-23 (In Chinese).

Wang, J. Y., Ma, Y. J., & Jiang, Z. P. (2005). Causes and developmental dynamic of land desertification in Zhanye,Gansu Province. *Journal of Desert Research*, 25(3), 427-432 (In Chinese).

Watson, I. W., (2003). *Preventing desertification and promoting rehabilitation in Western Australian semi-aridRangelands at the individual property scale. In J. Takahashi (Ed.), Global perspective in range rehabilitation and prevention of desertification, The 2003 Obihiro Asia and the pacific seminar on education for rural development (pp. 33–44), Hokkaido*, Japan: Obihiro University of Agriculture and Veterinary Medicine (August 26–September 2).
Wei, Y. D., & Fang, C. L. (2006). Geographical and structural constraints of regional development in Western China: A study of Gansu Province. *Issues & Studies*, 42(2), 131–170.
Williams, O. B., Suijdendrop, H., & Wilcox, D. G. (1980). The Gascoyne Basin. In M. R., Biswas & A. K., Biwas (Eds.), *Desertification: Associated case studies prepared for the United Nations conference on desertification* (pp. 3–106). Oxford: Pergamon Press.
Williams, O. B. (1978). Desertification in the pastoral rangelands of the Gascoyne Basin, Western Australia. *Search*, 9 (7), 257–261.
Worster, D. (2004). *Dust bowl: The southern plains in the 1930s*. New York: Oxford University Press.
Wu, Z. R. (2006, May 29). Neimenggu Aohan Qi 20 duo nian jianchi buxie zaolin lüjua baohu shengtai. *Economic Daily News*. Retrieved May 17, 2008, from http://news.xinhuanet.com/newscenter/2006-05/29/content_4614704.htm (In Chinese).
Wu, R., Xiao, L., & Yun, F. (2005). Lüran tianjiao shengdi—Yijinhuoluo linyeju fangsha zaolin lühua jishi. *Zhifu Zhiyou*, 9, 54-65 (In Chinese).
Wuzhong Dangjiang Wang. (2005, October 10). Zhisha gongchen—Bai Chunlan.. Retrieved May 19, 2008, from http://www.wzhdj.gov.cn/Article/ShowArticle.asp?ArticleID=136 (In Chinese).
Xia, Y. (2005). *Xiaogangcun Yu Dabaogan*. Retrieved March 4, 2006, from http://www.jsdj.com/luyou/lyzy/anxgc.htm (In Chinese).
Xinhua News Agency. (2007, November 7). 18.7b Yuan earmarked to fight desertification in Xinjiang. *Ministry of Environmental Protection the People's Republic of China*. Retrieved May 22, 2008, from http://english.sepa.gov.cn/News_service/media_news/200712/t20071217_114947.htm.
Xinjiang Uygur Autonomous Region government. (2007, October). People's Republic of China: Xinjiang Municipal Infrastructure and Environmental Improvement Project (Prepared for the Asian Development Bank, Project Number: 39,228). *Asian Development Bank*. Retrieved May 23, 2008, from http://www.adb.org/projects/project.asp?id=39228.
Xu, D. W. (1995). Naiman Qi guoyou linchang (pu) gaige jian chengxiao. *Journal of Inner Mongolia Forestry*, 8, 17-18 (In Chinese).
Xu, Z. (1848). *Mu-ling-shu* (In Chinese).
Xu, W, & Wang, J. 2007. Gan Wei "Lü Se" Zuo Gong Xian De Ren—Ji Nai Man Qi Xing Long Zhao Lin Chang Chang Zhang Li Zhe. *Journal of Inner Mongolia Forestry*, 12: 4 (In Chinese).
Xu, G. W., & Yang, Z. W. (1991). Linmu suo shalong huangmo bian lüzhou—Naiman Qi zhili shamohua jishi. *Journal of Inner Mongolia Forestry*, 8, 4-5 (In Chinese).
Yang L. (1999). Discussion on the long-lasting feudal society in China. *Yunnan Social Science*, 2, 69-77 (In Chinese).
Yang, Z. Z., Wu, R., Gao, C., & Ma, J. (2005). Qiantan Neimenggu Yijinhuoluo Qi tuigeng huailin jianshe yu houxu chanye fazhan. *Inner Mongolia Forestry Science & Technology*, 4, 35-36 (In Chinese).
Yang, S., & Zhang, P. C. (2005). Zhagen bianjiang de shuiwen ren—Ji quanguo laodong mofan, Xinjiang Hetian shuiwen shuiziyuan kanceju dangzu chengyuan, jijian zuzhang Wu Guiwu. *Supervision in China*, 16, 38-40 (In Chinese).
Yu J. (2001). *Yucun zhengzhi* (*Politics in yucun village*). Beijing: Shangwu Yinshuguan (In Chinese).
Yu, J. (2002). Xiangcun zizhi: Genju yu lujing (Village self-governance: Foundations and paths). *Zhanlue Yu Guanli* (*Strategy and Management*), 6: 117–120.
Yu, R. C., & Zhang, Y. X. (1984). Yijinhuoluo Qi gaizao jiechai gongzuo yijian chengxiao. *Modern Agriculture*, 6, 42 (In Chinese).

Zhan, Y. (2002, April 22). Zhongguo jiaqiang dui shachenbao de zhili. *Xinhua News*. Retrieved May 22, 2008, from http://news.xinhuanet.com/fortune/2002-04/22/content_368226.htm (In Chinese).

Zhang, J. L. (2005). *Barriers to water markets in China's Heihe River Basin* (Research report for Economy and Environment Program for Southeast Asia. No. 2005-RR8). Retrieved May 26, 2008, from http://www.crdi.ca/uploads/user-S/11500956601JunlianRR8.pdf.

Zhang, Y. Z, Chen, Z. Y., Zhu, B. Q., Luo, X. Y., Guan, Y. N., Guo, S., & Nie, Y. P. (2008, January 16). Land desertification monitoring and assessment in Yulin of Northwest China using remote sensing and geographic information systems (GIS). *Environmental Monitoring and Assessment*, 0167-6369 (Print) & 1573-2959 (Online).

Zhang, Z. (1991). *Zhongguo shishen—Guanyu qi zai 19 shiji zhongguo shehui zhong zuoyong de yanjiu*. Shanghai: Shanghai Shehui Kexueyuan Chubanshe (In Chinese).

Zhang, B., & Tan, M. Y. (1989). Naiman—Keerqin shadi de lüzhou. *Journal of Inner Mongolia Forestry*, 9, 8-9 (In Chinese).

Zhang, F. R., Wang, L. X., Niu, Z. G., Chen, H. W., & Li, B. G. (2002). Land use change and sustainable utilization in Ejin Horo Banner. *Journal of Desert Research*, 22(2), 166-171 (In Chinese).

Zhao, Y. P., Huang, F., Guo, J. J. (1999). Desertification research based on digital earth. *IEEE*, 646–648.

Zhao, H. L., Zhou, R. L., Zhang, T. H., Zhao, X. Y. (2006). Effects of desertification on soil and crop growth properties in Horqin sandy cropland of Inner Mongolia, north China. *Soil & Tillage Research*, 87, 175–185.

Zhi, Z. P. (2001, December 24). Neimenggu Zizhiqu Chifeng Shi zhili fengsha jingyan tanmi. *People's Daily*. Retrieved May 28, 2008, from http://www.fzepb.gov.cn/news/list.asp?fileno=12001000330 (In Chinese).

Zhou, H. Y. (2000, April 6). Xibu kaifa kan shengtai. *People's Daily*, P. A11 (In Chinese).

Zhou, L. Y. & Liang, X. Q. (2003, September 19). "Shahai yugong" Shi Shuzhu. *People's Daily*. Retrieved May 3, 2008, from http://www.people.com.cn/GB/shizheng/1026/2096917.html (In Chinese).

Zhu, L. (2007, April 5). Aohan Qi Bao Yongxin jia de zhisha linchang. *Zhongguang Wang*. Retrieved May 11, 2008, from http://www.cnr.cn/zhuanti1/scb/qxkb/200704/t20070405_504435799.html (In Chinese).

Zhou, H., Wang, J. H. (2006, April 13). Yanchi zaihuo quanguo zhisha xianjin xian. *Xinhua News*. Retrieved May 17, 2008, from http://www.nx.xinhuanet.com/xbst/2006-04/13/content_7002786.htm (In Chinese).

Zhuang, D. Y. (2006, August 21). Yanchi xian ruhe shixian shamohua nizhuan. *Guangming Daily*. Retrieved May 15, 2008, from http://www.gmw.cn/01gmrb/2006-08/21/content_467271.htm (In Chinese).

Zuo, L. L. (2002). Hetian qianghua caoyuan zhifa duiwu jianshe. *China Animal Husbandry Bulletin*, 1, 24 (In Chinese).

Bibliography

Anderies, J. M., Janssen, M. A. & Walker, B. N. (2002). Grazing management, resilience and the dynamics of a fire driven rangeland system. *Ecosystems,* 5(1), 23–44.

Castells, M. (2002). *The Internet Galaxy: Reflections on the Internet, Business and Society*. London: Oxford University Press.

Cohen, M. D., March, J. G., Olsen, J. P. (1972). A garbage can model of organizational choice. *Administrative Science Quarterly*, 17(1), 1–25.

Dye, T. R. (2004). *Understanding Public Policy* (10th ed.). Beijing: Chinese People University Press.

Easton, David. (1953). *The political system: An inquiry into the state of political science*. New York: Knopf.

Flacks, R. (1975). Review: [untitled] reviewed work(s): the American intellectual elite by Charles Kadushin. *Contemporary Sociology*, 4(5), 552–554.

Fukuyama, F. (1989). The end of history. *The National Interest*, 16(summer), 3–18.

George, A. L., & Bennett, A. (2005). *Case studies and theory development in the social sciences*. Cambridge, Massachusetts: Harvard University.

Hjern, B., & Hull, C. (1982). Implementation Research as Empirical Constitutionalism. *European Journal of Political Research*, 10(1), 105–115.

Hogwood, B. W., and Gunn, L. A. (1984). *Policy analysis for the real world*. London: Oxford University Press.

Jasnoff, S. (1990). *The fifth branch: Science advisers as policymakers*. Cambridge, Massachusetts: Harvard University Press.

Jenkins, W. (1978). *Policy analysis: A political and organisational perspective*. New York: St. Martin's Press.

John, P. (1998). *Analyzing public policy*. London and New York: Pinter.

Kadushin, C. (1974). *The American intellectual elite*. Boston, District of Columbia, USA: Little, Brown and Company,

Latham, E. 1956. The group basis of politics. In H. Eulau, S. J. Eldersveld, & M. Janowitz (Eds.), *Political Behavior* (pp. 232–245). New York: Free Press.

Lindblom, C. E., & Woodhouse, E. J. (1993). The policy-making process (3rd ed.). Englewood Cliffs, NJ: Prentice Hall.

Plummer, K. (2001). *Documents of life 2: An invitation to a critical humanism*. London, Thousand Oaks, and New Delhi: Sage Publications.

Prescott, Edward C, & Visscher, Michael. (1980). Organization capital. *Journal of Political Economy*, 88(3): 446–61.

Reyonlds, J. F. et al. Smith, D. M. S., Lambin, E. F., Turner, B. L. II, Mortimore, M., Batterbury, S. P. J., Downing, T. E., Dowlatabadi, H., Fernández, R. J., Herrick, J. E., Huber-Sannwald, E., Jiang, H., Leemans,R., Lynam,T., Maestre, F. T., Ayarza,M., Walker, B. (2007). Global desertification: Building a science for dryland Development, *Science*, 316 (5826), 847–851.

Sabatier, P. A. (1986). Top-down and bottom-up models of policy implementation: A critical and suggested synthesis. *Journal of Public Policy*, 6 (1), 21–48.

Simon, H. A. (1957). *Models of man*. New York: John Wiley & Sons, Inc.

Simon, H. A. (1983). *Reason in human affairs*. Oxford: Blackwell.

Wang, L., Wang, X., & Zhang, K. (Eds.). (1993). *The experiences of combating desertification in P. R. China*. Beijing: College of Soil and Water Conservation in the Beijing Forestry University.

Xu, Y. (1993, September 24). *Speaking at the National Desertification Conference in Chifeng, Inner Mongolia, as quoted in Xinhua News Agency (English report of Chinese address)* (Also available in FBIS-CHI-93-184.). Beijing: The National Desertification Conference in Chifeng, Inner Mongolia.

ZXBW, Zhongweixian Xianzhi Bianzuan Weiyuanhui. (2005). *Zhongwei xianzhi* (*County annals of Zhongwei*). Yinchuang: Ningxia Renmin Chubanshe.

MIX
Papier aus verantwortungsvollen Quellen
Paper from responsible sources
FSC® C105338

If you have any concerns about our products,
you can contact us on
ProductSafety@springernature.com

In case Publisher is established outside the EU,
the EU authorized representative is:
Springer Nature Customer Service Center GmbH
Europaplatz 3, 69115 Heidelberg, Germany

Printed by Libri Plureos GmbH
in Hamburg, Germany